Topological Riesz Spaces and Measure Theory

Topological Riesz Spaces and Measure Theory

D. H. FREMLIN

Lecturer in Mathematics, University of Essex

CAMBRIDGE
AT THE UNIVERSITY PRESS
1974

CAMBRIDGE UNIVERSITY PRESS
Cambridge, New York, Melbourne, Madrid, Cape Town, Singapore, São Paulo, Delhi

Cambridge University Press
The Edinburgh Building, Cambridge CB2 8RU, UK

Published in the United States of America by Cambridge University Press, New York

www.cambridge.org
Information on this title: www.cambridge.org/9780521201704

First published 1974
This digitally printed version 2008

A catalogue record for this publication is available from the British Library

Library of Congress Catalogue Card Number: 72–95410

ISBN 978-0-521-20170-4 hardback
ISBN 978-0-521-09031-5 paperback

Contents

Acknowledgements

The gestation of this book has extended over three appointments: as a lecturer at United College, The Chinese University of Hong Kong; as a Central Electricity Generating Board Junior Research Fellow at Churchill College, Cambridge; and as a lecturer at the University of Essex. I am grateful for the material support of all these institutions, and for the moral support of my colleagues there.

Almost every mathematician I have spoken to in the last six years has influenced the book in some way; but I should like to give special thanks to Professor W. A. J. Luxemburg, who introduced me to its subject matter; to Dr N. N. Chan, who suggested that it should be written; and to Dr F. Smithies, who gently moulded it into publishable form. Finally I must mention Miss M. Mitchell, who typed the bulk of the MS, and the editors of the Cambridge University Press, for their skill and patience in transforming an idiosyncratic manuscript into print.

Acknowledgements

Preface

This book is addressed to functional analysts who would like to understand better the application of their subject to the older discipline of measure theory. The relationship of the two subjects has not always been easy. Measure theory has been the source of many examples for functional analysis; and these examples have been leading cases for some of the most important developments of the general theory. Such a stimulation is, of course, entirely welcome. But there have in addition been several cases in which special results in measure theory have been applied to prove general theorems in analysis. The ordinary functional analyst feels inadequately prepared for these applications, and is exasperated by the intrusion of a large body of knowledge in an unfamiliar style into his own concerns.

My aim therefore is to identify those concepts in measure theory which have most affected functional analysis, and to integrate them into the latter subject in a way consistent with its own structure and habits of thought. The most powerful idea is undoubtedly that of Riesz space, or vector lattice. The principal Banach spaces which measure theory has contributed to functional analysis all have natural partial orderings which render them Riesz spaces. Many of their special properties can be related to the ways in which their orderings, their linear structures and their topologies are connected. For a clear understanding of the difference, for instance, between an L^1 space and an L^∞ space, there is no substitute for an abstract analysis of their properties as ordered linear topological spaces.

The other point at which measure theory has had an impact on functional analysis is in the representation of linear functionals as integrals, and the consequent deduction of surprising properties. I think that the analyst's instinctive rejection of such methods is a perfectly sound reaction. The difficulty is to provide techniques powerful enough to act as substitutes. Here again I believe that a thorough understanding of some quite simple topological Riesz spaces can see us through most of the difficulties.

Now as soon as we begin to look at measure theory with a sceptical and abstract eye, a number of peculiarities strike us. The first is the relative unimportance of measure spaces themselves. They are used to construct spaces of equivalence classes of functions which then take on a life of their own; and quite different measure spaces can give rise to isomorphic function spaces. So it is natural to ask just what it is about a measure space that determines the associated function spaces. The answer is readily to hand: it is a construction called the measure algebra. Two measure spaces with isomorphic measure algebras will have isomorphic L^1 spaces, isomorphic L^∞ spaces, and so on; and to a large extent the converse is true.

I think that the most convincing way of demonstrating these facts is to exhibit methods of constructing the function spaces (or, rather, isomorphic copies of them) directly from a measure algebra. These constructions are not particularly simple; in this book they take up Chapter 4 and the first half of Chapter 5. However, given some intuitive grasp of the nature of Riesz spaces, they are fairly straightforward, and provide invaluable insights.

For instance, another curiosity of traditional measure theory is the unimportance of any notion of homomorphism between measure spaces. This distinguishes it from all comparably abstract branches of mathematics. I believe that this deficiency occurs because the natural and important homomorphisms of the theory are between measure algebras and not between measure spaces at all. In §§45 and 54 I discuss such homomorphisms and their effects on the constructions I have set up.

The reader may be forgiven for wondering at this stage just how much he needs to know to cope with this book. A basic knowledge of functional analysis is essential. It would be possible to go a fair way with normed spaces alone; but this would shut off many of the most interesting ideas, and I have written on the assumption that I may call on the fundamental concepts of the theory of linear topological spaces. As for measure theory, in a formal sense I require none; I give every definition from that of measure space onwards. In an informal sense, I cannot pretend that this book is a genuine alternative to the traditional presentation of the elementary theory. §§61–3, while formally complete, are far too sketchy to be satisfactory as a first introduction. So if you are uncertain of your grounding in the subject, I refer you to the detailed advice in the Prerequisites section.

I have preferred, in ordering the material of this book, to arrange results by the contexts in which they apply; thus propositions which refer to arbitrary Riesz spaces go into §14 and those which refer only to Archimedean Riesz spaces go into §15. Now this is not, of course, a natural order for learning the subject. The beginner will certainly wish in the first instance to restrict his attention to Archimedean Riesz spaces, and leave any generalizations to one side for the moment. I think my decision is justified; it means that the hypotheses of successive theorems do not usually vary in arbitrary details, and are consequently much easier to recall precisely. But it makes a page-by-page progression inappropriate. To assist the reader to avoid spending too much time on material which will not be immediately essential, I have used asterisks to mark sections and propositions which can at first be passed over. They will be given references in the text when they come to be needed.

Most of this book is taken up with fulfilling the objects I described at the beginning of this preface. But I think that topological Riesz spaces are inherently fascinating, and I have laid Chapter 2 out as a survey of their elementary properties; and in Chapter 8 I try to take a special subject up to the area of present research work.

University of Essex D. H. FREMLIN
January 1973

Prerequisites

ZORN'S LEMMA, TRANSFINITE INDUCTION, AND THE AXIOM OF CHOICE The best introduction I know to this subject is in the appendix of KELLEY; HALMOS N.S.T. is also perfectly satisfactory. I shall apply the axiom of choice, and the principle of dependent choice for sequences, without comment, though when syntax allows I signal with such a word as 'choose' or 'choice'. Zorn's lemma is used most often in the form

> If $\mathscr{X}$ is any collection of sets such that for any subset $\mathscr{Y}$ of $\mathscr{X}$ which is totally ordered by ordinary inclusion, we have
> $$\bigcup \mathscr{Y} = \{t : \exists\ Y \in \mathscr{Y}, t \in Y\} \in \mathscr{X},$$
> then $\mathscr{X}$ has a maximal element.

(This is nearly what KELLEY calls the 'maximal principle'.) When employing this form, I shall generally say merely 'by Zorn's lemma, $\mathscr{X}$ has a maximal element', and leave it to the reader to verify that the hypotheses on $\mathscr{X}$ are satisfied – supposing, of course, that the verification is straightforward. But if the lemma is to be applied in the more general form

> If $\mathscr{X}$ is a partially ordered set such that every totally ordered subset has an upper bound in $\mathscr{X}$, then $\mathscr{X}$ has a maximal element,

then I shall give more details.

FUNCTIONAL ANALYSIS I can recommend BOURBAKI V, ROBERTSON & ROBERTSON, KELLEY & NAMIOKA, or SCHAEFER T.V.S.; any of these is adequate for the principal results. For details see §A1.

MEASURE THEORY As I explained in the Preface, I make no formal requirements in this field. But a knowledge of the elementary theory up to and including the Radon–Nikodým theorem would certainly be helpful. The first point is that although abstract measure theory is not particularly difficult in its early stages, the only easy examples of measure spaces are essentially trivial. There is no substitute for Lebesgue measure on the real line as a leading example. So at some stage the reader should make himself familiar with a proof that it exists.

Secondly, several of the most important ideas in this book appeared first in measure theory. Pre-eminent among these are the arguments leading up to the Radon–Nikodým theorem. I give this theorem in § 63 with a proof which refers to almost every preceding chapter. In fact my proof is very closely related to one of the traditional ones, but its component ideas have been scattered. I think that it is illuminating to see how they can be joined more closely together in a direct proof. I shall give a concordance in the proper place.

Meanwhile, I do not want the reader to feel that he must take a course in measure theory before proceeding further. Perhaps the books suggested below should be regarded as a starred section; optional on first reading, but likely to be useful later.

R. G. Bartle, *The Elements of Integration* (Wiley, 1966)

H. L. Royden, *Real Analysis* (Macmillan, 1963)

H. Widom, *Lectures on Measure and Integration* (van Nostrand Reinhold, 1969)

J. H. Williamson, *Lebesgue Integration* (Holt, Rinehart & Winston, 1962).

1. Riesz spaces

The prerequisites for this chapter are few; an acquaintance with linear spaces is the principal requirement. A Riesz space is simply a linear space over the field **R** of real numbers which has a special kind of partial ordering, and all we need to know about partial orderings will be covered in §§ 11 and 13. But the theory of Riesz spaces is already rich, and some of the work in §§ 16 and 17 is far from trivial. It does, however, have to be taken seriously. These are the basic results which will enable us to handle Riesz spaces with assurance and facility.

11 Partially ordered sets

This section is little more than a list of definitions. As such I suggest that it should be read carefully once, together with the associated examples; but that there is no need to consciously memorize anything. You will find the index perfectly reliable.

Actually the concepts here have applications to every branch of mathematics, and they will mostly be familiar in everything but name. I think it is amusing and instructive to seek such applications out and consciously appreciate them.

Now to work:

11A Definition **A partially ordered set** is a pair $(A, \leqslant)$ where A is a set and $\leqslant$ is a binary relation on A such that:

(i) $a \leqslant a$ for every $a \in A$;

(ii) if a and b belong to A and $a \leqslant b$ and $b \leqslant a$, then $a = b$;

(iii) if a, b and c belong to A and $a \leqslant b$ and $b \leqslant c$, then $a \leqslant c$.

In this context we write $a \geqslant b$ for $b \leqslant a$ (that is, $\geqslant$ will always be the inverse relation to $\leqslant$), and $a < b$ or $b > a$ for $a \leqslant b$ and $a \neq b$.

For examples see 1XA and 1XC.

11B Suprema and infima Let $(A, \leqslant)$ be a partially ordered set, and B a subset of A. There may, or may not, be an $a \in A$ with the property

$$\text{for every} \quad c \in A, \quad a \leqslant c \quad \text{iff} \quad b \leqslant c \quad \text{for every} \quad b \in B,$$

that is, $a \leqslant c$ iff c is an upper bound for B. Clearly there can be at most one such a [11A(ii)]. If a has this property, we say that 'B has a least upper bound, which is a', or '$\sup B$ exists, and $\sup B = a$', or simply '$\sup B = a$'.

Similarly, $\inf B$, if it exists, is that one element of A such that

$$\text{for every} \quad c \in A, \quad c \leqslant \inf B \quad \text{iff} \quad c \leqslant b \quad \text{for every} \quad b \in B.$$

We observe that

$$\inf B = \sup \{a : a \leqslant b \;\forall\; b \in B\},$$

$$\sup B = \inf \{a : b \leqslant a \;\forall\; b \in B\},$$

$$\inf A = \sup \varnothing, \quad \sup A = \inf \varnothing,$$

all these inequalities being true in the sense that if one side is defined, so is the other, and they are then equal.

For a warning of the dangers in this notation, see 1XB.

11C Directed sets Let $(A, \leqslant)$ be a partially ordered set. A set $B \subseteq A$ is **directed upwards,** $B\uparrow$, if for every pair a, b of elements of B there is a $c \in B$ such that $a \leqslant c$ and $b \leqslant c$.

I shall write $B \uparrow a$ to mean that $B\uparrow$ and that $\sup B = a$. Observe, for instance, that $\{a\} \uparrow a$ for every $a \in A$.

Associated with a directed set is a filter. If B is a non-empty set which is directed upwards, I shall write $\mathscr{F}(B\uparrow)$ for the filter on A with base

$$\{\{a : a \in B,\, a \geqslant b\} : b \in B\}.$$

$\mathscr{F}(B\uparrow)$ is sometimes called the **filter of sections** of the directed set B.

$B\downarrow$ ('B is directed downwards'), $B \downarrow a$ and $\mathscr{F}(B\downarrow)$ are defined in the same way, but upside down.

11D Functions Let $(A, \leqslant)$ and $(B, \preccurlyeq)$ be partially ordered sets. A function $f: A \to B$ is **increasing** if $a \leqslant b \Rightarrow fa \preccurlyeq fb$. When f is a sequence $\langle a_n \rangle_{n\in\mathbf{N}}$ say, I shall write $\langle a_n \rangle_{n\in\mathbf{N}}\uparrow$ to mean that it is an increasing sequence, that is, that $a_n \preccurlyeq a_{n+1}$ for every $n \in \mathbf{N}$. Similarly, $\langle a_n \rangle_{n\in\mathbf{N}}$ is a **decreasing** sequence, $\langle a_n \rangle_{n\in\mathbf{N}}\downarrow$, if $a_{n+1} \preccurlyeq a_n$ for every $n \in \mathbf{N}$. By analogy with 11C, I shall write $\langle a_n \rangle_{n\in\mathbf{N}} \uparrow a$ to mean that $\langle a_n \rangle_{n\in\mathbf{N}}\uparrow$ and that $\sup_{n\in\mathbf{N}} a_n = a$. $\langle a_n \rangle_{n\in\mathbf{N}} \downarrow a$ is defined similarly.

An increasing function $f: A \to B$ is **order-continuous** if

$$\sup f[C] = fa \quad \text{whenever } C \uparrow a \text{ and } C \text{ is not empty}$$

$$\inf f[C] = fa \quad \text{whenever } C \downarrow a \text{ and } C \text{ is not empty.}$$

(Of course, if f is increasing, then $f[C]$ is directed whenever C is.) Note that it is only directed sets which must have their sups and infs preserved in this way.

*There are two more concepts which will be useful to us in special circumstances. An increasing function $f: A \to B$ is **order-continuous on the left** if it satisfies the first half of the condition for order-continuity, that is, if $f[C] \uparrow fa$ whenever $C \uparrow a$ and $C \neq \varnothing$. And f is **sequentially order-continuous** if

$$\langle fa_n \rangle_{n \in \mathbf{N}} \uparrow fa \quad \text{whenever} \quad \langle a_n \rangle_{n \in \mathbf{N}} \uparrow a$$

and

$$\langle fb_n \rangle_{n \in \mathbf{N}} \downarrow fb \quad \text{whenever} \quad \langle b_n \rangle_{n \in \mathbf{N}} \downarrow b.$$

***11E Order-closed sets** If $(A, \leqslant)$ is a partially ordered set and $B \subseteq A$, I shall write

$$\mathscr{I}B = \{a: \exists\ C \subseteq B, \quad C \neq \varnothing, \quad C \uparrow a \quad \text{in} \quad A\},$$

$$\mathscr{D}B = \{a: \exists\ C \subseteq B, \quad C \neq \varnothing, \quad C \downarrow a \quad \text{in} \quad A\}.$$

Then $B \subseteq \mathscr{I}B$ and $B \subseteq \mathscr{D}B$.

B is **order-closed** if $\mathscr{I}B = B = \mathscr{D}B$.

***11F Order-bounded sets** Let $(A, \leqslant)$ be a partially ordered set. A set $B \subseteq A$ is **order-bounded** if it has both upper and lower bounds in A, i.e. if $B \subseteq [b, c] = \{a: b \leqslant a \leqslant c\}$ for some b and c in A.

***11G Products** Let $\langle (A_\iota, \leqslant_\iota) \rangle_{\iota \in I}$ be an indexed family of partially ordered sets. Let A be the cartesian product $\prod_{\iota \in I} A_\iota$, and define a relation $\leqslant$ on A by $a \leqslant b$ iff $a(\iota) \leqslant b(\iota)$ for every $\iota \in I$. Then $(A, \leqslant)$ is a partially ordered set, the **product** of the family $\langle (A_\iota, \leqslant_\iota) \rangle_{\iota \in I}$. If, for each $\iota \in I$, $\pi_\iota: A \to A_\iota$ is the canonical map, then π_ι is increasing and order-continuous.

***11H Exercises** (**a**) Let $(A, \leqslant)$ be a partially ordered set. Then there is a topology on A for which the closed sets are precisely the order-closed sets of A. [See 1XC.]

(**b**) Let $(A, \leqslant)$ and $(B, \leqslant)$ be partially ordered sets. Then an increasing function $f: A \to B$ is order-continuous iff $f^{-1}[C]$ is order-closed in A for every order-closed $C \subseteq B$.

Notes and comments We shall see a greatest lower bound or a least upper bound on every other page of this book; so it will be as well to grasp firmly the formal definitions in 11B. I generally take these formalities seriously; I want to know exactly what I mean by such expressions as inf $\varnothing$.

I cannot emphasize too strongly the fact that the least upper bound of a set depends on the partially ordered set in which it lies. There would be formal advantages in writing

$$\sup_{(A,\leqslant)} B$$

to mean 'the least upper bound of B taken in the partially ordered set $(A, \leqslant)$'. Of course this would be absurdly cumbersome; but the abbreviation '$\sup B$' must always be recognized as such. In 1XB I give a simple example of the danger. There's a more elaborate one in 4XF, and the distinctions become very important in Chapter 7. The same warning, of course, applies to the notations $\mathscr{I}$ and $\mathscr{D}$ in 11E.

There is a perfect symmetry in the notion of partially ordered set; if $\leqslant$ is a partial ordering, so is its inverse $\geqslant$. Consequently all the associated definitions are doubled, as in 11B and 11C. I want to call attention to the concept of the filter of sections $\mathscr{F}(B\uparrow)$, defined in 11C. I think that this filter is almost the most important thing associated with a directed set. You may have seen it already in the correspondence between net-convergence and filter-convergence in topological spaces [KELLEY 2L]. Note the technical point that $\mathscr{F}(B\uparrow)$ is a filter on the whole space A, not on B.

For another look at the material of this section, see BOURBAKI I, chapter III, § 1.

12 Partially ordered linear spaces

In this short section I discuss the simplest way in which a linear space structure and a partial ordering can be related. Although we shall very rarely have any reason to consider partially ordered linear spaces which are not Riesz spaces [§ 14], I think that the results here are clarified by being placed in their natural context.

Here, and everywhere in this book, all linear spaces have the real numbers for their underlying field. There do exist applications in which it is more convenient to have the complex numbers; but I think that these are best approached by way of the real case. Other fields are so far merely curiosities from the standpoint of this theory.

12A Definition A **partially ordered linear space** is a quadruple $(E, +, ., \leqslant)$ where $(E, +, .)$ is a linear space over the field $\mathbf{R}$ of real numbers and $\leqslant$ is a partial ordering on E such that

(i) if $x \leqslant y$, then $x+z \leqslant y+z$ for every $z \in E$;

(ii) If $x \geqslant 0$ in E, then $\alpha x \geqslant 0$ whenever $\alpha \geqslant 0$ in $\mathbf{R}$.

12B Positive cones From 12A(i), we see that $x \leqslant y \Leftrightarrow 0 \leqslant y-x$. So $\leqslant$ is determined entirely by $E^+ = \{x\colon x \in E, x \geqslant 0\}$, the **positive cone** of E. Given a linear space E over $\mathbf{R}$ and a set $P \subseteq E$, there is a partial ordering $\leqslant$ on E such that $(E, \leqslant)$ is a partially ordered linear space and $P = E^+$ iff

$$P \cap (-P) = \{0\} \quad \text{[for 11A(i) and (ii)]},$$

$$P+P \subseteq P \quad \text{[for 11A(iii)]},$$

$$\alpha P \subseteq P \quad \forall \ \alpha \geqslant 0 \quad \text{[for 12A(ii)]}.$$

In particular, $P = \{0\}$ will do [cf. 1XA].

12C Lemma Let E be a partially ordered linear space, $x \in E$, A, $B \subseteq E$. Then

(a) $\sup(x+A) = x+\sup A$ if either side exists.

(b) $\sup(-A) = -\inf A$ if either side exists.

(c) $\sup(A+B) = \sup A + \sup B$ if the right-hand side exists.

(d) If $\alpha \geqslant 0$, $\sup(\alpha A) = \alpha \sup A$ if the right-hand side exists.

Proof of (c) Apply (*a*) twice, as follows:

$$\begin{aligned}\sup A + \sup B &= \sup(A+\sup B)\\ &= \sup\{x+\sup B : x \in A\}\\ &= \sup\{\sup\{x+y : y \in B\} : x \in A\}\\ &= \sup\{x+y : x \in A, y \in B\}\\ &= \sup(A+B).\end{aligned}$$

Notes and comments The correspondence between an ordering and a positive cone [12B] is extremely important; it is one of the easiest ways of defining partial orderings on linear spaces.

In 12C I prove only (c), because the other parts are direct consequences of the definitions. In a partially ordered linear space E, the map $y \mapsto x+y : E \to E$ is an order-automorphism for every $x \in E$, and therefore must preserve suprema and infima. The same applies to the map $y \mapsto \alpha y : E \to E$ for every $\alpha > 0$. On the other hand, the map $y \mapsto -y : E \to E$ is order-reversing, that is, $x \leqslant y \Leftrightarrow -y \leqslant -x$, so it exchanges suprema for infima.

Although I shall have no space to discuss them, many of the ideas of this book, set out for Riesz spaces, have extensions to partially ordered linear spaces. Compatible topologies [§ 21] are an obvious example; Lebesgue topologies [§ 24] are another. Some of these developments may be found in PERESSINI.

13 Lattices

Once again we must have a section consisting mostly of definitions. I maintain a careful separation between general remarks on lattices and those concerning linear spaces because I wish to apply the former to Boolean lattices in § 41.

Although it is possible to regard a lattice as a set with two binary algebraic operations, governed by certain identities, I prefer to take it as a special kind of partially ordered set. This makes it easier to think of the suprema and infima of infinite sets, which is something we must do continually.

The important sections below are A–E, though G should be examined because it attaches a rather unusual meaning to the word 'disjoint'.

13A Definitions A **lattice** is a partially ordered set $(A, \leqslant)$ such that $\sup\{a, b\}$ and $\inf\{a, b\}$ exist for all elements a and b of A. It follows at once (by induction on the number of elements in B) that $\sup B$ and $\inf B$ exist for every non-empty finite set $B \subseteq A$.

In a lattice A, we write $a \vee b$ and $a \wedge b$ for $\sup\{a, b\}$ and $\inf\{a, b\}$ respectively. Thus $\vee$ and $\wedge$ may be thought of as binary operations on A, the **lattice operations.** Since

$$a \leqslant b \Leftrightarrow a \vee b = b \Leftrightarrow a \wedge b = a,$$

the lattice can be defined in terms of these operations. But apart from a few special instances [e.g. 22F below], I think it is easier to regard the partial ordering as fundamental. From this point of view, for instance, the identity.

$$(a \vee b) \vee c = a \vee (b \vee c)$$

is trivial, as both sides are equal to $\sup\{a, b, c\}$.

13B Dedekind completeness (a) A lattice A is **Dedekind complete,** or **order-complete,** if every non-empty $B \subseteq A$ which has an upper bound in A has a least upper bound.

(b) A lattice A is **Dedekind σ-complete,** or **sequentially Dedekind complete,** if every non-empty countable $B \subseteq A$ which has an upper bound or a lower bound has a least upper bound or greatest lower bound respectively.

13C Notes These definitions are extremely important. Observe:

(a) If A is a lattice, then A is Dedekind complete iff every non-empty subset of A which has a lower bound has a greatest lower bound. **P** Suppose that A is Dedekind complete and that $B \subseteq A$ has a lower bound and is not empty. Let

$$C = \{a : a \in A, \ a \text{ is a lower bound for } B\}.$$

Then C is a non-empty subset of A with an upper bound, so $\sup C$ exists. But $\sup C = \inf B$. The reverse implication is proved in exactly the same way. **Q**.

Consequently a Dedekind complete lattice is indeed Dedekind σ-complete. The definition of Dedekind σ-completeness must be given in terms of both suprema and infima because there is no way to get a countable C from a countable B.

(b) Let A be any lattice, and $B \subseteq A$. Let

$$B_1 = \{\sup C : C \subseteq B, \ C \text{ is finite and not empty}\}.$$

Then B_1 is closed under the lattice operation $\vee$ and has the same upper bounds as B. So 13Ba could be rewritten as: 'a lattice A is Dedekind complete if every non-empty set $B \subseteq A$, which is directed upwards and has an upper bound, has a least upper bound'. This will often be easier to prove [e.g. 16Db].

13D Distributive lattices A lattice A is **distributive** if

$$(a \vee b) \wedge c = (a \wedge c) \vee (b \wedge c), \quad (a \wedge b) \vee c = (a \vee c) \wedge (b \vee c)$$

for all $a, b, c \in A$; that is, if each of $\vee$ and $\wedge$ is distributive over the other.

All the lattices considered in this book are distributive, and I shall make no attempt to discuss any other kind. BIRKHOFF is still an excellent introduction to general lattice theory.

Exercise Show that each of the two halves of the condition above implies the other, so that either would be sufficient for the definition.

13E Lattice homomorphisms If A and B are lattices, a function $f: A \to B$ is a **lattice homomorphism** if

$$f(a \vee b) = fa \vee fb, \quad f(a \wedge b) = fa \wedge fb$$

for all $a, b \in A$. A lattice homomorphism must be increasing. Observe that an order-continuous lattice homomorphism [11D] is precisely a function which preserves all suprema and infima of non-empty sets; to see this we have to use the trick of 13Cb to transform an arbitrary supremum or infimum into the supremum or infimum of a directed set, so as to apply the hypothesis of order-continuity.

13F Sublattices Let A be a lattice. A **sublattice** of A is a subset of A closed under $\vee$ and $\wedge$. *A **σ-sublattice** is a sublattice B of A such that if $C \subseteq B$ is countable and not empty, and $\inf C$ or $\sup C$ exists in A, then it belongs to B.

13G Disjoint sets and functions Let A be a lattice with a least member a_0. I shall call a set $B \subseteq A$ **disjoint** (in A) if $b \wedge c = a_0$ whenever b and c are distinct members of B. (Observe that possibly $a_0 \in B$.) If I is any set, a function $f: I \to A$ is **disjoint** if $f\iota \wedge f\kappa = a_0$ whenever ι and κ are distinct members of I.

*I shall often use the following phrase: 'C is disjoint in E^+', where E is a Riesz space and E^+ is its positive cone. Now E^+ is a sublattice of the lattice E, and has a least member, viz. 0; so what I mean is that $x \wedge y = 0$ whenever x and y are distinct members of C.

***13H Products** Let $\langle (A_\iota, \leqslant_\iota) \rangle_{\iota \in I}$ be a family of partially ordered sets, and suppose that none of the A_ι is empty. Let $\leqslant$ be the product partial ordering defined on $A = \prod_{\iota \in I} A_\iota$ [11G]. Then

(*a*) $(A, \leqslant)$ is a lattice iff every $(A_\iota, \leqslant_\iota)$ is.

(*b*) $(A, \leqslant)$ is distributive, or Dedekind complete, or Dedekind σ-complete, iff every $(A_\iota, \leqslant_\iota)$ is.

(*c*) The canonical maps from A to each A_ι are always lattice homomorphisms.

Notes and comments As I have rigorously expunged from this section everything which will not be used later, it is not an adequate introduction to lattice theory. The only two kinds of lattice which we shall be thinking about, Riesz spaces and Boolean lattices, share many special properties; to begin with, they are distributive, which makes them rather uninteresting as lattices. The distributive property is so natural that one is liable to use it unconsciously; which is dangerous if there are any nondistributive lattices about. But I think that for now we can accept these simplifications cheerfully.

Since most of the objects of discussion of this book are lattices, I shall not trouble with many examples at this point. The obvious ones to glance at, before continuing to the next section, are 1XB, 1XC and 4XA.

14 Riesz spaces

As soon as we join §§ 12 and 13 together, and consider partially ordered linear spaces which are lattices, the character of our study changes completely. Riesz spaces have a remarkable wealth of properties.

Some of the most elementary, identities relating the lattice and linear space operations, are in 14B. These are so numerous that one naturally seeks an alternative to memorization; and such an alternative exists in the '**R**-test', which runs 'if an identity is true in **R**, it is true in all Riesz spaces'. Actually, there is a perfectly valid metatheorem along these lines. But the metatheorem is a good deal more complex then the work it eliminates; besides, direct proofs are more illuminating of associated structures such as lattice-ordered groups. So I give the identities most commonly required in an order which makes each an easy consequence of the preceding ones.

The really surprising thing about Riesz spaces is the strong distributive law which they obey [14D]. After this, the rest of the section is fairly straightforward, with the expected relationships between products, subspaces, quotients and homomorphisms.

14A Definitions A **Riesz space,** or **vector lattice,** is a partially ordered linear space $(\boldsymbol{E}, +, ., \leqslant)$ such that $(\boldsymbol{E}, \leqslant)$ is a lattice.

If $\boldsymbol{E}$ is a Riesz space, we write

$$x^+ = x \vee 0, \quad x^- = (-x) \vee 0, \quad |x| = x \vee (-x)$$

for any $x \in \boldsymbol{E}$.

Exercise Let $\boldsymbol{E}$ be a partially ordered linear space such that $\sup\{x, 0\}$ exists for every $x \in \boldsymbol{E}$. Then $\boldsymbol{E}$ is a Riesz space. [Use 12Ca and 12Cb.]

14B Elementary identities Let E be a Riesz space, x, y and z members of E, and α and β real numbers. Then

(**a**) $(x \vee y)+z = (x+z) \vee (y+z)$ [special case of 12Ca].

(**b**) $(x \wedge y)+z = (x+z) \wedge (y+z)$ [similarly].

(**c**) $-(x \wedge y) = (-x) \vee (-y)$ [special case of 12Cb].

(**d**) If $\alpha \geqslant 0$, $\alpha x \vee \alpha y = \alpha(x \vee y)$ [special case of 12Cd].

(**e**) If $\alpha \geqslant 0$, $\alpha x \wedge \alpha y = \alpha(x \wedge y)$ [similarly].

(**f**) $2(x \vee y) = 2x \vee 2y = x+y+|x-y|$ [definition of $|\ |$ and (a) above].

(**g**) $2(x \wedge y) = x+y-|x-y|$ [similarly, using (b) and (c)].

(**h**) $x+y = x \vee y + x \wedge y$ [adding (f) and (g)].

(**i**) $x = x^+ - x^-$ [apply (a) to $x+x^-$].

(**j**) $|x| = x^+ + x^-$ [apply (d) and (a) to $x+2x^-$].

(**k**) $x^+ \vee x^- = |x|$ [from the definitions, since $|x| \geqslant 0$ by (j)].

(**l**) $x^+ \wedge x^- = 0$ [by (j), (k) and (h)].

(**m**) $|\alpha x| = |\alpha|\,|x|$ [consider $\alpha \geqslant 0$ and $\alpha \leqslant 0$ separately].

(**n**) $|x+y| \leqslant |x|+|y|$ [since $x+y \leqslant x+|y| \leqslant |x|+|y|$; and similarly $-x-y \leqslant |x|+|y|$].

14C Definitions Let E be a Riesz space. A set $A \subseteq E$ is **solid** if $y \in A$ whenever there is an $x \in A$ such that $|y| \leqslant |x|$. The intersection of any set of solid sets is solid.

If A is any subset of E, the set $\{y : \exists\, x \in A, |y| \leqslant |x|\}$ is solid; it is the smallest solid set including A, and is called the **solid hull** of A.

14D The distributive law Let E be a Riesz space and A a subset of E such that $\sup A$ exists. Then, for any $x \in E$, $\sup\{x \wedge y : y \in A\}$ exists and is equal to $x \wedge \sup A$.

Proof Plainly $x \wedge \sup A$ is an upper bound for $\{x \wedge y : y \in A\}$. Conversely, suppose that $z \geqslant x \wedge y$ for every $y \in A$. Then

$$z \geqslant (x-y) \wedge 0 + y$$

$$\geqslant (x - \sup A) \wedge 0 + y, \quad \text{for every } y \in A.$$

So $$z \geqslant (x - \sup A) \wedge 0 + \sup A = x \wedge \sup A.$$

Similarly $x \vee \inf A = \inf\{x \vee y : y \in A\}$ if the left-hand side exists.

Corollary E is a distributive lattice.

14E Linear maps We shall be concerned with several classes of operators between Riesz spaces; the important ones are:

(**a**) Increasing linear maps. A linear map T is increasing iff $Tx \geqslant 0$ whenever $x \geqslant 0$. (For this reason, increasing linear maps are often called **positive**.)

(**b**) **Riesz homomorphisms.** A Riesz homomorphism is a linear map which is also a lattice homomorphism. A linear map T is a Riesz homomorphism

$$\text{iff} \quad (Tx)^+ = T(x^+) \quad \text{for every } x;$$

$$\text{iff} \quad |Tx| = T|x| \quad \text{for every } x;$$

$$\text{iff} \quad Tx \wedge Ty = 0 \quad \text{whenever } x \wedge y = 0.$$

(Use 14Bf, g and j. For the last condition, observe that in this case

$$\begin{aligned}(Tx)^+ &= (T(x^+) - T(x^-)) \vee 0 \\ &= T(x^+) \vee T(x^-) - T(x^-) \\ &= T(x^+) + T(x^-) - T(x^+) \wedge T(x^-) - T(x^-) \\ &= T(x^+),\end{aligned}$$

using 14Bi, a, h and l.)

(**c**) Order-continuous increasing linear maps. A linear map T is increasing and order-continuous

$$\text{iff} \quad A \downarrow 0 \quad \text{and} \quad A \neq \varnothing \;\Rightarrow\; T[A] \downarrow 0;$$

$$\text{iff} \quad \varnothing \subset A \subseteq E^+ \quad \text{and} \quad A \uparrow x \;\Rightarrow\; T[A] \uparrow Tx.$$

[Use 12Ca and 12Cb. Note that in the second condition we consider only sets A which are bounded below by 0. These are enough, for any set which is directed upwards can be truncated by one of its own members; I mean that if $A \uparrow$ and $A \neq \varnothing$, we can consider

$$A' = \{x : x \in A,\, x \geqslant y\}$$

for some $y \in A$; and now A' is bounded below and has the same upper bounds as A, so the action of T on A' will tell us all we need to know. And by a simple translation we can move y to 0.]

*(**d**) Sequentially order-continuous increasing linear maps.

(e) Order-continuous (or sequentially order-continuous) Riesz homomorphisms.

(f) Maps expressible as the difference of two increasing linear maps [see 16*B*].

(g) Maps expressible as the difference of two order-continuous increasing linear maps [see 16G].

Much of the richness of the more advanced theory of Riesz spaces derives from this multiplicity; each kind of operator is associated with a different class of results. The following chart shows the relationships between these seven classes:

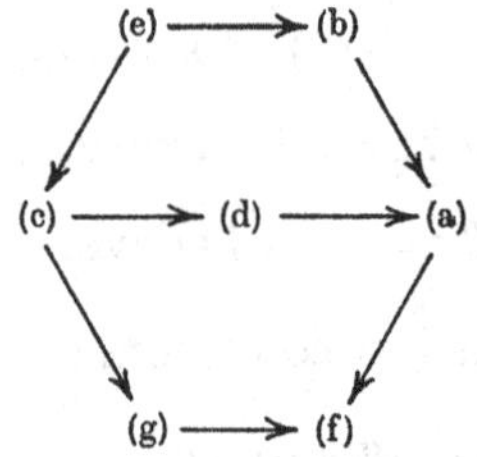

where the arrow $(e) \rightarrow (b)$ signifies that every operator of type (e) is also of type (b).

14F Subspaces Let $\boldsymbol{E}$ be a Riesz space. A **Riesz subspace** of $\boldsymbol{E}$ is a linear subspace which is also a sublattice [13F]. A solid linear subspace is always a Riesz subspace.

A Riesz subspace $\boldsymbol{F}$ of $\boldsymbol{E}$ is **order-dense** if, for every $x \in \boldsymbol{E}^+$,

$$x = \sup\{y : 0 \leqslant y \leqslant x,\ y \in \boldsymbol{F}\}.$$

$\boldsymbol{F}$ is **super-order-dense** if for every $x \in \boldsymbol{E}^+$ there is a sequence $\langle y_n \rangle_{n \in \mathbf{N}}$ in $\boldsymbol{F}^+$ such that $\langle y_n \rangle_{n \in \mathbf{N}} \uparrow x$ in $\boldsymbol{E}$. $\boldsymbol{F}$ is **locally order-dense** if it is order-dense in its solid hull (which must be a solid linear subspace of $\boldsymbol{E}$); i.e. if

$$0 \leqslant x \leqslant x_0 \in \boldsymbol{F} \Rightarrow x = \sup\{y : y \in \boldsymbol{F},\ 0 \leqslant y \leqslant x\}.$$

A solid linear subspace is always locally order-dense.

A **band** or **normal subspace** of $\boldsymbol{E}$ is a solid linear subspace $\boldsymbol{F}$ such that whenever $A \subseteq \boldsymbol{F}$ and $\sup A$ exists in $\boldsymbol{E}$, then $\sup A \in \boldsymbol{F}$; i.e. an order-closed solid linear subspace [11E].

The intersection of any set of Riesz subspaces is a Riesz subspace; the intersection of any set of solid linear subspaces is a solid linear subspace; the intersection of any set of bands is a band.

The range of a Riesz homomorphism is a Riesz subspace; the kernel of a Riesz homomorphism is a solid linear subspace; the kernel of an order-continuous Riesz homomorphism is a band.

14G Quotient spaces: proposition Let E be a Riesz space and F a solid linear subspace of E. Then the linear space quotient E/F can be given a partial ordering such that it is a Riesz space and the canonical map $\phi: E \to E/F$ is a Riesz homomorphism.

Proof Let $P = \{\phi x : x \in E^+\} \subseteq E/F$.

(a) Given $x \in E$, $\phi x \in P \Leftrightarrow x^- \in F$. **P** If $x^- \in F$, then

$$\phi x = \phi(x^+) - \phi(x^-) = \phi x^+ \in P.$$

Conversely, if $\phi x \in P$, there is a $y \in E^+$ such that $\phi y = \phi x$, i.e. $y - x \in F$. Now $-x \leqslant y - x$, so

$$0 \leqslant x^- = (-x) \vee 0 \leqslant (y-x) \vee 0 = (y-x)^+ \leqslant |y-x|,$$

and $x^- \in F$ as F is solid. **Q**

(b) Now P satisfies the conditions of 12B. **P** If $\phi x \in P \cap (-P)$, then $x^- \in F$ by (a) above. But also $\phi(-x) \in P$, so $x^+ = (-x)^- \in F$, and $x = x^+ - x^- \in F$. Thus $P \cap (-P) = \{0\}$. But the other conditions are trivially satisfied. **Q**

(c) So E/F can be given a partial ordering under which it is a partially ordered linear space and $P = (E/F)^+$. Now if $x \in E$, $\sup\{\phi x, 0\}$ exists in E/F and is equal to $\phi(x^+)$. **P** Of course $\phi x^+ = \phi x + \phi x^- \geqslant \phi x$, and $\phi x^+ \geqslant 0$. On the other hand, if $\phi y \geqslant \phi x$ and $\phi y \geqslant 0$, then $y^- \in F$ and $(y-x)^- \in F$, by (*a*) above. So

$$\begin{aligned}(y - x^+)^- &= (x \vee 0 - y) \vee 0 = (x-y) \vee (-y) \vee 0 \\ &= (y-x)^- \vee y^- \in F,\end{aligned}$$

as a solid linear subspace is always a sublattice. So $\phi y \geqslant \phi x^+$. As y is arbitrary, $\phi(x^+) = \sup\{\phi x, 0\}$. **Q**

(d) By the exercise in 14A, E/F is a Riesz space, and ϕ is a Riesz homomorphism by 14Eb.

Remarks Thus not only is the kernel of every Riesz homomorphism a solid linear subspace, but every solid linear subspace is the kernel of some Riesz homomorphism. For this reason some authors call solid linear subspaces 'ideals'.

Although it will not be used in this book, the result of 14Lb is very important in understanding the nature and significance of bands.

***14H Products** Let $\langle E_\iota \rangle_{\iota \in I}$ be an indexed family of Riesz spaces. On $\prod_{\iota \in I} E_\iota$ there are natural linear space and lattice structures [13H] which render it a Riesz space. The canonical projections to each E_ι are all order-continuous Riesz homomorphisms.

***14I Order units** Two concepts we shall want to use later are the following. Let E be a Riesz space. An **order unit** of E (sometimes called a **strong order unit**) is an $e \in E^+$ such that for every $x \in E$ there is an $n \in \mathbf{N}$ such that $x \leqslant ne$. A **weak order unit** of E is an $e \in E^+$ such that

$$x = \sup_{n \in \mathbf{N}} x \wedge ne \quad \forall\ x \in E.$$

***14J** At various points of the book we shall want to use the following results.

Lemma Let E be a Riesz space.

(**a**) If x, y and z belong to E^+,

$$z \wedge (x+y) \leqslant z \wedge x + z \wedge y.$$

(**b**) If $\langle x_i \rangle_{i<n}$ is a finite sequence in E^+ and $|y| \leqslant \sum_{i<n} x_i$, then there is a finite sequence $\langle y_i \rangle_{i<n}$ in E such that $y = \sum_{i<n} y_i$ and $|y_i| \leqslant x_i$ for every $i < n$.

(**c**) If $\langle x_i \rangle_{i<n}$ and $\langle y_j \rangle_{j<m}$ are finite sequences in E^+ such that $\sum_{i<n} x_i = \sum_{j<m} y_j$, then there is a double sequence $\langle z_{ij} \rangle_{i<n, j<m}$ in E^+ such that $x_i = \sum_{k<m} z_{ik}$ and $y_j = \sum_{k<n} z_{kj}$ for every $i < n$ and $j < m$.

(**d**) The convex hull of a solid set is solid.

Proofs (a) Set $w = z \wedge (x+y) - z \wedge x$. Then by 14Bh

$$\begin{aligned} w &= z + x + y - z \vee (x+y) - z - x + z \vee x \\ &= y + z \vee x - z \vee (x+y) \leqslant y. \end{aligned}$$

But also $w \leqslant z$, so $w \leqslant y \wedge z$, as required.

(**b**) (**c**) We prove these by induction on m and n. Let us begin with (c), with $m = 2$, so that $y_0 + y_1 = \sum_{i<n} x_i$. If $n = 1$, set $z_{00} = y_0$ and $z_{01} = y_1$. For the inductive step $n \to n+1$, when we have

$$y_0 + y_1 = \sum_{i \leqslant n} x_i,$$

set $z_{n0} = x_n \wedge y_0$, $z_{n1} = x_n - z_{n0}$. Now

$$x_n = x_n \wedge (y_0 + y_1) \leqslant x_n \wedge y_0 + x_n \wedge y_1$$

by (a) above, so $z_{n1} \leqslant x_n \wedge y_1 \leqslant y_1$. So if we set $w_0 = y_0 - z_{n0}$ and $w_1 = y_1 - z_{n1}$, we shall have $w_0 + w_1 = \sum_{i<n} x_i$, and by the inductive

hypothesis we can choose the z_{i0} and z_{i1}, for $i < n$, such that $z_{i0}+z_{i1} = x_i$ for every $i < n$ and $\sum_{i<n} z_{i0} = w_0 = y_0 - z_{n0}$, $\sum_{i<n} z_{i1} = w_1 = y_1 - z_{n1}$. Thus the induction continues.

The proof of (c) now concludes with a simple induction on m.

As for (b), we apply (c) to the sum

$$\sum_{i<n} x_i = y^+ + y^- + (\sum_{i<n} x_i - |y|).$$

If $\langle z_{ij}\rangle_{i<n, j<3}$ is a double sequence in E^+ such that

$$z_{i0}+z_{i1}+z_{i2} = x_i \quad \forall\ i < n,$$
$$\sum_{i<n} z_{i0} = y^+,$$
$$\sum_{i<n} z_{i1} = y^-,$$

set $y_i = z_{i0} - z_{i1}$ for each $i < n$. Then $\sum_{i<n} y_i = y$ and

$$|y_i| \leqslant z_{i0}+z_{i1} \leqslant x_i \quad \forall\ i < n.$$

(d) Let $A \subseteq E$ be a solid set. Suppose that x belongs to the convex hull of A and that $|y| \leqslant |x|$. Then x can be expressed as $\sum_{i<n} \alpha_i x_i$ where $x_i \in A$ and $\alpha_i \in \mathbf{R}^+$ for each $i < n$, and $\sum_{i<n} \alpha_i = 1$. So

$$|x| \leqslant \sum_{i<n} |\alpha_i x_i| = \sum_{i<n} \alpha_i |x_i|$$

by 14Bm and n. Now by (b) above there is a sequence $\langle y_i\rangle_{i<n}$ such that $y = \sum_{i<n} y_i$ and $|y_i| \leqslant \alpha_i |x_i|$ for every $i < n$. Set $z_i = \alpha_i^{-1} y_i$ if $\alpha_i \neq 0$, $z_i = 0$ if $\alpha_i = 0$. Then $|z_i| \leqslant |x_i|$ so $z_i \in A$ for each $i < n$. But $y = \sum_{i<n} \alpha_i z_i$ is now in the convex hull of A, as required.

***14K Lemma** Let E be a Riesz space.

(a) If $|x| \wedge |y| = 0$ in E, then $|x+y| = |x| + |y|$.

(b) If $x \wedge y = 0$ in E, then $\alpha x \wedge \beta y = 0$ for every $\alpha, \beta \in \mathbf{R}^+$.

(c) If $\langle x_i\rangle_{i<n}$ is a sequence in E such that $|x_i| \wedge |x_j| = 0$ whenever $i \neq j$, and supposing that $n \neq 0$, then

$$|\sum_{i<n} \alpha_i x_i| = \sum_{i<n} |\alpha_i|\,|x_i| = \sup\{|\alpha_i|\,|x_i| : i < n\}$$

for any sequence $\langle \alpha_i\rangle_{i<n}$ in $\mathbf{R}$.

Proof (a) We know that

$$x^+ \wedge x^- = x^+ \wedge y^- = y^+ \wedge x^- = y^+ \wedge y^- = 0.$$

So by 14Ja $u \wedge v = 0$, where $u = x^+ + y^+$ and $v = x^- + y^-$. Now

$$(x+y)^+ = (u-v)^+ = (u-v) \vee 0 = u \vee v - v$$
$$= u + v - u \wedge v - v = u.$$

Similarly $(x+y)^- = v$. So $|x+y| = u+v = |x|+|y|$.

(**b**) $\alpha x \wedge \beta y \leqslant (\alpha+\beta)x \wedge (\alpha+\beta)y = (\alpha+\beta)(x \wedge y) = 0$.

(**c**) For $n = 2$, $|\alpha_0 x_0| \wedge |\alpha_1 x_1| = |\alpha_0|\,|x_0| \wedge |\alpha_1|\,|x_1| = 0$ by (b). So $|\alpha_0 x_0 + \alpha_1 x_1| = |\alpha_0 x_0| + |\alpha_1 x_1|$ by (a), and also

$$|\alpha_0 x_0| + |\alpha_1 x_1| = |\alpha_0 x_0| \vee |\alpha_1 x_1| + |\alpha_0 x_0| \wedge |\alpha_1 x_1|$$
$$= |\alpha_0 x_0| \vee |\alpha_1 x_1|.$$

The inductive step to general n is now easy, using 14Ja to see that

$$|\alpha_n x_n| \wedge \textstyle\sum_{i<n} |\alpha_i x_i| = 0.$$

14L Exercises (**a**) Let $\boldsymbol{E}$ be a Riesz space and $\boldsymbol{F}$ a linear subspace. Then these are equivalent: (i) $x^+ \in \boldsymbol{F}$ for every $x \in \boldsymbol{F}$ (ii) $|x| \in \boldsymbol{F}$ for every $x \in \boldsymbol{F}$ (iii) $\boldsymbol{F}$ is a Riesz subspace. [Use 14Bf, g and j.]

*(**b**) Let $\boldsymbol{E}$ be a Riesz space and $\boldsymbol{F}$ a solid linear subspace. Let $\phi: \boldsymbol{E} \to \boldsymbol{E}/\boldsymbol{F}$ be the canonical Riesz homomorphism [14G]. Then ϕ is order-continuous iff $\boldsymbol{F}$ is a band, and ϕ is sequentially order-continuous [definition: 11D] iff $\boldsymbol{F}$ is a σ-sublattice [definition: 13F].

*(**c**) If $\boldsymbol{E}$ is any Riesz space and $x \in \boldsymbol{E}^+$, the band in $\boldsymbol{E}$ generated by x is

$$\{y : y \in \boldsymbol{E},\ |y| = \sup_{n\in\mathbf{N}} |y| \wedge nx\}.$$

Notes and comments A result due to AMEMIYA G.S.T. [theorem 10.1] states that every Riesz space is isomorphic to a quotient of a subspace of a product of copies of **R**. This is the basis of the result that any identity involving only addition, subtraction, scalar multiplication, and the lattice operations, which is true in **R**, must be true in every Riesz space. Amemiya's representation actually depends on the axiom of choice; but the result can be proved without it. BAKER gives a slightly stronger metatheorem of the same kind.

The actual examples of Riesz spaces we deal with will often be either Riesz subspaces of products of copies of **R** (as in § A2) or quotients of such subspaces (like those in § 62). But we need the general theory to cope with spaces generated in other ways, as in § 16. For the moment I think it is enough to consider $\mathbf{R}^X$ itself [1XD], some of the familiar subspaces of $\mathbf{R}^{\mathbf{N}}$ [2XA–2XC] and of $C(X)$ [§ A2], and, for variety, 1XF.

The distributive law [14D] is enormously important, and I shall tend to employ it tacitly. BIRKHOFF gives a result [chapter XIII § 4 theorem 4] which shows something of what lies beneath it. All the concepts introduced in 14E–14I will be needed sometime; but the only one of much interest at this stage is the quotient space construction [14G]. Although such a construction is an obvious thing to look for, it is far from clear that it will appear so simply.

15 Archimedean Riesz spaces

All the most familiar Riesz spaces, and all those with which we shall be concerned in this book, possess the 'Archimedean' property. A Riesz space is Archimedean if, as in the real numbers, $\{nx : n \in \mathbf{N}\}$ is unbounded above for any strictly positive x. This property is so common that one's intuitive idea of Riesz spaces is liable inadvertently to include it. In our present inquiry such an unconscious assumption will do little harm, as all spaces considered are Archimedean. However, for the benefit of readers who may wish later to examine non-Archimedean spaces, I have segregated here the elementary special consequences of the Archimedean property; and elsewhere in the book I try to include the hypothesis 'Archimedean' only when necessary.

The short paragraphs up to 15E are technical, but extremely useful. 15F gives a characterization of bands in Archimedean Riesz spaces, which is more important for a general understanding than for the special needs of this book. 15G sorts out one-dimensional Riesz spaces, and will be needed only in Chapter 8.

15A Definition A Riesz space E is **Archimedean** if, for x and y belonging to E,

$$x \geqslant ny \quad \forall \; n \in \mathbf{N} \Rightarrow y \leqslant 0.$$

Clearly, a Riesz subspace of an Archimedean Riesz space is Archimedean.

15B Proposition A Dedekind σ-complete Riesz space is Archimedean.

Proof If $x \geqslant ny$ for every $n \in \mathbf{N}$, set $z = \sup\{ny : n \in \mathbf{N}\}$. Then

$$y + z = \sup\{(n+1)y : n \in \mathbf{N}\} \leqslant z$$

[using 12Ca], so $y \leqslant 0$.

15C Lemma Let E be an Archimedean Riesz space. Let A be a non-empty subset of E and suppose that the set B of upper bounds of A is also non-empty. Then $\inf(B - A) = 0$.

Proof z is a lower bound for $B-A$ iff $z \leqslant y-x$ for every $y \in B$ and $x \in A$, i.e. iff $y-z \in B$ for every $y \in B$. Consequently, if $y_0 \in B$ and $x_0 \in A$, $y_0 - nz \geqslant x_0$ for every $n \in \mathbf{N}$. Hence $z \leqslant 0$.

15D Lemma Let $\boldsymbol{E}$ be an Archimedean Riesz space. Then for any $x \in \boldsymbol{E}^+$ the map $\alpha \mapsto \alpha x$: $\mathbf{R} \to \boldsymbol{E}$ is order-continuous, so that if $A \uparrow \alpha$ in $\mathbf{R}$, $\{\beta x : \beta \in A\} \uparrow \alpha x$ in $\boldsymbol{E}$.

Proof The point is that if $y \geqslant \beta x$ for every $\beta \in A$,

$$(\alpha - n^{-1})x \leqslant y \quad \forall \; n \geqslant 1,$$

i.e. $n(\alpha x - y) \leqslant x$ for every $n \geqslant 1$; so $\alpha x \leqslant y$.

15E Lemma Let $\boldsymbol{E}$ be an Archimedean Riesz space and $\boldsymbol{F}$ a Riesz subspace of $\boldsymbol{E}$ such that for every non-zero $x \in \boldsymbol{E}^+$ there is some $y \in \boldsymbol{F}$ such that $0 < y \leqslant x$. Then $\boldsymbol{F}$ is order-dense.

Proof Let $x \in \boldsymbol{E}^+$, and let $A = \{y : y \in \boldsymbol{F},\ 0 \leqslant y \leqslant x\}$. **?** Suppose, if possible, that x is not the least upper bound of A, that is, that there is an upper bound x_1 of A such that $x_1 \ngeqslant x$. Then $0 < x - x \wedge x_1$, so there is a $y_0 \in \boldsymbol{F}$ such that $0 < y_0 \leqslant x - x \wedge x_1$. But now

$$y + y_0 \leqslant x \wedge x_1 + y_0 \leqslant x$$

for every $y \in A$. Inducing on n, we see that $ny_0 \in A$ for every $n \in \mathbf{N}$; which is impossible. **X**

***15F Bands in Archimedean Riesz spaces** Let $\boldsymbol{E}$ be a Riesz space, and $A \subseteq \boldsymbol{E}$. Let us write

$$A^d = \{x : x \in \boldsymbol{E},\ |x| \wedge |y| = 0 \quad \forall \; y \in A\}.$$

Then (a) A^d is a band in $\boldsymbol{E}$ (b) if $\boldsymbol{E}$ is Archimedean, and $\boldsymbol{F}$ is a band in $\boldsymbol{E}$, then $\boldsymbol{F} = (\boldsymbol{F}^d)^d$.

Proof **(a)** By 14Bn and 14Ja, A^d is closed under addition; by 14Bm and 14Kb it is a linear subspace; clearly it is solid; now by 14D it is a band.

(b) Of course $\boldsymbol{F}^{dd} \supseteq \boldsymbol{F}$. On the other hand, suppose that $x \in \boldsymbol{E} \backslash \boldsymbol{F}$. As $\boldsymbol{F}$ is solid, $|x| \notin \boldsymbol{F}$; so $|x|$ cannot be the least upper bound of

$$B = \{y : y \in \boldsymbol{F},\ 0 \leqslant y \leqslant |x|\},$$

that is, there is an upper bound x_1 of B in $\boldsymbol{E}$ such that $x_1 \not\geq |x|$. Let $z = |x| - |x| \wedge x_1 > 0$. Now for any $y \in B$, $y + z \leq |x| \wedge x_1 + z = |x|$. But for any $y \in \boldsymbol{F}$ and $n \in \mathbf{N}$

$$(n+1)(|y| \wedge z) \leq n(|y| \wedge z) + z,$$

and $(n+1)(|y| \wedge z) \in \boldsymbol{F}$. So, inducing on n, $n(|y| \wedge z) \in B$ for every $n \in \mathbf{N}$; that is, $n(|y| \wedge z) \leq |x|$ for every $n \in \mathbf{N}$. Because $\boldsymbol{E}$ is Archimedean, $|y| \wedge z$ must be 0. But y is arbitrary, so $z \in \boldsymbol{F}^d$; and

$$|x| \wedge z = z > 0,$$

so $x \notin \boldsymbol{F}^{dd}$. This shows that $\boldsymbol{F}^{dd} \subseteq \boldsymbol{F}$; so $\boldsymbol{F}^{dd} = \boldsymbol{F}$, as required.

***15G Lemma** Let $\boldsymbol{E}$ be a Riesz space. Then the following are equivalent:

(i) $\boldsymbol{E} = \{0\}$ or $\boldsymbol{E} \cong \mathbf{R}$;

(ii) $\boldsymbol{E}$ is Archimedean and totally ordered;

(iii) $\boldsymbol{E}$ is Archimedean and, whenever $x, y \in \boldsymbol{E}^+ \backslash \{0\}$, $x \wedge y > 0$;

(iv) $\boldsymbol{E}$ has no non-trivial bands;

(v) $\boldsymbol{E}$ has no non-trivial solid linear subspaces.

Proof (a) (i) $\Rightarrow$ (v) $\Rightarrow$ (iv) are elementary.

(b) (iv) $\Rightarrow$ (iii) Suppose that $\boldsymbol{E}$ satisfies (iv).

(α) Let $x > 0$ in $\boldsymbol{E}$. Then it is easy to see that

$$\{y : |y| \leq \alpha x \quad \forall\ \alpha > 0\}$$

is a band in $\boldsymbol{E}$, not containing x; so it is $\{0\}$. It follows that $\boldsymbol{E}$ is Archimedean.

(β) If $x, y \in \boldsymbol{E}$ and $x \wedge y = 0$, then

$$\boldsymbol{F} = \{z : |z| \wedge x = 0\}$$

is a band in $\boldsymbol{E}$ [15Fa]. If $\boldsymbol{F} = \{0\}$, then $y = 0$; if $\boldsymbol{F} = \boldsymbol{E}$, then $x = 0$.

(c) (iii) $\Rightarrow$ (ii) Suppose that $\boldsymbol{E}$ satisfies (iii). If x, $y \in \boldsymbol{E}$, then $(x-y)^+ \wedge (y-x)^+ = 0$ [14Bl], so either $(x-y)^+ = 0$ or $(y-x)^+ = 0$ i.e. either $x \leq y$ or $y \leq x$.

(d) (ii) $\Rightarrow$ (i) Suppose that $\boldsymbol{E}$ satisfies (ii), and is not $\{0\}$. Let $x_0 \in \boldsymbol{E}^+ \backslash \{0\}$. Given $y \in \boldsymbol{E}^+$, let $\alpha_0 = \sup\{\alpha : \alpha \in \mathbf{R}, \alpha x_0 \leq y\} < \infty$ as $\boldsymbol{E}$ is Archimedean. Then $\alpha_0 x_0 \leq y$ [15D]. Also, $\alpha > \alpha_0 \Rightarrow \alpha x_0 \not\leq y \Rightarrow \alpha x_0 > y$; so $\alpha_0 x_0 \geq y$ [15D again].

Thus $\boldsymbol{E} = \boldsymbol{E}^+ - \boldsymbol{E}^+$ [14Bi] $= \{\alpha x_0 : \alpha \in \mathbf{R}\}$, and the map $\alpha \mapsto \alpha x_0$ is the required isomorphism between $\mathbf{R}$ and $\boldsymbol{E}$.

***15H Exercises** (a) A product of Archimedean Riesz spaces is Archimedean.

(b) If E is a finite-dimensional Archimedean Riesz space, $E \cong \mathbf{R}^n$ [1XD] for some $n \in \mathbf{N}$.

Notes and comments As we shall not have any need to consider non-Archimedean spaces, I give only one example, the lexicographic ordering of $\mathbf{R}^2$ [1XF]. I think that what is important about 15C–15E is the fact that they fail for spaces of this type (indeed, with a little thought one can see that each of these results is a necessary and sufficient condition for the Archimedean property). This failure gives perhaps a salutary shock.

Because $\mathbf{R}^X$ is Archimedean [1XD], all its Riesz subspaces are Archimedean, that is, all true function spaces are Archimedean. The spaces of linear maps in the next section, which will loom large in this book, are plainly Archimedean because they are defined from Archimedean spaces. However, many of the Riesz spaces of measure theory are defined as quotient spaces, and quotient spaces do not in general have to be Archimedean (see 1XE, and recall the theorem of Amemiya quoted at the end of § 14, which can be used to show that every Riesz space can be represented as a quotient of an Archimedean Riesz space). I think that a reason why the important examples should be Archimedean is to be found in 22Eb; they are defined so as to be normed spaces, and only Archimedean spaces can carry norms of the right type.

It is the case that every Archimedean Riesz space is isomorphic to a Riesz subspace of a Dedekind complete Riesz space; by 15B, this is another necessary and sufficient condition. From this stems Ogasawara's representation of an arbitrary Archimedean Riesz space as a space of continuous functions on a suitable topological space, each function being permitted to take the values $\pm\infty$ on a nowhere dense closed set. I have no space to discuss this further; the results are to be found in VULIKH, theorems IV.11.1 and V.7.1, and LUXEMBURG & ZAANEN R.S., theorem 50.8.

16 Linear maps between Riesz spaces

In 14E we glanced briefly at a number of classes of operators between Riesz spaces. The natural ones were those of (a)–(e), for they preserve aspects of the order and linear structures of the spaces. Thus Riesz homomorphisms [14Eb], which preserve both the linear and the lattice structures, are associated with an effective notion of quotient space [14G]. However, the set of all Riesz homomorphisms between two Riesz spaces does not ordinarily have a simple structure of its own. (Consider, for example, the set of Riesz homomorphisms from $\mathbf{R}^2$ to $\mathbf{R}$, which is not closed under ordinary addition.) Similarly, the set of order-continuous increasing linear maps, though closed under addition [see 16H below], is not closed under ordinary multiplication by a negative scalar.

If therefore we seek a set of linear maps which will be even a linear space, we must go outside these classes. The set of increasing linear maps, for instance, is closed under addition and multiplication by non-negative scalars; consequently, the set of functions expressible as the difference of two increasing linear maps will be a linear space. Now it happens that if the range space is Dedekind complete, this space of linear maps is a Riesz space. Since much the most important range space is of course $\mathbf{R}$, this strong condition is acceptable. The first six paragraphs of this section are accordingly devoted to proving this result, and providing some tools for handling the Riesz space we have constructed; it turns out itself to be Dedekind complete.

This space which I shall call $L^{\sim}(E; F)$, is not actually the most exciting space of linear maps from E to F. More interesting is $L^{\times}(E; F)$, the space of differences of order-continuous increasing linear maps. This also is a Dedekind complete Riesz space, being in fact a band in $L^{\sim}(E; F)$ [16H]; and $L^{\sim}(E; F)$ will be important to us principally because of this relationship, which is a valuable source of results about $L^{\times}(E; F)$.

In the applications in this book, the range space F will usually, though not always, be the real numbers; but the proofs are made very little different by the generalization to arbitrary Dedekind complete Riesz spaces.

16A Lemma Let E and F be Riesz spaces, and $T: E^+ \to F$ a function. Then T extends to a linear map from E to F iff

$$T(x+y) = Tx+Ty \quad \forall\, x, y \in E^+,$$

$$T(\alpha x) = \alpha Tx \quad \forall\; x \in E^+, \alpha \in \mathbf{R}^+.$$

[For by 14Bi $E = E^+ - E^+$.]

16B Definition Let E and F be Riesz spaces. Let $L^{\sim}(E; F)$ be the set of all linear maps from E to F which can be expressed as the difference of two increasing linear maps.

Then $L^{\sim}(E; F)$ has a natural linear structure, being a linear subspace of the set of all linear maps from E to F, and in addition a partial ordering given by

$$S \leqslant T \quad \text{iff } Sx \leqslant Tx \quad \forall x \in E^+,$$

which renders it a partially ordered linear space; its positive cone being just the set of increasing linear maps.

I shall write $E^{\sim}$ for $L^{\sim}(E; \mathbf{R})$.

16C Theorem Let E be any Riesz space and F a Dedekind complete Riesz space. Then a linear map $T: E \to F$ is in $L^{\sim}(E; F)$ iff

$$\{Ty: 0 \leqslant y \leqslant x\}$$

is bounded above in F for every $x \in E^+$.

Proof (a) If $T = T_1 - T_2$, where T_1 and T_2 are increasing, then $\{Ty : 0 \leqslant y \leqslant x\}$ is bounded above by $T_1 x$ for every $x \in E^+$.

(b) Conversely, if the condition is satisfied, define $S: E^+ \to F$ by

$$Sx = \sup\{Ty : 0 \leqslant y \leqslant x\} \quad \forall\ x \in E^+,$$

which always exists because F is supposed Dekekind complete.

Clearly $S(\alpha x) = \alpha Sx$ whenever $x \in E^+$ and $\alpha \in \mathbf{R}^+$. Now if $x_1, x_2 \in E^+$,

$$\begin{aligned} Sx_1 + Sx_2 &= \sup\{T(y_1+y_2) : 0 \leqslant y_1 \leqslant x_1,\ 0 \leqslant y_2 \leqslant x_2\} \\ &\leqslant \sup\{Tz : 0 \leqslant z \leqslant x_1 + x_2\} \\ &= \sup\{T(z \wedge x_1) + T(z - z \wedge x_1) : 0 \leqslant z \leqslant x_1 + x_2\} \\ &\leqslant Sx_1 + Sx_2. \end{aligned}$$

(Because if $0 \leqslant z \leqslant x_1 + x_2$, $0 \leqslant z - z \wedge x_1 \leqslant x_2$; cf. 14Ja prf. Note also that in the first line I used 12Cc.) So S extends to a linear map from E to F [16A]. Now S and $S - T$ are increasing, so $T \in L^{\sim}(E; F)$.

16D Theorem Let E be a Riesz space and F a Dedekind complete Riesz space. Then $L^{\sim}(E; F)$ is a Dedekind complete Riesz space and

(a) if $T \in L^{\sim}(E; F)$ then T^+ is given by

$$T^+ x = \sup\{Ty : 0 \leqslant y \leqslant x\} \quad \forall\ x \in E^+;$$

(b) if A is a non-empty set in $L^{\sim}(E; F)$ and $A \uparrow$, then $\sup A$ exists iff $\sup\{Tx : T \in A\}$ exists for every $x \in E^+$, and in this case

$$(\sup A)(x) = \sup\{Tx : T \in A\} \quad \forall\ x \in E^+.$$

Proof (a) We know that $L^{\sim}(E; F)$ is a partially ordered linear space. But we have seen in the course of proving 16C that if $T \in L^{\sim}(E; F)$ there is a linear map $S: E \to F$ such that

$$Sx = \sup\{Ty : 0 \leqslant y \leqslant x\} \quad \forall\ x \in E^+,$$

and now it is obvious that $S = \sup\{T, 0\}$ in $L^{\sim}(E; F)$. So $L^{\sim}(E; F)$ is a Riesz space and $S = T^+$.

(b) Suppose that $A \uparrow$ in $L^{\sim}(E; F)$ and that $Sx = \sup\{Tx : T \in A\}$ exists for every $x \in E^+$. Because A is directed upwards, S satisfies the conditions of 16A, and extends to a linear map $S_0: E \to F$. A cannot be empty (unless $F = \{0\}$), and, if $T_0 \in A$, $S_0 - T_0$ is increasing; so in any case $S_0 \in L^{\sim}(E; F)$, and plainly $S_0 = \sup A$.

It follows at once, by the device of 13Cb, that $L^{\sim}(E; F)$ is Dedekind complete; for if A is bounded above by T_0, then $\{Tx: T \in A\}$ is bounded above by $T_0 x$, and therefore has a least upper bound, for every $x \in E^+$.

***16E Corollary** Let E and F satisfy the conditions of 16D.

(a) If $T \in L^{\sim}(E; F)$, then $|T|$ is given by

$$|T|(x) = \sup\{Ty : |y| \leqslant x\} \quad \forall\ x \in E^+.$$

(b) If $T_1, T_2 \in L^{\sim}(E; F)$, then $T_1 \vee T_2$ and $T_1 \wedge T_2$ are given by

$$(T_1 \vee T_2)(x) = \sup\{T_1 x_1 + T_2 x_2 : x_1, x_2 \in E^+, x_1 + x_2 = x\},$$

$$(T_1 \wedge T_2)(x) = \inf\{T_1 x_1 + T_2 x_2 : x_1, x_2 \in E^+, x_1 + x_2 = x\}$$

for every $x \in E^+$.

(c) If A is a non-empty subset of $L^{\sim}(E; F)$ and $A \downarrow$, then

$$A \downarrow 0 \text{ iff } \inf\{Tx : T \in A\} = 0 \text{ for every } x \in E^+.$$

Proofs (a) and (b) follow from the formula in 16Da by applying the identities

$$|T| = (2T)^+ - T,\ T_1 \vee T_2 = T_2 + (T_1 - T_2)^+,\ T_1 \wedge T_2 = T_1 + T_2 - T_1 \vee T_2;$$

(c) follows by applying 16Db to the set $-A$. [See also 16I.]

***16F Lemma** Let E be a Riesz space and F a Dedekind complete Riesz space. Let $T: E \to F$ be a linear map. For each $x \in E^+$, set

$$A_x = \{\textstyle\sum_{i<n} |Tx_i| : \langle x_i \rangle_{i<n} \text{ is a finite sequence in } E^+ \text{ such that } \sum_{i<n} x_i = x\}.$$

Then (a) $A_x \uparrow$ for each $x \in E^+$; (b) $T \in L^\sim(E; F)$ iff A_x is bounded above in F for each $x \in E^+$, and in this case

$$|T|(x) = \sup A_x \quad \forall\ x \in E^+.$$

Proof (a) If $\langle x_i \rangle_{i<n}$ and $\langle y_j \rangle_{j<m}$ are finite sequences in E^+ such that $\sum_{i<n} x_i = \sum_{j<m} y_j = x$, then by 14Jc there is a double sequence $\langle z_{ij} \rangle_{i<n, j<m}$ in E^+ such that $\sum_{j<m} z_{ij} = x_i$ for every $i < n$ and

$$\textstyle\sum_{i<n} z_{ij} = y_j$$

for every $j < m$. Now it follows that

$$\textstyle\sum_{i<n} |Tx_i| = \sum_{i<n} |\sum_{j<m} T(z_{ij})| \leqslant \sum_{i<n, j<m} |T(z_{ij})|$$

by 14Bn. So $\sum_{i<n, j<m} |T(z_{ij})|$ is a member of A_x greater than or equal to both $\sum_{i<n} |Tx_i|$ and $\sum_{j<m} |Ty_j|$. Thus $A_x \uparrow$.

(b) Of course, if $T \in L^\sim(E; F)$, then A_x must be bounded above by $|T|(x)$ for every $x \in E^+$, for

$$\textstyle\sum_{i<n} |Tx_i| \leqslant \sum_{i<n} |T|(x_i) = |T|(\sum_{i<n} x_i)$$

for any sequence $\langle x_i \rangle_{i<n}$ in E^+. On the other hand, if A_x is bounded above for every $x \in E^+$, then fix on any $x \in E^+$ and consider, for $|y| \leqslant x$,

$$Ty = Ty^+ - Ty^- \leqslant |Ty^+| + |Ty^-| + |T(x - |y|)|,$$

which belongs to A_x. Thus $\{Ty : 0 \leqslant y \leqslant x\}$ is bounded above by the upper bounds of A_x, and $T \in L^\sim(E; F)$. Also, by 16Ea,

$$|T|(x) = \sup \{Ty : |y| \leqslant x\} \leqslant \sup A_x,$$

for any $x \in E^+$; so $|T|(x) = \sup A_x$.

16G Definition Let E and F be Riesz spaces. Then $L^\times(E; F)$ is the set of linear maps from E to F which are expressible as the difference of two order-continuous increasing linear maps. I shall write $E^\times$ for $L^\times(E; \mathbf{R})$.

16H Theorem Let E be a Riesz space and F a Dedekind complete Riesz space. Then $L^\times(E; F)$ is a band in $L^\sim(E; F)$.

Proof (a) Suppose that T_1 and T_2 are increasing order-continuous linear maps from E to F. Then so is $T_1 + T_2$. **P** Of course $T_1 + T_2$ is linear and increasing. Suppose that A is a non-empty set in E such that $A \downarrow 0$. Then $(T_1 + T_2)[A] \downarrow$. But

$$\inf \{T_1 x + T_2 x : x \in A\} = \inf \{T_1 x + T_2 y : x, y \in A\}$$

[because A is directed downwards and T_1 and T_2 are increasing]

$$= \inf(T_1[A]+T_2[A])$$
$$= \inf T_1[A]+\inf T_2[A] \quad \text{[as in 12Cc]}$$
$$= 0$$

because T_1 and T_2 are both order-continuous. Because T_1+T_2 is linear, this is enough to show that it is order-continuous [14Ec]. **Q**

(b) So let P be the set of all order-continuous increasing linear maps from $\boldsymbol{E}$ to $\boldsymbol{F}$, a subset of $\boldsymbol{L}^{\sim}(\boldsymbol{E};\boldsymbol{F})^+$. Then $P+P \subseteq P$. Also, it is clear that $\alpha P \subseteq P$ for every $\alpha \in \mathbf{R}^+$; so that $P-P = \boldsymbol{L}^{\times}(\boldsymbol{E};\boldsymbol{F})$ is a linear subspace of $\boldsymbol{L}^{\sim}(\boldsymbol{E};\boldsymbol{F})$. Next, $0 \leqslant S \leqslant T \in P \Rightarrow S \in P$, again using the first condition in 14Ec. So $\boldsymbol{L}^{\times}(\boldsymbol{E};\boldsymbol{F})$ is solid. **P** If $|S| \leqslant |T_1-T_2|$, then $0 \leqslant S^+ \leqslant |S| \leqslant |T_1|+|T_2| \in P$, so $S^+ \in P$. Similarly $S^- \in P$ so

$$S = S^+ - S^- \in \boldsymbol{L}^{\times}(\boldsymbol{E};\boldsymbol{F}). \quad \mathbf{Q}$$

In particular, $\boldsymbol{L}^{\times}(\boldsymbol{E};\boldsymbol{F})$ is a Riesz subspace of $\boldsymbol{L}^{\sim}(\boldsymbol{E};\boldsymbol{F})$.

(c) Finally, suppose that A is a non-empty subset of $\boldsymbol{L}^{\times}(\boldsymbol{E};\boldsymbol{F})$ and that $S = \sup A$ exists in $\boldsymbol{L}^{\sim}(\boldsymbol{E};\boldsymbol{F})$. Fix $T_0 \in A$. Following 13Cb, set

$$B = \{\sup(C \cup \{T_0\}) : C \subseteq A,\ C \text{ finite}\}.$$

Thus $B \subseteq \boldsymbol{L}^{\times}(\boldsymbol{E};\boldsymbol{F})$, B is bounded below by T_0, and $B \uparrow S$. We know that $S-T_0$ is increasing; I shall show that it is order-continuous with the second criterion of 14Ec. For if $\varnothing \subset D \subseteq \boldsymbol{E}^+$ and $D \uparrow x$, then

$$(S-T_0)(x) = \sup_{T \in B} Tx - T_0 x \qquad \text{[16Db]}$$
$$= \sup_{T\in B}(T-T_0)(x) \qquad \text{[12Ca]}$$
$$= \sup_{T\in B}\sup_{y\in D}(T-T_0)(y)$$

[as $T-T_0$ is order-continuous for every $T \in B$]

$$= \sup_{y\in D}\sup_{T\in B}(T-T_0)(y)$$
$$= \sup_{y\in D}(S-T_0)(y)$$

(repeating the arguments of the first two lines); and this is the required result. Consequently, $S = (S-T_0)+T_0 \in \boldsymbol{L}^{\times}(\boldsymbol{E};\boldsymbol{F})$. As A is arbitrary, $\boldsymbol{L}^{\times}(\boldsymbol{E};\boldsymbol{F})$ is a band in $\boldsymbol{L}^{\sim}(\boldsymbol{E};\boldsymbol{F})$.

***16I Exercise** Let $\boldsymbol{E}$ be a Riesz space and $\boldsymbol{F}$ a Dedekind complete Riesz space. Let A be a non-empty subset of $\boldsymbol{L}^{\sim}(\boldsymbol{E};\boldsymbol{F})$. Then A is bounded above in $\boldsymbol{L}^{\sim}(\boldsymbol{E};\boldsymbol{F})$ iff

$$B_x = \{\textstyle\sum_{i<n} T_i x_i : x_i \geqslant 0 \quad \forall\ i<n,\ T_i \in A \quad \forall\ i<n,\ \sum_{i<n} x_i = x\}$$

is bounded above in F for every $x \in E^+$, and in this case

$$(\sup A)(x) = \sup B_x$$

for every $x \in E^+$. Similarly, $(\inf A)(x) = \inf B_x$ provided the right-hand side is always defined.

Notes and comments The results of this section are the first justification we see for the serious study of abstract Riesz spaces. The spaces $L^\sim(E; F)$ and $L^\times(E; F)$ are defined in such a way that their structure is not immediately accessible, and we must deduce as much as possible from the theorems 16D and 16H.

It is in this section also that I begin to assume familiarity with the basic properties of Riesz spaces laid out in §§ 12 and 14. Consequently the proofs above contain a number of elisions in which elementary but non-trivial results have to be applied. It is a valuable exercise to search out these applications.

An outstanding problem in the subject is that of salvaging something from the above work if F is not Dedekind complete. It seems that within the methods and concepts I have introduced, not much can be done; but in a suitably enlarged context of partially ordered linear spaces there ought to be interesting extensions.

17 Order-dense Riesz subspaces

A number of important results are associated with the concepts of order-dense or locally order-dense Riesz subspace. [For definitions, see 14F.] The ones here are those which have applications to the work of this book. In particular, they can be thought of as the core of the Radon–Nikodým theorem; but this will become apparent much later.

Since the results of this section are of interest primarily because of their applications, I have starred all the paragraphs except the first two. 17A gives a valuable sufficient condition for greatest lower bounds (and, of course, least upper bounds) to be the same whether taken in the whole space or in a subspace; you will recall that this cannot be taken for granted [1XB]. 17B is an important general theorem on the extension of order-continuous increasing linear maps.

17A Lemma If E is a Riesz space and F is a locally order-dense Riesz subspace of E, then whenever A is a non-empty subset of F and x is the greatest lower bound of A in F, x is the greatest lower bound of A in E.

Proof ? For otherwise A has a lower bound x_1 in E such that $x_1 \nleqslant x$, i.e. $x \vee x_1 > x$. Let y be any member of A; then

$$0 < x \vee x_1 - x \leqslant y - x \in F.$$

As F is locally order-dense, there is a $z \in F$ such that $0 < z \leqslant x \vee x_1 - x$. But now $x \vee x_1 \geqslant x + z$, so $x + z$ is a lower bound for A in F strictly greater than x; which is impossible. **X**

Remark The same result will be true of least upper bounds. This is sometimes expressed by saying that F is **regularly embedded** in E.

17B Theorem Let E and G be Riesz spaces, and F an order-dense Riesz subspace of E. Suppose that $T: F \to G$ is an order-continuous increasing linear map such that

$$Ux = \sup\{Ty : y \in F, 0 \leqslant y \leqslant x\}$$

exists in G for every $x \in E^+$. Then T has a unique extension to an order-continuous increasing linear map from E to G.

Proof **(a)** I show first that if A is a non-empty set in F^+, and $A \uparrow x$ in E, then $\sup T[A]$ exists and is equal to Ux.

P Of course $Ty \leqslant Ux$ for every $y \in A$. On the other hand, suppose that w is any upper bound for $T[A]$, and suppose that $z \in F$ and that $0 \leqslant z \leqslant x$. Then

$$B = \{z \wedge y : y \in A\} \uparrow z \wedge x = z$$

in E, by 14D. *A fortiori*, $B \uparrow z$ in F. So, as T is order-continuous, $T[B] \uparrow Tz$. But we know that for any $y \in A$,

$$T(z \wedge y) \leqslant Ty \leqslant w.$$

So w is an upper bound for $T[B]$, and $w \geqslant Tz$.

As z is arbitrary, $w \geqslant Ux$; which proves the result. **Q**

(b) Now it follows that

$$U(x_1 + x_2) = Ux_1 + Ux_2 \quad \forall \; x_1, x_2 \in E^+.$$

P For let
$$A_i = \{y : y \in F, 0 \leqslant y \leqslant x_i\}$$

for $i = 1, 2$. Because F is order-dense, $A_i \uparrow x_i$ in E for both i. So $A_1 + A_2 \uparrow x_1 + x_2$ [12Cc]. Now by (a) above,

$$\begin{aligned} U(x_1 + x_2) &= \sup T[A_1 + A_2] = \sup (T[A_1] + T[A_2]) \\ &= \sup T[A_1] + \sup T[A_2] = Ux_1 + Ux_2, \end{aligned}$$

using 12Cc again. **Q**

(c) Since it is easy to see that $U(\alpha x) = \alpha Ux$ whenever $\alpha \geqslant 0$ and $x \in \boldsymbol{E}^+$, U has an extension to a linear map $U: \boldsymbol{E} \to \boldsymbol{G}$ [16A]. Clearly $Ux = Tx$ for every $x \in \boldsymbol{E}^+$, so U is an extension of T. Finally, U is order-continuous. **P** For suppose that $\varnothing \subset A \subseteq \boldsymbol{E}^+$ and that $A \uparrow x$. Then

$$B = \{z : z \in \boldsymbol{F}, \exists\, y \in A, 0 \leqslant z \leqslant y\} \uparrow x,$$

since clearly A and B have the same upper bounds. So $Ux = \sup T[B]$ by (a). But any upper bound of $U[A]$ is an upper bound of $T[B]$, so $\sup U[A]$ exists and is equal to Ux. As A is arbitrary, U is order-continuous. **Q**

(d) Of course the extension of T is unique, because it must agree with U on $\boldsymbol{E}^+$ and therefore on $\boldsymbol{E}$.

***17C Proposition** Let $\boldsymbol{E}$ and $\boldsymbol{G}$ be Riesz spaces, and $\boldsymbol{F}$ an order-dense Riesz subspace of $\boldsymbol{E}$. Let $U: \boldsymbol{E} \to \boldsymbol{G}$ be an order-continuous increasing linear map, and let T be the restriction of U to $\boldsymbol{F}$. (For example, U might have been constructed from T as in 17B above.) Then:

(a) if T is a Riesz homomorphism, so is U;

(b) if T is a one-to-one Riesz homomorphism, so is U.

Proof (a) Suppose that T is a Riesz homomorphism, and that $x \wedge y = 0$ in $\boldsymbol{E}$. Set

$$A = \{z : z \in \boldsymbol{F}, 0 \leqslant z \leqslant x\} \quad \text{and} \quad B = \{w : w \in \boldsymbol{F}, 0 \leqslant w \leqslant y\}.$$

Then $A \uparrow x$ and $B \uparrow y$ [because $\boldsymbol{F}$ is order-dense], so, because U is order-continuous,

$$\begin{aligned} Ux \wedge Uy &= (\sup_{z \in A} Uz) \wedge Uy \\ &= \sup_{z \in A} (Uz \wedge Uy) \\ &= \sup_{z \in A} (Uz \wedge (\sup_{w \in B} Uw)) \\ &= \sup_{z \in A} \sup_{w \in B} (Uz \wedge Uw), \end{aligned}$$

using 14D twice. But if $z \in A$ and $w \in B$, $0 \leqslant z \wedge w \leqslant x \wedge y = 0$, so $Uz \wedge Uw = Tz \wedge Tw = T(z \wedge w) = 0$. Thus $Ux \wedge Uy = 0$. As x and y are arbitrary, U is a Riesz homomorphism, by 14Eb.

(b) If now T is one-to-one, then for any non-zero $x \in \boldsymbol{E}$ there is a $z \in \boldsymbol{F}$ such that $0 < z \leqslant |x|$. So

$$|Ux| = U(|x|) \geqslant Tz > 0.$$

Thus $Ux \neq 0$. As x is arbitrary, U is one-to-one.

***17D Lemma** Let E be a Dedekind complete Riesz space, F any Riesz space, and $T: E \to F$ an order-continuous Riesz homomorphism such that $T[E]$ is locally order-dense in F. Then $T[E]$ is a solid linear subspace of F.

Proof As T is a Riesz homomorphism, $T[E]$ is a Riesz subspace of F. It is therefore enough to show that if $w \in T[E]$ and $0 \leqslant y \leqslant w$, then $y \in T[E]$. Let $x \in E$ be such that $Tx = w$, and consider

$$A = \{z \wedge |x| : z \in E, Tz \leqslant y\}.$$

A is bounded above by $|x|$ and $0 \in A$, so $\sup A$ exists; also, it is clear that $A \uparrow$, so, as T is order-continuous,

$$\begin{aligned} T(\sup A) &= \sup T[A] \\ &= \sup\{T(z \wedge |x|) : z \in E, Tz \leqslant y\} \\ &= \sup\{Tz \wedge |Tx| : z \in E, Tz \leqslant y\} \\ &= \sup\{Tz : z \in E, Tz \leqslant y\} \\ &= y, \end{aligned}$$

because $T[E]$ is locally order-dense.

***17E Lemma** If E and F are Riesz spaces and $T: E \to F$ is a one-to-one Riesz homomorphism such that $T[E]$ is locally order-dense in F, then T is order-continuous.

Proof Since T is a one-to-one Riesz homomorphism, it is a Riesz space isomorphism between E and $T[E]$. Now if $\varnothing \subset A \downarrow 0$ in E, $\inf T[A] = 0$ in $T[E]$. By 17A, $\inf T[A] = 0$ in F. So by 14Ec T is order-continuous.

***17F Corollary** If E is a Dedekind complete Riesz space, F is any Riesz space, and $T: E \to F$ is a one-to-one Riesz homomorphism such that $T[E]$ is locally order-dense in F, then $T[E]$ is solid in F.

Proof By 17E, T is order-continuous; now by 17D $T[E]$ is solid.

17G Exercises (a) Suppose in 17B, that F is locally order-dense but not order-dense in E. Show that T has an extension, as in the theorem, but that the extension may not be unique.

(b) Let E be a Riesz space and F an order-dense Riesz subspace of E

such that every member of E is dominated by a member of F (that is, $\forall\ x \in E\ \exists\ y \in F$ such that $|x| \leqslant y$). Show that $E^\times$ and $F^\times$ may be identified as Riesz spaces.

*(c) Let E and G be Riesz spaces, and F a super-order-dense Riesz subspace of E. Suppose that $T: F \to G$ is a sequentially order-continuous increasing linear map such that $\sup_{n \in \mathbf{N}} Ty_n$ exists in G whenever $\langle y_n \rangle_{n \in \mathbf{N}}$ is an increasing sequence in F^+ with a least upper bound in E. Then T has a unique extension to a sequentially order-continuous increasing linear map from E to G. [Hint: show that if $\langle y_n \rangle_{n \in \mathbf{N}}$ is a sequence in F^+ such that $\langle y_n \rangle_{n \in \mathbf{N}} \uparrow x$ in E, then

$$\sup_{n \in \mathbf{N}} Ty_n = \sup\{Ty : y \in F,\ 0 \leqslant y \leqslant x\}.$$

Now use the method of 17B.]

(d) If E is any Riesz space and F is a locally order-dense Riesz subspace of E which is (in itself) Dedekind complete, then F is solid. [Use 17F, or otherwise.]

*(e) If E is any Riesz space, F is a locally order-dense Riesz subspace of E, and G is a locally order-dense Riesz subspace of F, then G is locally order-dense in E.

Notes and comments The extension theorem 17B is used in various places in Chapter 6. Like many results concerning order-continuous maps it has a 'sequential' form referring to sequentially order-continuous maps [17Gc]; for some applications in measure theory, this is really the natural one to use [see 62K/62Me]. Note that the sequential form of the concept 'order-dense' is 'super-order-dense'.

*The hypotheses of 17D and 17E should be examined carefully; they are not quite natural, and applications often fulfil them with little to spare. On the other hand, it is not clear that they can be relaxed in any useful way. These two results, of course, tend to be employed together, as in 17F.

*18 The countable sup property

A large proportion of the elementary examples of Riesz spaces share the special property that every set with a least upper bound has a countable subset with the same least upper bound. (It follows, of course, that the same will be true for greatest lower bounds.) While this is not vitally important to us, it is of considerable interest, especially in view of the elegant characterization of Archimedean Riesz spaces of this type in 18D below.

18A Definition A Riesz space $\boldsymbol{E}$ has the **countable sup property** if, whenever A is a subset of $\boldsymbol{E}$ with a least upper bound, there is a countable $B \subseteq A$ such that $\sup B = \sup A$.

18B Lemma Let $\boldsymbol{E}$ be an Archimedean Riesz space with the countable sup property. Let $A \subseteq \boldsymbol{E}$ be bounded above. Then there is a countable set $B \subseteq A$ with the same upper bounds as A.

Proof If $A = \varnothing$, set $B = \varnothing$. Otherwise, let C be the set of upper bounds of A. As $\boldsymbol{E}$ is Archimedean, $\inf(C-A) = 0$ [15C]. So there is a countable set $D \subseteq C-A$ such that $\inf D = 0$. Now we can choose a countable $B \subseteq A$ such that $D \subseteq C-B$, so $\inf(C-B) = 0$.

Suppose that $x \in A$ and that w is an upper bound for B. Then for any $z \in C$ and $y \in B$,

$$z-y \geqslant x-w,$$

so $x-w \leqslant \inf(C-B)$ i.e. $x \leqslant w$. Thus every upper bound for B is an upper bound for A; but the converse is trivial, so the result is proved.

18C Corollary If $\boldsymbol{E}$ is an Archimedean Riesz space with the countable sup property, every Riesz subspace of $\boldsymbol{E}$ has the countable sup property.

18D Theorem Let $\boldsymbol{E}$ be an Archimedean Riesz space. Then $\boldsymbol{E}$ has the countable sup property iff every disjoint set in $\boldsymbol{E}^+$ which is bounded above is countable.

Proof (a) Suppose that $\boldsymbol{E}$ has the countable sup property, and that A is a disjoint set in $\boldsymbol{E}^+$ bounded above by x_0. (Recall that by 'disjoint' I mean that $x \wedge y = 0$ whenever x and y are distinct members of A; 13G.) Then there is a countable set $B \subseteq A$ with the same upper bounds as A[18B]. But $A \subseteq B \cup \{0\}$. **P** **?** If y is a non-zero member of $A \backslash B$, then $y \wedge z = 0$ for every $z \in B$. Suppose that x is any upper bound for A; then for any $z \in B$

$$z = z \wedge x \leqslant z \wedge (x-y) + z \wedge y = z \wedge (x-y)$$

[using 14Ja], so $x-y$ is also an upper bound for B and therefore for A. Inducing on n, $x-ny \geqslant y$ for every $n \in \mathbf{N}$. **X** **Q** So A is countable.

(b) Suppose that $\boldsymbol{E}$ satisfies the condition, and that $A \subseteq \boldsymbol{E}^+$ is a non-empty set with a least upper bound x. Let $0 \leqslant \alpha < 1$; then there is a countable $A_\alpha \subseteq A$ such that

$$\alpha x = \sup\{y \wedge \alpha x : y \in A_\alpha\}.$$

P Set

$$B = \{z : \exists\ y \in A,\ 0 \leqslant z \leqslant (y - \alpha x)^+\},$$

and let $\mathscr{X}$ be the family of all subsets of B which are disjoint in $\boldsymbol{E}^+$. By Zorn's lemma, $\mathscr{X}$ has a maximal element C say; as C is bounded above by $(1-\alpha)x$, it is countable, and we can choose a non-empty countable set $A_\alpha \subseteq A$ such that

$$\forall\ z \in C \quad \exists\ y \in A_\alpha \quad \text{such that} \quad 0 \leqslant z \leqslant (y - \alpha x)^+.$$

? Suppose that there is a $w > 0$ such that

$$\alpha x - w \geqslant y \wedge \alpha x \quad \forall\ y \in A_\alpha,$$

i.e.
$$0 < w \leqslant (\alpha x - y)^+ \quad \forall\ y \in A_\alpha.$$

Then $w \wedge z = 0$ for every $z \in C$, so (C being maximal) $w \wedge z = 0$ for every $z \in B$ i.e.

$$w \wedge (y - \alpha x)^+ = 0 \quad \forall\ y \in A.$$

But now

$$w \wedge (1-\alpha)x = w \wedge (x - \alpha x)^+ = \sup_{y \in A} w \wedge (y - \alpha x)^+ = 0$$

[using 12Ca and 14D], and as $1 - \alpha > 0$,

$$w \wedge \alpha x = 0$$

[14Kb]. But $0 < w \leqslant \alpha x$. **X**

So $\alpha x = \sup\{y \wedge \alpha x : y \in A_\alpha\}$. **Q**

Now let $\alpha(n) = 1 - 2^{-n}$ and $D = \bigcup_{n \in \mathbf{N}} A_{\alpha(n)}$. Then

$$\begin{aligned} x &= \sup\{\alpha(n)x : n \in \mathbf{N}\} \quad [15\text{D}] \\ &= \sup\{y \wedge \alpha(n)x : n \in \mathbf{N},\ y \in A_{\alpha(n)}\} \\ &= \sup D. \end{aligned}$$

Finally, suppose that A is an arbitrary non-empty subset of $\boldsymbol{E}$ with a least upper bound x. Fix $x_0 \in A$ and set $A' = \{y \vee x_0 : y \in A\}$. Then $\sup A' = x$ and A' has a lower bound. It is clear that, applying the result above to $A' - x_0$, A' has a countable subset B' with least upper bound x. Now choose a countable $B \subseteq A$ such that

$$B' \subseteq \{y \vee x_0 : y \in B\},$$

and see that $x = \sup(\{x_0\} \cup B)$, which is a countable subset of A.

18E Exercises *(a) Show that an Archimedean Riesz space $\boldsymbol{E}$ has the countable sup property iff whenever $\boldsymbol{F}$ is an Archimedean

Riesz space and $T: E \to F$ is increasing, linear and sequentially order-continuous, then T is order-continuous.

(b) Let E be a Riesz space with an order unit [definition: 14I], and F a Dedekind complete Riesz space with the countable sup property. Then $L^{\sim}(E; F)$ has the countable sup property.

(c) Let E be a Riesz space with a weak order unit [definition: 14I], and F a Dedekind complete Riesz space with the countable sup property. Then $L^{\times}(E; F)$ has the countable sup property.

Notes and comments There are simple examples of Riesz spaces with the countable sup property in 2XA–2XC; see also 25Lc. Theorem 18D is clearly a very powerful method of identifying such spaces, as well as being an important property of them. From this theorem we can see that the countable sup property is intimately related to the 'countable chain condition' on topological spaces and Boolean lattices.

It is important, however, to appreciate that the countable sup property is not the same as the property: 'every disjoint set in the positive cone is countable', which is shared by many of the elementary examples. The latter is of course much stronger.

1X Examples for Chapter 1

Because the structures dealt with in this chapter will recur throughout the book, I shall give very few examples at this stage, and most of them will be slightly off the main line of thought. Thus 1XA–1XC are designed to illustrate technical aspects of the concepts I have introduced, rather than their practical applications. In contrast, 1XD and 1XE discuss a fundamental class of Riesz spaces, the spaces $\mathbf{R}^X$, which also have important topological properties.

1XA A partially ordered set Let A be any set (even $\varnothing$); let $\leqslant$ be the relation on A given by

$$a \leqslant b \Leftrightarrow a = b.$$

Then $(A, \leqslant)$ is a partially ordered set [11A]. If $B \subseteq A$, then $\sup B$ exists iff $\inf B$ exists iff either B has just one element or B is empty and A has just one element.

If $(C, \leqslant)$ is any other partially ordered set, every function from A to C is increasing and order-continuous [11D]. Conversely, every increasing function from C to A is order-continuous.

1XB A lattice Let X be any set, and let $\leqslant$ be the restriction of $\subseteq$ to $\mathscr{P}X$, the set of subsets of X. Then $(\mathscr{P}X, \leqslant)$ is a Dedekind complete distributive lattice; if A is a non-empty subset of $\mathscr{P}X$,

$$\sup A = \bigcup A \quad \text{and} \quad \inf A = \bigcap A.$$

[See also 4XA.]

Now let $\mathfrak{T}$ be any topology on X. Then $\mathfrak{T}$ is a sublattice of $\mathscr{P}X$ [13F]. Let $\preccurlyeq$ be the restriction of $\leqslant$ or $\subseteq$ to $\mathfrak{T}$, so that $(\mathfrak{T}, \preccurlyeq)$ is a lattice in its own right. The embedding of $\mathfrak{T}$ in $\mathscr{P}X$ is a lattice homomorphism (because $\mathfrak{T}$ is a sublattice of $\mathscr{P}X$); and it is order-continuous on the left [11D], because if $A \subseteq \mathfrak{T}$, then $\bigcup A$, which is $\sup A$ in $\mathscr{P}X$, belongs to $\mathfrak{T}$, and therefore must be sup A in $\mathfrak{T}$. We see also from this that $\mathfrak{T}$ is a Dedekind complete lattice [13B]. But suppose that A is a non-empty subset of $\mathfrak{T}$. Then $\inf A$ in $\mathfrak{T}$ is $\operatorname{int}(\bigcap A)$, while $\inf A$ in $\mathscr{P}X$ is $\bigcap A$ itself. Of course these are not in general the same. Thus we have here an example in which the meaning of the expression $\inf A$ depends on which lattice we consider A as a subset of.

***1XC Totally ordered sets** Let $(A, \leqslant)$ be a totally ordered set, i.e. a partially ordered set such that for every a and b belonging to A, either $a \leqslant b$ or $b \leqslant a$. Then A must be a lattice. On A the **order topology** is the topology generated by all sets of the forms $\{a : a > b\}$ and $\{a : a < b\}$ as b runs through A. This topology is always Hausdorff and regular. A subset of A is closed for the order topology iff it is order-closed in the sense of 11E. *The order topology is the coarsest topology which is 'compatible' in the sense of the note at the end of § 21; and it is the finest topology which is 'Lebesgue' in the sense of the introduction to § 24.

A is Dedekind complete iff every closed interval

$$[b, c] = \{a : b \leqslant a \leqslant c\}$$

is compact in the order topology. (In the special case $A = \mathbf{R}$, this is the Heine–Borel theorem; cf. also 24Lb.) A is compact in the order topology iff either it is empty or it is Dedekind complete and has greatest and least elements.

1XD The space $\mathbf{R}^X$. Let X be any set. Then $\mathbf{R}^X$, the set of all functions from X to $\mathbf{R}$, has a natural Riesz space structure given by

$$(f+g)(t) = ft + gt \quad \forall\ t \in X,$$
$$(\alpha f)(t) = \alpha ft \quad \forall\ t \in X,$$
$$f \geqslant g \Leftrightarrow ft \geqslant gt \quad \forall\ t \in X;$$

that is to say, $\mathbf{R}^X$ is regarded as a product of copies of the Riesz space $\mathbf{R}$ [14H]. $\mathbf{R}^X$ is Dedekind complete [13Hb, or directly].

Regarded as a product of copies of $\mathbf{R}$, $\mathbf{R}^X$ has a natural topology, which is of course a complete Hausdorff locally convex linear space topology. This topology is Fatou [23A], Levi [23I] and Lebesgue [24A]; so the completeness of $\mathbf{R}^X$ is an example of Nakano's theorem [23K]. The linear topological space dual $(\mathbf{R}^X)'$ is equal to $(\mathbf{R}^X)^\times$.

1XE Quotients of $\mathbf{R}^X$. Let X be a non-empty set, and $\mathscr{F}$ a filter on X. Let E be $\mathbf{R}^X$ with its usual Riesz space structure, and set

$$F = \{x : x \in E, \{t : x(t) = 0\} \in \mathscr{F}\}.$$

Then F is a solid linear subspace of E. Now the quotient Riesz space E/F [14G] is Archimedean iff $\mathscr{F}$ is closed under countable intersections, and in this case the canonical map from E to E/F is sequentially order-continuous [11D, 14Lb]. F is a band iff $\mathscr{F}$ is generated by a single set Y, and in this case $E/F \cong \mathbf{R}^Y$.

For example, let $X = \mathbf{N}$ and let $\mathscr{F}$ be any filter on X including the Fréchet filter, i.e. containing the complement of every finite set. Then E/F cannot be Archimedean.

***1XF A non-Archimedean Riesz space** Give $\mathbf{R}^2$ its usual linear space structure, and define the **lexicographic ordering** $\leqslant$ on $\mathbf{R}^2$ by

$$(\alpha, \beta) \leqslant (\gamma, \delta) \quad \text{iff either} \quad \alpha < \gamma \quad \text{or} \quad \alpha = \gamma \quad \text{and} \quad \beta \leqslant \delta.$$

As this is a total ordering, $\mathbf{R}^2$ is a lattice; clearly it is a partially ordered linear space; therefore it is a Riesz space.

If $x = (1, 0)$ and $y = (0, 1)$, then $ny \leqslant x$ for every $n \in \mathbf{N}$, but $y > 0$; so $\mathbf{R}^2$ is not Archimedean [15A].

The usual linear space topology on $\mathbf{R}^2$ is Hausdorff, Lebesgue and Levi, but not of course compatible [21Ba].

Further reading for Chapter 1 For an elegant elementary discussion of partially ordered sets, see BOURBAKI I, chapter III, § 1. For an alternative introduction to the theory of Riesz spaces, see BOURBAKI VI, chapter II; there is a deeper approach in BIRKHOFF, chapter XV. A large part of the work done for Riesz spaces in this chapter can be applied to lattice-ordered groups; for hints see BIRKHOFF, chapter XIII. For the more advanced theory of Riesz spaces, see LUXEMBURG & ZAANEN R.S.

2. Topological Riesz spaces

This chapter and the next are an exposition of the elementary theory of Riesz spaces with linear space topologies on them; they form a natural confluence of the Riesz space theory of Chapter 1 with ordinary abstract functional analysis. The present chapter proceeds by studying a series of properties that Riesz space topologies can have (most of them being, of course, relations between the topology and the order structure). The properties are all chosen to be ones possessed by important special cases; the theory never gets far from the applications that will be made of it, except perhaps in the work leading up to Nakano's theorem in § 23. In § 26 the conditions imposed are so strong that they become significantly less abstract; this section almost forms a layer intermediate between the first five sections of this chapter and the concrete examples from measure theory that will follow later.

Throughout this chapter I avoid the hypothesis of local convexity as far as possible. This is for both positive and negative reasons. The negative justification is that it can be done without an excessive amount of extra work. On the positive side, the most important examples of non-locally-convex spaces in analysis are Riesz spaces and are particularly accessible by the methods of this chapter [see, for example, 63K]. But of course it is quite possible to read this chapter with only locally convex spaces in mind.

Although we shall not study dual spaces thoroughly until the next chapter, I shall occasionally use the symbol E' for the space of continuous linear functionals on a linear topological space E, as well as $E^{\sim}$ and $E^{\times}$, defined in 16B and 16G respectively.

21 Compatible topologies

The simplest requirement we can make of a linear space topology on a Riesz space is that the positive cone be closed. Every Hausdorff topology we study will be of this type. It is perhaps slightly surprising that this already has non-trivial consequences.

21 A Definition If E is a Riesz space, a linear space topology on E is said to be **compatible** with the order structure of E if E^+ is closed.

21B Proposition Let E be a Riesz space and $\mathfrak{T}$ a compatible linear space topology on E. Then:

(a) E is Archimedean;

(b) $\{x : x \leqslant x_0\}$ and $\{x : x \geqslant x_0\}$ are closed for every $x_0 \in E$;

(c) $\mathfrak{T}$ is Hausdorff;

*(d) Let $\varnothing \subset A\uparrow$ (i.e. let A be a non-empty set which is directed upwards) in E, and suppose that $\mathscr{F}(A\uparrow)$ [definition: 11C] has a cluster point x_0. Then $x_0 = \sup A$.

*(e) In particular, if A is a non-empty relatively compact subset of E which is directed upwards, then $\sup A$ exists and is $\lim \mathscr{F}(A\uparrow)$.

Proofs (a) If $x \geqslant 2^n y$ for every $n \in \mathrm{N}$ i.e. $2^{-n}x - y \geqslant 0$ for every $n \in \mathrm{N}$, then $-y = \lim_{n\to\infty} 2^{-n}x - y \geqslant 0$ i.e. $y \leqslant 0$.

(b) and (c) are easy.

*(d) (i) If $x \in A$, then $\{y : y \geqslant x\}$ is a closed set belonging to $\mathscr{F}(A\uparrow)$, so contains x_0; thus x_0 is an upper bound for A. (ii) Now if x is any upper bound for A, $\{y : y \leqslant x\}$ is a closed set belonging to $\mathscr{F}(A\uparrow)$, so contains x_0; thus x_0 is the least upper bound of A.

*(e) is now easy.

***21C Exercise** Let E be a Riesz space and $\mathfrak{T}$ a locally convex compatible linear space topology on E. Then the weak topology $\mathfrak{T}_s(E, E')$ is also compatible, and if $x \in E \backslash E^+$ there is an increasing continuous linear functional f on E such that $fx < 0$.

***Notes and comments** It is clear that (with a suitable definition of 'Archimedean' space) everything in this section can be applied to general partially ordered linear spaces. Indeed, 21Bb provides a definition of 'compatible' topology on any partially ordered set, for which 21Bd and 21Be will still be true [cf. 1XC].

22 Locally solid topologies

A linear space topology on a Riesz space is 'locally solid' if the solid neighbourhoods of 0 form a basis. This is true of most of the 'strong' topologies which interest us, though not of 'weak' topologies. The results of this section are mostly straightforward. Locally solid topologies are really very natural and easy to work with, largely because the technical result 22C enables us to use pseudo-norms effectively.

22A Definitions If E is a Riesz space, a linear space topology on E is **locally solid** if 0 has a neighbourhood basis consisting of solid sets.

If E is a Riesz space, a **Riesz pseudo-norm** on E is a function $\rho: E \to \mathbf{R}$ such that

(i) $\rho x \geqslant 0$,

(ii) $\rho(x+y) \leqslant \rho x + \rho y$,

(iii) $|x| \leqslant |y| \Rightarrow \rho x \leqslant \rho y$,

(iv) $\lim_{\alpha \to 0} \rho(\alpha x) = 0$

for all $x, y \in E$. Observe that in this case

$$|\alpha| \leqslant 1 \Rightarrow |\alpha x| \leqslant |x| \Rightarrow \rho(\alpha x) \leqslant \rho(x) \quad \forall\ x \in E.$$

A **Riesz norm** or **Riesz seminorm** on E is a Riesz pseudo-norm which is a norm or seminorm respectively.

22B Proposition Let E be a Riesz space with a locally solid linear space topology. Then:

(a) the function $x \mapsto |x|: E \to E$ is uniformly continuous;

(b) the lattice operations $\vee$ and $\wedge$ on E are uniformly continuous;

(c) the solid hull of a bounded set is bounded;

(d) order-bounded sets [definition: 11F] are bounded;

*(e) the closure of a solid set is solid.

Proofs (a) This is because

$$||x|-|y|| \leqslant |x-y| \quad \forall\ x, y \in E,$$

since by 14Bn $|x|-|y| \leqslant |x-y|$ and $|y|-|x| \leqslant |y-x| = |x-y|$. So if U is a solid neighbourhood of 0 and $x-y \in U$, $|x|-|y| \in U$.

(**b**) now follows from 14Bf and 14Bg.

(**c**) Suppose $A \subseteq E$ is bounded, i.e. is absorbed by every neighbourhood of 0. Let B be the solid hull of A. If U is any neighbourhood of 0, there is a solid neighbourhood V of 0 such that $V \subseteq U$; as A is bounded, there is an $\alpha \geqslant 0$ such that $A \subseteq \alpha V$. But now αV is solid, so

$$B \subseteq \alpha V \subseteq \alpha U.$$

As U is arbitrary, B is bounded.

(**d**) follows, since

$$[x, y] \subseteq x + (\text{solid hull of } \{y-x\}).$$

*(e) Let $A \subseteq E$ be a solid set. Suppose that $x \in \bar{A}$ and that $|y| \leqslant |x|$. Then the map $\phi: E \to E$ given by

$$\phi z = (y \wedge |z|) \vee (-|z|)$$

is continuous [(a) and (b) above]. Now $\phi x = y$, so

$$y \in \phi[\bar{A}] \subseteq \overline{\phi[A]}.$$

But $\phi[A] \subseteq A$ (because $|\phi z| \leqslant |z|$ for every $z \in E$), so $y \in \bar{A}$, as required.

22C Proposition Let E be a Riesz space and $\mathfrak{T}$ a locally solid linear space topology on E. Then $\mathfrak{T}$ is defined by the continuous Riesz pseudo-norms on E, that is, for every neighbourhood U of 0 there is a continuous Riesz pseudo-norm ρ on E such that

$$\{x : \rho x < 1\} \subseteq U.$$

Proof Choose a sequence $\langle V_n \rangle_{n \in \mathbf{N}}$ of solid neighbourhoods of 0 such that $V_0 \subseteq U$ and $V_{n+1} + V_{n+1} + V_{n+1} \subseteq V_n$ for every $n \in \mathbf{N}$. Define $d: E \to \mathbf{R}$ by:

$$d(x) = 1 \quad \text{if } x \notin V_0;$$

$$d(x) = 2^{-n} \quad \text{if } x \in V_n \backslash V_{n+1};$$

$$d(x) = 0 \quad \text{if } x \in \textstyle\bigcap_{n \in \mathbf{N}} V_n.$$

Then $|x| \leqslant |y| \Rightarrow d(x) \leqslant d(y)$, because every V_n is solid.

Define $\rho: E \to \mathbf{R}$ by

$$\rho(x) = \inf\{\textstyle\sum_{i<n} d(x_i) : x = \sum_{i<n} x_i\}.$$

Now

(i) plainly $\rho(x) \geqslant 0$ for every $x \in E$;

(ii) and $\rho(x+y) \leqslant \rho(x) + \rho(y)$ for every $x, y \in E$;

(iii) and $|x| \leqslant |y| \Rightarrow \rho(x) \leqslant \rho(y)$. **P** For if $y = \sum_{i<n} y_i$, then

$$|x| \leqslant |y| \leqslant \textstyle\sum_{i<n} |y_i|;$$

so by 14Jb there exists a sequence $\langle x_i \rangle_{i<n}$ such that $x = \sum_{i<n} x_i$ and $|x_i| \leqslant |y_i|$ for every $i < n$. Now

$$\rho(x) \leqslant \textstyle\sum_{i<n} d(x_i) \leqslant \sum_{i<n} d(y_i),$$

which is arbitrarily close to $\rho(y)$. **Q**

(iv) Also, $\rho(x) \leqslant d(x)$ for every $x \in E$, so that $\rho(x) \leqslant 2^{-n}$ whenever $x \in V_n$. Hence [using (ii) above] ρ is continuous. As $\rho(0) = 0$,

$$\lim_{\alpha \to 0} \rho(\alpha x) = 0 \quad \forall\ x \in E,$$

and ρ is a continuous Riesz pseudo-norm.

Finally, observe that for every finite sequence $\langle x_i \rangle_{i<n}$ in E and $m \in \mathbf{N}$,

$$\sum_{i<n} d(x_i) < 2^{-m} \Rightarrow \sum_{i<n} x_i \in V_m.$$

P Induce on n simultaneously for all m. For $n = 1$, we have an x_0 such that $d(x_0) < 2^{-m}$, so that $x_0 \in V_{m+1} \subseteq V_m$. Inductive step: If $n \geqslant 1$, there must be an r such that $0 \leqslant r \leqslant n-1$ and

$$\sum_{i=0}^{r-1} d(x_i) < 2^{-m-1}, \quad \sum_{i=r+1}^{n-1} d(x_i) < 2^{-m-1},$$

with the convention that $\sum_{i=0}^{-1}$ and $\sum_{i=n}^{n-1}$ are both zero. Now by the inductive hypothesis

$$\sum_{i=0}^{r-1} x_i \in V_{m+1}, \quad \sum_{i=r+1}^{n-1} x_i \in V_{m+1},$$

and of course $x_r \in V_{m+1}$ as $d(x_r) < 2^{-m}$. So

$$\sum_{i=0}^{n-1} x_i \in V_{m+1} + V_{m+1} + V_{m+1} \subseteq V_m. \quad \mathbf{Q}$$

In particular, $\sum_{i<n} d(x_i) < 1 \Rightarrow \sum_{i<n} x_i \in V_0 \subseteq U$. But it follows at once that $\rho(x) < 1 \Rightarrow x \in U$.

22D Proposition Let E be a Riesz space with a locally solid linear space topology. Then its dual E' is a solid linear subspace of $E^\sim$, and the solid hull of an equicontinuous set in E' is equicontinuous.

Proof From the condition in 16C it is obvious that a linear functional on E is in $E^\sim$ iff it is bounded on order-bounded sets; so 22Bd above is enough to show that $E' \subseteq E^\sim$. As for the equicontinuous sets in E', consider the sets

$$U^0 = \{f : f \in E^\sim, |fx| \leqslant 1 \quad \forall\ x \in U\}$$

for solid neighbourhoods U of 0 in E. If $f \in U^0$ and $|g| \leqslant |f|$ in $E^\sim$, then

$$|gx| \leqslant |g|(|x|) \leqslant |f|(|x|) = \sup\{fy : |y| \leqslant |x|\}$$
$$\leqslant 1 \quad \forall\ x \in U$$

[using 16Ea]. So U^0 is a solid subset of $E^\sim$. Now (i) E' is a union of solid sets, so is solid (ii) every equicontinuous set is included in a solid equicontinuous set.

[See also 26Hc.]

22E Proposition Let E be a Riesz space and $\mathfrak{T}$ a locally solid Hausdorff linear space topology on E. Then

(a) $\mathfrak{T}$ is compatible with the order structure of E [because $E^+ = \{x : x \wedge 0 = 0\}$];

(b) E is Archimedean [21Ba];

(c) Bands in E are closed [15Fb];

*(d) If $A \subseteq E$ and $x \in \bar{A}$, then

$$x \in \overline{\{x \wedge y : y \in A\}},$$

so $x = \sup_{y \in A} x \wedge y = \inf_{y \in A} x \vee y$.

22F Proposition Let E be a Riesz space with a Hausdorff locally solid linear space topology $\mathfrak{T}$. Then the uniform space completion $E^\wedge$ of E has a natural Riesz space structure, and its topology $\mathfrak{T}^\wedge$ is a locally solid Hausdorff linear space topology; and the embedding $E \to E^\wedge$ is a Riesz homomorphism.

Proof We use 22Bb. (**a**) Because the operation $\vee : E \times E \to E$ is uniformly continuous, it has a (unique) continuous extension to an operator $\vee : E^\wedge \times E^\wedge \to E^\wedge$. The following identities are true for all $x, y, z \in E$:

(i) $x \vee y = y \vee x$,

(ii) $x \vee (y \vee z) = (x \vee y) \vee z$,

(iii) $x \vee x = x$,

(iv) $x + (y \vee z) = (x+y) \vee (x+z)$,

(v) $\alpha(x \vee y) = \alpha x \vee \alpha y \quad \forall\ \alpha \geqslant 0$,

and therefore, by continuity, for all $x, y, z \in E^\wedge$.

(**b**) These are enough to make $\vee$ define a Riesz space structure on $E^\wedge$, as follows. Say that $x \leqslant y$ in $E^\wedge$ iff $x \vee y = y$. Now

$$x \leqslant y \quad \text{and} \quad y \leqslant z \Rightarrow x \vee z = x \vee (y \vee z) = (x \vee y) \vee z = y \vee z = z$$
$$\Rightarrow x \leqslant z;$$

$$x \leqslant y \quad \text{and} \quad y \leqslant x \Rightarrow x = y;$$

$$x \in E^\wedge \Rightarrow x \vee x = x \Rightarrow x \leqslant x;$$

thus (ii) and (iii) show that $\leqslant$ is a partial ordering. Next,

$$x \leqslant y \Rightarrow x \vee y = y \Rightarrow (x+z) \vee (y+z) = y+z$$
$$\Rightarrow x+z \leqslant y+z \quad \forall\ z \in E^\wedge;$$

$$0 \leqslant x \Rightarrow 0 \vee x = x \Rightarrow 0 \vee \alpha x = \alpha x \Rightarrow 0 \leqslant \alpha x \quad \forall\ \alpha \geqslant 0;$$

thus (iv) and (v) show that $(E^\wedge, \leqslant)$ is a partially ordered linear space. Finally,

$$y \leqslant x \text{ and } z \leqslant x \Rightarrow (y \vee z) \vee x = y \vee (z \vee x) = y \vee x = x$$

$$\Rightarrow y \vee z \leqslant x,$$

and of course

$$y \vee (y \vee z) = (y \vee y) \vee z = y \vee z,$$

$$z \vee (y \vee z) = z \vee (z \vee y) = z \vee y = y \vee z,$$

from which $y \vee z = \sup\{y, z\}$ in $E^\wedge$; as y and z are arbitrary, $E^\wedge$ is a Riesz space [14A, exercise].

(c) Any of the conditions in 14Eb can be used to show that the identity map $E \to E^\wedge$ is a Riesz homomorphism.

(d) Finally, let W be any closed neighbourhood of 0 in $E^\wedge$. Then $W \cap E$ is a neighbourhood of 0 in E, so by 22C there is a continuous Riesz pseudo-norm ρ on E such that $\{x : x \in E, \rho x < 1\} \subseteq W \cap E$. As ρ is uniformly continuous, it has a (unique) continuous extension $\hat{\rho}$ on $E^\wedge$. Now $\hat{\rho}$ is still a Riesz pseudo-norm. **P** Consider the conditions of 22A. We can rewrite (iii) as

$$\rho(x) = \rho(x \vee (-x)) \leqslant \rho(x \vee (-x) \vee y) \quad \forall\ x, y \in E;$$

which, like (i) and (ii), will be satisfied by $\hat{\rho}$ if satisfied by ρ. And condition (iv) is a consequence of the continuity of $\hat{\rho}$. **Q** But it is easy to see that

$$\{x : x \in E^\wedge, \hat{\rho}(x) < 1\} \subseteq \overline{W \cap E} \subseteq W.$$

So $\mathfrak{T}^\wedge$ is defined by the continuous Riesz pseudo-norms on E and is locally solid.

Remarks Observe that the conditions (i)–(v) of the proof above are necessary and sufficient conditions for a binary operation $\vee$ to define a Riesz space structure on a given linear space. An alternative proof of the proposition in the locally convex case can be got from 31C in the next chapter. Note also 22Gb; the positive cone of $E^\wedge$ is precisely the closure in $E^\wedge$ of the positive cone of E.

22G Exercises **(a)** Let E be a Riesz space and $\| \ \|$ a Riesz norm on E. Then the dual norm on E' is also a Riesz norm.

(b) Let E be a Riesz space with a locally solid linear space topology, and F a Riesz subspace of E. Then $\bar{F}$ is also a Riesz subspace of E, and the positive cone of $\bar{F}$ is precisely the closure in E of the positive cone of F.

***(c)** Let E be a Riesz space with a locally convex locally solid linear space topology $\mathfrak{T}$. Then $\mathfrak{T}$ is defined by the continuous Riesz seminorms on E. [Hint: 14Jd.]

***(d)** Let E be a Riesz space. Then there is a finest linear space topology on E for which all order-bounded sets are bounded, and this topology is locally solid. [Hint: 25F, proof.]

(e) Let E be a Riesz space with a metrizable linear space topology $\mathfrak{T}$ such that the solid hull of a bounded set is always bounded. Then $\mathfrak{T}$ is locally solid.

Notes and comments It is so natural to require a topology on a Riesz space to be locally solid that some authors use the phrase 'topological Riesz space' to mean a Riesz space endowed with a locally solid linear space topology. But we shall also be interested in weak topologies, which are hardly ever locally solid; consider, for instance, the examples 2XA–2XC.

It is well known that any linear space topology can be defined by pseudo-norms; a simple adaptation of this result gives 22C. Actually we really need this for the proof of 23B below rather than for itself. For the time being, though, we know just that locally solid linear space topologies are precisely those which can be defined by Riesz pseudo-norms. Note that, as hinted in 22Gc, the locally convex case is rather simpler, though it also relies on 14Jb (via 14Jd).

Clearly it is 22Ba/b that give rise to 22F, as well as 22Be and 22E; these results suggest that it might be interesting to study topologies for which the lattice operations are continuous or uniformly continuous. Another generalization, which can be applied to any partially ordered linear space, has been extensively studied; this is the notion of a 'normal' positive cone, treated in PERESSINI and SCHAEFER T.V.S.

The essential idea of 22D and 22Ga is that the polar in $E^{\sim}$ of a solid set in E is solid; this recurs in other contexts.

23 Fatou topologies

All the locally solid topologies in which we are interested actually satisfy a further condition: 0 has a base of order-closed neighbourhoods. This leads to some much deeper results, the outstanding one being

Nakano's theorem [23K], which gives a powerful sufficient condition for a topological Riesz space to be complete. Most of the section is devoted to proving this result through a series of lemmas, many of which have other applications; but as the theorem itself will not be applied in its full strength except in 64E, this work is all starred. The unstarred parts are Theorem 23B, which is an essential tool for coping with Fatou topologies, and the definition of 'Levi' topology [23I], which will be used later in other contexts.

23A Definitions Let E be a Riesz space. A **Fatou** topology on E is a locally solid linear space topology such that 0 has a neighbourhood basis consisting of sets U such that $\mathscr{I}U \subseteq U$ [definition: 11E].

A **Fatou pseudo-norm** on E is a Riesz pseudo-norm ρ such that

$$\varnothing \subset A \uparrow x \text{ in } E^+ \Rightarrow \rho x = \sup\{\rho y : y \in A\};$$

that is, ρ is order-continuous on the left on E^+ [11D]. Following 22A, a **Fatou seminorm** or **Fatou norm** is a Fatou pseudo-norm which is a seminorm or norm respectively.

23B Theorem Let E be a Riesz space and $\mathfrak{T}$ a Fatou topology on E. Then $\mathfrak{T}$ is defined by the continuous Fatou pseudo-norms on E.

Proof (a) Let U be any neighbourhood of 0 for $\mathfrak{T}$, let V be a solid neighbourhood of 0 such that $V \subseteq U$, and let W be a neighbourhood of 0 such that $\mathscr{I}W \subseteq W \subseteq V$. By 22C, there is a continuous Riesz pseudo-norm θ_1 on E such that

$$\theta_1(x) < 1 \Rightarrow x \in W.$$

Let Φ be the set of all functions ρ: $E^+ \to \mathbf{R}^+$ such that (α) $0 \leqslant \rho x \leqslant \theta_1 x$ for every $x \in E^+$ and (β) if $\varnothing \subset A \uparrow x$ in E^+ then $\rho x \leqslant \sup_{y \in A} \rho y$. Then the function ρ_0 given by

$$\rho_0 x = 0 \quad \forall\ x \in E^+ \cap W, \quad \rho_0 x = 1 \quad \forall\ x \in E^+ \backslash W$$

belongs to Φ, so $\Phi \neq \varnothing$. Define θ_0: $E^+ \to \mathbf{R}^+$ by

$$\theta_0 x = \sup\{\rho x : \rho \in \Phi\} \quad \forall\ x \in E^+.$$

Then it is plain that $\theta_0 \in \Phi$ and that $\theta_0 x < 1 \Rightarrow \rho_0 x < 1 \Rightarrow x \in W$.

(b) θ_0 is increasing. **P** Given $x \in E^+$, set $\rho y = \theta_0(x \wedge y)$ for every $y \in E^+$; then a simple computation shows that $\rho \in \Phi$, so that for every $y \geqslant x$

$$\theta_0 y \geqslant \rho y = \theta_0(x \wedge y) = \theta_0 x. \textbf{ Q}$$

Next, $\theta_0(x+y) \leqslant \theta_0 x + \theta_0 y$ for every $x, y \in E^+$.

P (i) Let $z \in E^+$. Define $\rho: E^+ \to \mathbf{R}^+$ by

$$\rho u = \max(0, \theta_0(z+u) - \theta_1 z) \quad \forall\ u \in E^+.$$

Then $\rho \in \Phi$ so $\rho y \leqslant \theta_0 y$, i.e. $\theta_0(z+y) \leqslant \theta_1 z + \theta_0 y$; and this is true for every $z \in E^+$.

(ii) So now define $\rho: E^+ \to \mathbf{R}^+$ by

$$\rho z = \theta_0(z+y) - \theta_0 y \quad \forall\ z \in E^+;$$

the remarks so far are just what we need to see that $0 \leqslant \rho z \leqslant \theta_1 z$ for every $z \in E^+$, so $\rho \in \Phi$ and $\rho x \leqslant \theta_0 x$, i.e. $\theta_0(x+y) \leqslant \theta_0 x + \theta_0 y$, as required. **Q**

(c) Now define $\theta_0: E \backslash E^+ \to \mathbf{R}^+$ by $\theta_0 x = \theta_0(|x|)$. Then

$$\theta_0(x+y) = \theta_0(|x+y|) \leqslant \theta_0(|x|+|y|)$$
$$\leqslant \theta_0(|x|) + \theta_0(|y|) = \theta_0 x + \theta_0 y$$

for all $x, y \in E$. As moreover $\theta_0 x \leqslant \theta_1 x$ for every $x \in E$, θ_0 is continuous; hence, it is a Riesz pseudo-norm; being order-continuous on the left on E^+, it is a Fatou pseudo-norm. Finally

$$\theta_0 x < 1 \Rightarrow \theta_0(|x|) < 1 \Rightarrow |x| \in W \Rightarrow x \in V \subseteq U.$$

[See also 23Na.]

***23C Lemma** Let E be a Riesz space and ρ a Fatou pseudo-norm on E. Suppose that $\varnothing \subset A \uparrow x$ in E; then $\rho x \leqslant \sup_{y \in A} \rho y$.

Proof Consider $B = \{y^+ + x^- : y \in A\}$. Then $B \uparrow$ because $A \uparrow$, and

$$\sup B = x^- + \sup\{y^+ : y \in A\} = x^- + (\sup A)^+ = |x|.$$

So $$\rho x = \rho(|x|) = \sup_{z \in B} \rho z = \sup\{\rho(y^+ + x^-) : y \in A\}.$$

But $$y \in A \Rightarrow y \leqslant x \Rightarrow y^- \geqslant x^- \Rightarrow y^+ + x^- \leqslant |y|,$$

so $$\rho x \leqslant \sup\{\rho(|y|) : y \in A\} = \sup\{\rho y : y \in A\}.$$

***23D Lemma** Let E be a Dedekind σ-complete Riesz space and ρ a Fatou pseudo-norm on E. Let $\langle x_n \rangle_{n \in \mathbf{N}}$ be an order-bounded sequence in E such that $\rho(x_n - x_{n+1}) \leqslant 2^{-n}$ for every $n \in \mathbf{N}$. Then

$$\rho(x_n - \sup_{m \in \mathbf{N}} \inf_{i \geqslant m} x_i) \leqslant 2^{-n+1} \quad \forall\ n \in \mathbf{N}.$$

Proof $\rho(x_n - \sup_{m\in\mathbf{N}} \inf_{i\geqslant m} x_i) = \rho(\sup_{m\in\mathbf{N}} \inf_{i\geqslant m} x_i - x_n)$

$$= \rho(\sup_{m\geqslant n} \inf_{i\geqslant m} x_i - x_n)$$

$$= \rho(\sup_{m\geqslant n} (\inf_{i\geqslant m} x_i - x_n))$$

$$\leqslant \sup_{m\geqslant n} \rho(\inf_{i\geqslant m} x_i - x_n)$$

[using 23C for the first time]

$$= \sup_{m\geqslant n} \rho(x_n - \inf_{i\geqslant m} x_i)$$

$$= \sup_{m\geqslant n} \rho(\sup_{k\geqslant m}(x_n - \inf_{k\geqslant i\geqslant m} x_i))$$

$$\leqslant \sup_{k\geqslant m\geqslant n} \rho(x_n - \inf_{k\geqslant i\geqslant m} x_i)$$

[using 23C for the second time]. Now however, if $k \geqslant m \geqslant n$,

$$x_n - \inf_{k\geqslant i\geqslant m} x_i = \sum_{i=n}^{m-1} (x_i - x_{i+1}) + \sum_{i=m}^{k-1} (\inf_{i\geqslant j\geqslant m} x_j - \inf_{i+1\geqslant j\geqslant m} x_j)$$

$$= \sum_{i=n}^{m-1} (x_i - x_{i+1}) + \sum_{i=m}^{k-1} (\inf_{i\geqslant j\geqslant m} x_i - x_{i+1})^+,$$

so
$$|x_n - \inf_{k\geqslant i\geqslant m} x_i| \leqslant \sum_{i=n}^{k-1} |x_{i+1} - x_i|,$$

and
$$\rho(x_n - \inf_{k\geqslant i\geqslant m} x_i) \leqslant \sum_{i=n}^{k-1} \rho(x_{i+1} - x_i) \leqslant 2^{-n+1},$$

whence the result.

***23E Notation** In the rest of this section we shall have occasion to use some manipulations of filters which, while fairly straightforward, seem to have no generally accepted description.

The principle I use is the following. If X is a (non-empty) set, $\mathscr{F}$ a filter on X, Y another set, and $f: X \to Y$ a function, then $f\mathscr{F}$ is the filter on Y

$$\{A : A \subseteq Y, f^{-1}[A] \in \mathscr{F}\},$$

i.e. the filter on Y with base

$$\{f[A] : A \in \mathscr{F}\}.$$

If X and Y are topological spaces and f is continuous, then $f\mathscr{F} \to ft$ whenever $\mathscr{F} \to t$. If X and Y are uniform spaces and f is uniformly continuous, then $f\mathscr{F}$ is Cauchy whenever $\mathscr{F}$ is.

Two special cases we shall be concerned with are:

(i) Let E be a linear space, $\rho: E \to \mathbf{R}$ a function, $u \in E$, and $\mathscr{F}$ a filter on E. Then $\rho(\mathscr{F} - u)$ is the filter on $\mathbf{R}$

$$\{A : \{x : \rho(x-u) \in A\} \in \mathscr{F}\},$$

so that $\rho(\mathscr{F} - u) \to 0$ iff

$$\forall\ \epsilon > 0 \quad \exists\ A \in \mathscr{F}, \quad |\rho(x-u)| \leqslant \epsilon \quad \forall\ x \in A.$$

(ii) Let E be a Riesz space, $u \in E$, and $\mathscr{F}$ a filter on E. Then $\mathscr{F} \vee u$ is the filter on E

$$\{A : \{x : x \vee u \in A\} \in \mathscr{F}\}.$$

$\mathscr{F} \wedge u$ is defined similarly.

***23F Lemma** Let E be a Riesz space and ρ a Fatou pseudo-norm on E. Then

(a) $F = \{u : \rho u = 0\}$ is a band in E.

(b) If $x \in E$, $\{u : \rho(x-u) = 0\}$ is an order-closed [definition: 11E] sublattice of E.

(c) If $\mathscr{F}$ is a filter on E, $A = \{u : \rho(\mathscr{F} - u) \to 0\}$ is an order-closed sublattice of E (perhaps $\varnothing$).

Proof (a) is a simple computation; F is solid and closed under addition, so is a solid linear subspace; now 23C shows that it is a band. Any translate of a band is an order-closed sublattice, which proves (b). As for (c), it is easy to see that if $x \in A$, then $A = x + F$, reducing it to (b).

***23G Lemma** Let E be a Dedekind σ-complete Riesz space and ρ a Fatou pseudo-norm on E. Let $\mathscr{F}$ be a filter on E which is 'ρ-Cauchy', that is,

$$\inf_{A \in \mathscr{F}} \sup_{u, v \in A} \rho(u-v) = 0.$$

Suppose that A is a member of $\mathscr{F}$ which is order-bounded and closed under the lattice operation $\wedge$. Then there is a $u \in \mathscr{I}\mathscr{D}A$ such that $\rho(\mathscr{F} - u) \to 0$.

Proof Choose a sequence $\langle A_n \rangle_{n \in \mathbf{N}}$ in $\mathscr{F}$ such that

$$\rho(x-y) \leqslant 2^{-n} \quad \forall\ x, y \in A_n \quad \forall\ n \in \mathbf{N}.$$

For each $n \in \mathbf{N}$, choose an element

$$u_n \in A \cap \bigcap_{i \leqslant n} A_i.$$

Then $\{u_n : n \in \mathbf{N}\}$ is order-bounded and $\rho(u_n - u_{n+1}) \leqslant 2^{-n}$ for every $n \in \mathbf{N}$. So if

$$u = \sup_{m\in\mathbf{N}} \inf_{i\geqslant m} u_i,$$

then $\rho(u_n - u) \leqslant 2^{-n+1}$ for every $n \in \mathbf{N}$ by 23D, and

$$\rho(x-u) \leqslant 3.2^{-n} \quad \forall\ x \in A_n \quad \forall\ n \in \mathbf{N},$$

so $\rho(\mathscr{F} - u) \to 0$.

But as A is closed under $\wedge$,

$$\inf_{i\geqslant m} u_i = \inf_{k\geqslant m} \inf_{k\geqslant i\geqslant m} u_i \in \mathscr{D}A \quad \forall\ m \in \mathbf{N},$$

and $u \in \mathscr{I}\mathscr{D}A$.

***23H Lemma** Let $\boldsymbol{E}$ be a Dedekind complete Riesz space with a Fatou topology. Let $B \subseteq \boldsymbol{E}$ be order-bounded and closed under the lattice operation $\wedge$. Let $\mathscr{F}$ be a filter on $\boldsymbol{E}$ such that for every continuous Fatou pseudo-norm ρ on $\boldsymbol{E}$

$$C(\rho) = \{u : u \in B,\ \rho(\mathscr{F} - u) \to 0\} \neq \varnothing.$$

Then there is an $x_0 \in \mathscr{I}\mathscr{D}B$ such that $\mathscr{F} \to x_0$.

Proof If ρ is a continuous Fatou pseudo-norm on $\boldsymbol{E}$, $C(\rho)$ is closed under $\wedge$ [23Fc]. So if $w(\rho) = \inf C(\rho)$, $C(\rho) \downarrow w(\rho)$ and $w(\rho) \in \mathscr{D}B$. But also, again by 23Fc,

$$\rho(\mathscr{F} - w(\rho)) \to 0.$$

Now if ρ_1 and ρ_2 are both continuous Fatou pseudo-norms on $\boldsymbol{E}$, $\rho_1 + \rho_2$ is also a continuous Fatou pseudo-norm on $\boldsymbol{E}$, and

$$C(\rho_1) \supseteq C(\rho_1 + \rho_2),$$

so $w(\rho_1) \leqslant w(\rho_1 + \rho_2)$. So

$$D = \{w(\rho) : \rho \text{ is a continuous Fatou pseudo-norm on } \boldsymbol{E}\}\uparrow;$$

as $D \subseteq \mathscr{D}B$ and B is bounded above, $D \uparrow x_0$ say, and $x_0 \in \mathscr{I}\mathscr{D}B$.

Moreover, if ρ_0 is any continuous Fatou pseudo-norm on $\boldsymbol{E}$,

$$D_0 = \{w(\rho + \rho_0) : \rho \text{ is a continuous Fatou pseudo-norm on } \boldsymbol{E}\}$$

is cofinal with D, i.e. for every $w \in D$ there is a $v \in D_0$ such that $v \geqslant w$, so $D_0 \uparrow x_0$; and

$$D_0 \subseteq \{u : \rho_0(\mathscr{F} - u) \to 0\}$$

because $(\rho + \rho_0)\,(\mathscr{F} - w(\rho + \rho_0)) \to 0$ for every continuous Fatou pseudo-norm ρ. So by 23Fc again, $\rho_0(\mathscr{F} - x_0) \to 0$; as ρ_0 is arbitrary, $\mathscr{F} \to x_0$ [using 23B].

23I Definition Let E be a Riesz space. A linear space topology on E is **Levi** if every set $A \subseteq E$ which is bounded (in the linear topological space sense) and directed upwards, has an upper bound (in the ordering of E).

***23J Lemma** Let E be a Dedekind complete Riesz space with a Levi Hausdorff Fatou topology. Suppose that $\mathscr{F}$ is a Cauchy filter on E such that $\mathscr{F} \wedge z$ [notation: 23E(ii)] converges for each $z \in E$. Then $\mathscr{F}$ converges.

Proof For each $z \in E$, let $x(z)$ be the limit of $\mathscr{F} \wedge z$.

(a) If $z_1 \leqslant z_2$, $x(z_1) \leqslant x(z_2)$. **P** $\mathscr{F} \wedge z_1 = \mathscr{F} \wedge (z_2 \wedge z_1) = (\mathscr{F} \wedge z_2) \wedge z_1$ and the function $u \mapsto u \wedge z_1$ is continuous; so

$$x(z_1) = \lim (\mathscr{F} \wedge z_2 \wedge z_1) = z_1 \wedge \lim (\mathscr{F} \wedge z_2) = z_1 \wedge x(z_2). \text{ Q}$$

(b) Set $A = \{x(z) : z \in E^+\}$. Then $A \uparrow$, by (a) above. But also A is bounded. **P** Let U be any solid neighbourhood of 0 in E. Let V be a balanced neighbourhood of 0 such that $V + V \subseteq U$, and let $B \in \mathscr{F}$ be such that $B - B \subseteq V$. Let $u \in B$; there is an $\alpha \in \mathbf{R}^+$ such that $u \in \alpha V$ and then $B \subseteq u + V \subseteq \alpha V + V \subseteq (\alpha+1)(V+V) \subseteq (\alpha+1)U$. But for any $z \geqslant 0$

$$\begin{aligned} x(z) &\in \overline{\{y \wedge z : y \in B\}} \\ &\subseteq \overline{\{y \wedge z : y \in (\alpha+1)U\}} \\ &\subseteq \overline{(\alpha+1)U} \end{aligned}$$

as U is solid. So A is absorbed by $\overline{U}$. As U is arbitrary, A is bounded. **Q**

So A is bounded above and $x_0 = \sup A$ exists.

(c) Now $\mathscr{F} \to x_0$. **P** Let ρ be a continuous Fatou pseudo-norm on E, and let $\epsilon > 0$. Let $B \in \mathscr{F}$ be such that

$$\rho(u-v) \leqslant \epsilon \quad \forall\ u, \quad v \in B,$$

and fix $u_0 \in B$. Now we know that $A \uparrow x_0$, so

$$\{x(z) - u_0 : z \geqslant u_0^+\} \uparrow x_0 - u_0$$

[using (a) above], and by 23C there is a $z_1 \geqslant u_0^+$ such that

$$\rho(x_0 - u_0) \leqslant \epsilon + \rho(x(z_1) - u_0).$$

But we know that $\quad x(z_1) \in \overline{\{u \wedge z_1 : u \in B\}}$,

so there is a $u_1 \in B$ such that

$$\rho(x(z_1) - u_1 \wedge z_1) \leqslant \epsilon.$$

Now $$|u_1 \wedge z_1 - u_0| = |u_1 \wedge z_1 - u_0 \wedge z_1| \leqslant |u_1 - u_0|,$$

so $\rho(u_1 \wedge z_1 - u_0) \leqslant \rho(u_1 - u_0) \leqslant \epsilon$, and

$$\begin{aligned}\rho(x_0 - u_0) &\leqslant \epsilon + \rho(x(z_1) - u_0)\\ &\leqslant \epsilon + \rho(x(z_1) - u_1 \wedge z_1) + \rho(u_1 \wedge z_1 - u_0)\\ &\leqslant 3\epsilon.\end{aligned}$$

So $$\rho(x_0 - u) \leqslant 4\epsilon \quad \forall\ u \in B;$$

as ϵ and ρ are arbitrary, $\mathscr{F} \to x_0$ [using 23B again]. **Q**

***23K Nakano's theorem** Let E be a Dedekind complete Riesz space with a Levi Hausdorff Fatou topology. Then E is complete (as a uniform space).

Proof Let $\mathscr{F}$ be a Cauchy filter on E. Then for any $y, z \in E$, $(\mathscr{F} \vee y) \wedge z$ is Cauchy and contains the order-bounded set $[y \wedge z, z]$, so it converges [23G, 23H, setting $A = B = [y \wedge z, z]$]. So $\mathscr{F} \vee y$ converges [23J]. So $\mathscr{F}$ converges [23J, upside down].

***23L Proposition** Let E be a Riesz space with a Hausdorff Fatou topology. Let $A \subseteq E$ be a solid set such that $\mathscr{I}A \subseteq A$. Then A is closed.

Proof (a) Let Φ be the set of continuous Fatou pseudo-norms on E. For $\rho \in \Phi$, set

$$F(\rho) = \{z : z \in E, \rho z = 0\}.$$

Then $F(\rho)$ is a band in E [23Fa]. Set

$$F = \bigcup \{F(\rho)^d : \rho \in \Phi\},$$

where for each $\rho \in \Phi$

$$F(\rho)^d = \{u : u \in E, |u| \wedge |z| = 0 \quad \forall\ z \in F(\rho)\}$$

as in 15F. We recall that each $F(\rho)^d$ is a band [15Fa]. Also, if ρ_1 and ρ_2 belong to Φ, so does $\rho_1 + \rho_2$, and

$$F(\rho_1 + \rho_2)^d = (F(\rho_1) \cap F(\rho_2))^d \supseteq F(\rho_1)^d \cup F(\rho_2)^d,$$

from which it follows that F is a solid linear subspace of E.

(b) In fact, F is order-dense. **P** Suppose that $w > 0$ in E. As the topology on E is Hausdorff, there is a neighbourhood U of 0 such that $w \notin U$; by 23B, there is a $\rho \in \Phi$ such that $\rho w > 0$. So $w \notin F(\rho) = F(\rho)^{dd}$ [15Fb; recalling that by 22Eb E is Archimedean]. Let $z \in F(\rho)^d$ be such that $u = |z| \wedge w > 0$. Then $u \in F$ and $0 < u \leqslant w$. As w is arbitrary, F is order-dense by the criterion of 15E. **Q**

(c) Let x be any member of $\bar{A}$. Suppose that $\rho \in \Phi$, $y \in F(\rho)^d$, and that $0 \leqslant y \leqslant |x|$. We know that $\bar{A}$ is solid [22Be], so $y \in \bar{A}$.

Choose a sequence $\langle x_n \rangle_{n \in \mathbf{N}}$ in A such that $\rho(y - x_n) \leqslant 2^{-n}$ for every $n \in \mathbf{N}$. Set $y_n = y \wedge |x_n|$, so that $0 \leqslant y_n \leqslant y$, $y_n \in A$, and $\rho(y - y_n) \leqslant 2^{-n}$ for every $n \in \mathbf{N}$. Let

$$B = \{z : z \in E^+, \quad \exists\ n \in \mathbf{N}, \quad z \leqslant y_i \quad \forall\ i \geqslant n\},$$

so that $B \uparrow$, $B \subseteq A$ and y is an upper bound for B.

? Suppose, if possible, that $y \neq \sup B$. Then there is a $w < y$ which is also an upper bound for B. As $y - w > 0$ and $y - w \in F(\rho)^d$, $y - w \notin F(\rho)$ i.e. $\rho(y - w) > 0$. Let $n \in \mathbf{N}$ be such that $2^{-n} < \frac{1}{2}\rho(y - w)$.

Since every lower bound of $\{y_i : i \geqslant n\}$ is less than or equal to w,

$$0 \leqslant \inf_{m \geqslant n}(y_m - w)^+$$
$$\leqslant \inf\{y_m - z : m \geqslant n, \quad z \leqslant y_i \quad \forall\ i \geqslant n\} = 0$$

by 15C. So $\inf_{m \geqslant n}(y_m - w)^+ = 0$, i.e.

$$w = \inf_{m \geqslant n}(y_m \vee w).$$

So $\quad y - w = \sup_{m \geqslant n}(y - y_m \vee w) = \sup_{m \geqslant n} \sup_{m \geqslant i \geqslant n}(y - y_i \vee w),$

and

$$\rho(y - w) = \sup_{m \geqslant n} \rho(\sup_{m \geqslant i \geqslant n}(y - y_i \vee w))$$
$$\leqslant \sup_{m \geqslant n} \rho(\textstyle\sum_{m \geqslant i \geqslant n}(y - y_i))$$
$$\leqslant \textstyle\sum_{i \geqslant n} \rho(y - y_i) \leqslant 2 \,.\, 2^{-n} < \rho(y - w). \ \mathbf{X}$$

So indeed $B \uparrow y$ and $y \in \mathscr{I}A \subseteq A$.

(d) Thus

$$D = \{y : y \in F, 0 \leqslant y \leqslant |x|\} \subseteq A.$$

But as F is order-dense, $D \uparrow |x|$ so $|x| \in A$. As A is solid, $x \in A$; which proves the proposition.

***23M Lemma** Let E be any Riesz space. Then $\mathfrak{T}_s(E^\sim, E)$ and $\mathfrak{T}_s(E^\times, E)$ are Levi topologies on $E^\sim$ and $E^\times$ respectively.

Proof Suppose that A is a non-empty subset of $E^\sim$, directed upwards and bounded for $\mathfrak{T}_s(E^\sim, E)$. Then $\sup\{fx : f \in A\} < \infty$ for every $x \in E^+$. By 16Db, $\sup A$ exists in $E^\sim$, and A is bounded above in $E^\sim$. If $A \subseteq E^\times$, then $\sup A \in E^\times$ because $E^\times$ is a band in $E^\sim$; so A is bounded above in $E^\times$.

***23N Exercises** **(a)** Let E be a Riesz space with a locally convex Fatou topology $\mathfrak{T}$. Then $\mathfrak{T}$ is defined by the continuous Fatou seminorms on E.

(b) Let E be a Riesz space with a Fatou topology. Let A be a closed totally bounded set in E. Then A is order-closed.

(c) Let E be a Riesz space with a Fatou topology. Then 0 has a neighbourhood basis consisting of solid order-closed sets.

(d) A locally solid Levi linear space topology on an Archimedean Riesz space is Hausdorff.

(e) Let E be a Riesz space with a Levi linear space topology such that order-bounded sets are bounded. Then E has the 'boundedness property' i.e.

If $B \subseteq E^+$ is such that for every sequence $\langle x_n \rangle_{n\in\mathbf{N}}$ in B and every sequence $\langle \alpha_n \rangle_{n\in\mathbf{N}}$ decreasing to 0 in $\mathbf{R}$, the set $\{\alpha_n x_n : n \in \mathbf{N}\}$ is bounded above, then B is bounded above.

(f) Conversely, if E has the boundedness property, then any sequentially complete compatible linear space topology on E must be Levi.

(g) Let E be a Dedekind complete Riesz space and F a band in E. Let $\mathfrak{T}$ be a Levi linear space topology on E. Then the topology on F induced by $\mathfrak{T}$ is Levi.

(h) Let E be a Riesz space with a locally solid topology. Then $\mathfrak{T}_b(E', E)$ is Fatou. In particular, if E has a Riesz norm, the dual norm on E' is Fatou.

Notes and comments Fatou topologies are conspicuous for two reasons; first, they are very common (see the examples in 24C, § 26, 2XA–2XC, 2XF–G, 23Nh above, 42Re, 63K); second, one can prove deep results about them. We are now beginning to see some really interesting interactions between topological, algebraic and order properties. The results of § 22, while sometimes requiring a moment's thought, are all more or less to be expected. But in this section even the simplest arguments bear watching.

*The name 'Levi topology' is suggested by B. Levi's theorem [26Hj/63Ma]. The term 'Fatou norm' is taken from LUXEMBURG & ZAANEN B.F.S. [note XIII, § 41], and the phrase 'Fatou topology' seems to follow naturally. PERESSINI [chapter 4, § 1, pp. 138 *et seq.*] uses both concepts, but only in Dedekind complete spaces; a space with a Fatou topology is called **locally order complete,** whereas one with a Levi topology is called **boundedly order complete** [PERESSINI, p. 139].

The best tool we have for handling Fatou topologies is 23B. (Actually the proof seems odder than the result.) The same method, starting from 22Gc instead of 22C, produces 23Na. But one of the difficulties with Fatou topologies is that there is no effective dual characterization known in the locally convex case. Consequently we are denied the techniques which normally make locally convex topologies easier to handle.

The lemmas leading up to Nakano's theorem, particularly 23G and 23H, are slightly more precise than is necessary for the arguments here; but I have left them in these forms because there exist applications in which the precision is useful.

Nakano's theorem itself has a rather curious position. It is a very striking result and is pleasing for the way in which it relates order-completeness to Cauchy completeness. Its conditions are satisfied by a great variety of important cases, including most of the examples cited above. But in essentially all of these much easier proofs are available. Suppose, for instance, that the topology on E is metrizable; then a simple adaptation of 23D proves the result; and in fact we can do rather better [25Nb]. And 81C is another example of a special case in which a quite different method can be applied.

Indeed, general theorems about Fatou topologies are often of this type. Even 23B is usually bypassed by particular cases, because most Fatou topologies are specified directly by Fatou pseudo-norms. Or consider 23L. This is interesting in various ways, not least because of the construction of the order-dense solid linear subspace F in parts (a)–(b) of the proof, which is often useful in analysing Fatou topologies. But it will actually be applied only to a Dedekind complete normed space [65D], in which case 23D again gives the result at once.

24 Lebesgue topologies

The last condition of this kind which we study is in some ways the most interesting. A topology on a partially ordered space is 'Lebesgue' if every closed set is order-closed in the sense of 11D. The consequences of this property are so important that in the rest of this book I shall tend to classify topological Riesz spaces according to whether or not they possess it.

I begin by running through the elementary general properties of Lebesgue linear space topologies on Riesz spaces [up to 24G]. These are subtle but not difficult. The rest of the section hinges on an effective necessary and sufficient condition for a topology on an Archimedean

Riesz space to be Lebesgue [24I]; 'effective' in the sense that its corollaries 24J and 24K are useful.

24A Definition Let E be a Riesz space. A linear space topology on E is **Lebesgue** if $\varnothing \subset A \downarrow 0 \Rightarrow 0 \in \bar{A}$.

24B Theorem Let E be a Riesz space with a Lebesgue linear space topology.

(a) If $A \subseteq E$, $\mathscr{I}A \subseteq \bar{A}$ and $\mathscr{D}A \subseteq \bar{A}$.

(b) 0 has a base of neighbourhoods U such that $\mathscr{I}U \subseteq U$.

(c) If $\varnothing \subset A \downarrow 0$ in E, and U is a neighbourhood of 0, then there is an $x \in A$ such that $[-x, x] \subseteq U$.

(d) If $\varnothing \subset A \downarrow 0$ in E, then $\mathscr{F}(A\downarrow) \to 0$.

Proofs (a) and (b) are easy.

(c) Let V be a closed neighbourhood of 0 such that $V - V \subseteq U$. Then there is an $x \in A$ such that $[0, x] \subseteq V$. **P** **?** Otherwise, for each $x \in A$, set $A_x = A \cap [0, x]$ and

$$B_x = \{z : \exists\, y \in A_x,\ y \leqslant z \leqslant x\}.$$

Now $A_x \downarrow 0$, so for each $w \in [0, x]$, $\{w \vee y : y \in A_x\} \downarrow w$, and $w \in \mathscr{D}B_x \subseteq \overline{B_x}$. Thus $[0, x] \subseteq \overline{B_x}$. But we are supposing that $[0, x] \nsubseteq V$, and V is closed, so $B_x \nsubseteq V$.

Set

$$C = \{z : z \in E \backslash V,\quad \exists\ x,\quad y \in A,\quad y \leqslant z \leqslant x\}$$
$$= \textstyle\bigcup_{x \in A}(B_x \backslash V).$$

Now

(i) $\forall\ x \in A \quad \exists\ z \in C,\ z \leqslant x$,

(ii) $\forall\ z \in C \quad \exists\ y \in A,\ y \leqslant z$,

(iii) $A \downarrow 0$.

It follows that $C \downarrow 0$. But $0 \notin \bar{C}$. **XQ**

Now $[-x, x] = [0, x] - [0, x] \subseteq V - V \subseteq U$, as required.

(d) follows at once.

24C Corollary A Lebesgue locally solid linear space topology is Fatou.

24D Theorem Let $\boldsymbol{E}$ be an Archimedean Riesz space with a Lebesgue linear space topology.

(a) If A is a non-empty subset of $\boldsymbol{E}$ which is directed upwards and bounded above, $\mathscr{F}(A\uparrow)$ is Cauchy.

(b) Order-bounded sets in $\boldsymbol{E}$ are bounded.

Proof (a) Let B be the set of upper bounds of A in $\boldsymbol{E}$. Because $A\uparrow$, $(B-A)\downarrow$; as $\boldsymbol{E}$ is Archimedean, $\inf(B-A) = 0$ [15C]. So, given any neighbourhood U of 0 in $\boldsymbol{E}$, there is an $x_0 \in A$ and a $y_0 \in B$ such that

$$|z| \leqslant y_0 - x_0 \Rightarrow z \in U$$

[24Bc]. Now suppose that $x_1, x_2 \in A$ and that $x_1 \geqslant x_0$, $x_2 \geqslant x_0$. Then $|x_1 - x_2| \leqslant y_0 - x_0$, so $x_1 - x_2 \in U$. As U is arbitrary, $\mathscr{F}(A\uparrow)$ is Cauchy.

(b) Suppose that $B \subseteq \boldsymbol{E}$ is order-bounded; that is, there exist $x \leqslant y$ such that $B \subseteq [x, y]$. We know that $\{\alpha(y-x) : \alpha > 0\}\downarrow 0$ in $\boldsymbol{E}$ because $\boldsymbol{E}$ is Archimedean [15D]. So for any neighbourhood U of 0 there is an $\alpha > 0$ such that $[0, \alpha(y-x)] \subseteq U$, i.e. $[0, y-x] \subseteq \alpha^{-1}U$ [24Bc]. Thus $[0, y-x]$ is bounded. So $B \subseteq x + [0, y-x]$ must be bounded.

24E Corollary If $\boldsymbol{E}$ is a Riesz space with a compatible Lebesgue linear space topology under which it is complete as a uniform space, it is Dedekind complete.

Proof 21Ba, 24Da, 21Bd.

24F Corollary If $\boldsymbol{E}$ is an Archimedean Riesz space with a Lebesgue linear space topology, then $\boldsymbol{E}' \subseteq \boldsymbol{E}^{\times}$.

Proof Let $f \in \boldsymbol{E}'$. Then f is bounded on order-bounded sets [24Db], so $f \in \boldsymbol{E}^{\sim}$[16C]. Consider $|f|$ in $\boldsymbol{E}^{\sim}$. If $\varnothing \subset A\downarrow 0$ in $\boldsymbol{E}$, and $\epsilon > 0$, then $U = \{y : y \in \boldsymbol{E}, |fy| \leqslant \epsilon\}$ is a neighbourhood of 0, so there is an $x \in A$ such that $[-x, x] \subseteq U$ [24Bc]. Now $|f|(x) \leqslant \epsilon$ [16Ea]. As ϵ is arbitrary, $\inf_{x \in A} |f|(x) = 0$; as A is arbitrary, $|f|$ is order-continuous [14Ec], i.e. $|f| \in \boldsymbol{E}^{\times}$. So $f \in \boldsymbol{E}^{\times}$, as required [16H].

24G Proposition Let $\boldsymbol{E}$ be a Riesz space and $\mathfrak{T}$ a locally convex locally solid linear space topology on $\boldsymbol{E}$ such that $\boldsymbol{E}' \subseteq \boldsymbol{E}^{\times}$. Then $\mathfrak{T}$ is Lebesgue.

Proof Suppose that $\varnothing \subset A\downarrow 0$ in $\boldsymbol{E}$. Then 0 belongs to the closure of A for the weak topology $\mathfrak{T}_s(\boldsymbol{E}, \boldsymbol{E}')$. **P** Suppose $\langle f_i \rangle_{i<n}$ is a finite

sequence in E', and $\epsilon > 0$. Then $f = \sum_{i<n} |f_i|$ is defined in $E^\times$; so there is an $x \in A$ such that $fx \leqslant \epsilon$, and now $|f_i x| \leqslant \epsilon$ for every $i \leqslant n$. **Q** Consequently 0 must belong to the closed convex hull of A for $\mathfrak{T}$. Now let U be any solid neighbourhood of 0 for $\mathfrak{T}$; U meets the convex hull of A, so there exist finite sequences $\langle x_i \rangle_{i<n}$ in A and $\langle \alpha_i \rangle_{i<n}$ in $\mathbf{R}^+$ such that $\sum_{i<n} \alpha_i = 1$ and $\sum_{i<n} \alpha_i x_i \in U$. Let x be a member of A such that $x \leqslant \inf_{i<n} x_i$. Then

$$0 \leqslant x = \textstyle\sum_{i<n} \alpha_i x \leqslant \sum_{i<n} \alpha_i x_i \in U,$$

so $x \in U$. So A meets every neighbourhood of 0 for $\mathfrak{T}$. As A is arbitrary, $\mathfrak{T}$ is Lebesgue.

***24H** The following remarkable result has various applications in the analysis of Lebesgue topologies. Apart from the corollaries below, it will be employed in § 65 and again in Chapter 8.

Proposition Let E be a Riesz space and $\mathfrak{T}$ a linear space topology on E such that order-bounded sets are bounded. Then the following are equivalent:

(i) whenever $\langle x_n \rangle_{n \in \mathbf{N}}$ is a disjoint sequence in E^+ which is bounded above, $\langle x_n \rangle_{n \in \mathbf{N}} \to 0$;

(ii) whenever $\varnothing \subset A \uparrow$ in E, and A is bounded above, $\mathscr{F}(A\uparrow)$ is Cauchy.

Proof (a) Suppose that condition (i) holds. Let us say that a sequence $\langle x_i \rangle_{i \in \mathbf{N}}$ has *property* P_r if it is bounded above and $\inf_{i \in J} x_i = 0$ whenever $J \subseteq \mathbf{N}$ has $r+1$ members.

Then the sequences with property P_r are convergent to 0. **P** Induce on r. For $r = 0$, only the constant zero sequence has property P_r; so the result is trivial. **?** So suppose, if possible, that $\langle x_i \rangle_{i \in \mathbf{N}}$ is a sequence in E^+, bounded above by x say, which has property P_{r+1} but is not convergent to zero. Then there is a neighbourhood U of 0 and a subsequence $\langle y_i \rangle_{i \in \mathbf{N}}$ of $\langle x_i \rangle_{i \in \mathbf{N}}$ such that $y_i \notin U$ for every $i \in \mathbf{N}$. Note that $\langle y_i \rangle_{i \in \mathbf{N}}$ still has property P_{r+1}. Choose inductively a sequence $\langle V_i \rangle_{i \in \mathbf{N}}$ of neighbourhoods of 0 such that $V_0 + V_0 \subseteq U$ and $V_{i+1} + V_{i+1} \subseteq V_i$ for every $i \in \mathbf{N}$.

I shall construct inductively sequences $\langle w_i \rangle_{i \in \mathbf{N}}$ and $\langle z_{ij} \rangle_{i,j \in \mathbf{N}}$ in such a way that:

(α) $\langle z_{ij} \rangle_{j \in \mathbf{N}}$ has property P_{r+1} and is bounded above by x for every $i \in \mathbf{N}$;

(β) $z_{ij} \notin V_0 + V_i$ for every $i, j \in \mathbf{N}$;

(γ) $0 \leqslant w_i \leqslant z_{i0}$ for every $i \in \mathbf{N}$;

(δ) $w_i \notin V_0$ for every $i \in \mathbf{N}$;

(ϵ) $w_k \wedge z_{ij} = 0$ whenever $j \in \mathbf{N}$ and $k < i \in \mathbf{N}$.

Construction Set $z_{0j} = y_j$ for every $j \in \mathbf{N}$. Then $\langle z_{0j} \rangle_{j \in \mathbf{N}}$ has properties (α) and (β) because $V_0 + V_0 \subseteq U$. Now suppose $\langle z_{ij} \rangle_{j \in \mathbf{N}}$ and $\langle w_k \rangle_{k<i}$ have been found. Because $[0, x]$ is bounded, there is a $\beta > 0$ such that $\beta[0, x] \subseteq V_i$. Set $w_i = (z_{i0} - \beta x)^+$. Then $z_{i0} - w_i = z_{i0} \wedge \beta x \in V_i$, so $w_i \notin V_0$ because $z_{i0} \notin V_0 + V_i$. Now consider the sequence

$$\langle z_{ij} \wedge \beta^{-1} z_{i0} \rangle_{j \geqslant 1}.$$

Because $\langle z_{ij} \rangle_{j \in \mathbf{N}}$ has property P_{r+1}, this sequence has property P_r, and by the inductive hypothesis is convergent to 0. So there is an $m \in \mathbf{N}$ such that $z_{ij} \wedge \beta^{-1} z_{i0} \in V_{i+1}$ for every $j \geqslant m$. Set

$$z_{i+1,j} = (z_{i,j+m} - \beta^{-1} z_{i0})^+ \quad \forall\ j \in \mathbf{N}.$$

Then clearly $\langle z_{i+1,j} \rangle_{j \in \mathbf{N}}$ has property P_{r+1}. Next,

$$z_{i,j+m} - z_{i+1,j} \in V_{i+1},$$

and $z_{i,j+m} \notin V_0 + V_{i+1} + V_{i+1}$, so

$$z_{i+1,j} \notin V_0 + V_{i+1} \quad \forall\ j \in \mathbf{N}.$$

If $k < i$ and $j \in \mathbf{N}$,

$$w_k \wedge z_{i+1,j} \leqslant w_k \wedge z_{i,j+m} = 0;$$

while finally

$$\begin{aligned} w_i \wedge \beta z_{i+1,j} &= (z_{i0} - \beta x)^+ \wedge \beta(z_{i,j+m} - \beta^{-1} z_{i0}) \\ &\leqslant (z_{i0} - \beta x)^+ \wedge (\beta x - z_{i0})^+ = 0, \end{aligned}$$

so by 14Kb $w_i \wedge z_{i+1,j} = 0$. Thus the process continues.

But now, if $k < i$,

$$0 \leqslant w_k \wedge w_i \leqslant w_k \wedge z_{i0} = 0;$$

so $\langle w_i \rangle_{i \in \mathbf{N}}$ is a disjoint sequence in E^+, bounded above by x; but it is not convergent to 0. **X**

Thus every sequence with property P_{r+1} is convergent to 0; and the induction continues. **Q**

(b) (i) ⇒ (ii) **?** Suppose, if possible, that (i) is true but that there is a non-empty set A, directed upwards and bounded above, such that $\mathscr{F}(A\uparrow)$ is not Cauchy. Then there is a neighbourhood U of 0 such that

$$\{y : y \in A, y \geqslant z\} \nsubseteq z + U$$

for any $z \in A$. Choose a sequence $\langle z_n \rangle_{n \in \mathbf{N}}$ in A such that, for each $n \in \mathbf{N}$, $z_{n+1} \geqslant z_n$ and $z_{n+1} - z_n \notin U$. Let u be an upper bound for A.

Let V be a neighbourhood of 0 such that $V + V \subseteq U$. Since $[0, u - z_0]$ is bounded, there is an integer $r \in \mathbf{N}$ such that

$$[0, u - z_0] \subseteq (r+1) V.$$

Now define

$$x_i = (z_{i+1} - z_i - (r+1)^{-1}(u - z_0))^+.$$

Then $\langle x_i \rangle_{i \in \mathbf{N}}$ has property P_r. **P** Suppose that $J \subseteq \mathbf{N}$ has $r+1$ members. Set $y = \inf_{i \in J} x_i$. Then

$$y \wedge ((r+1)^{-1}(u - z_0) + z_i - z_{i+1})^+ = 0 \quad \forall\ i \in J,$$

so $\quad 0 = y \wedge \sum_{i \in J}((r+1)^{-1}(u - z_0) + z_i - z_{i+1})^+$

[using 14Ja]

$$= y \wedge \sum_{i \in J}((r+1)^{-1}(u - z_0) + z_i - z_{i+1} + x_i)$$

$$= y \wedge ((u - z_0) - \sum_{i \in J}(z_{i+1} - z_i) + \sum_{i \in J} x_i)$$

$$\geqslant y \wedge \sum_{i \in J} x_i = y. \textbf{ Q}$$

So by (a) above $\langle x_i \rangle_{i \in \mathbf{N}} \to 0$ and there is an $i \in \mathbf{N}$ such that $x_i \in V$. But now

$$z_{i+1} - z_i = (z_{i+1} - z_i) \wedge (r+1)^{-1}(u - z_0) + x_i$$

$$\in V + V \subseteq U.$$

contradicting the choice of $\langle z_i \rangle_{i \in \mathbf{N}}$. **X**

(c) **(ii) ⇒ (i)** If $\langle x_n \rangle_{n \in \mathbf{N}}$ is a disjoint sequence in $\boldsymbol{E}^+$, bounded above by x, set

$$A = \{\textstyle\sum_{i<n} x_i : n \in \mathbf{N}\}.$$

Then $\varnothing \subset A \uparrow$ and A is also bounded above by x [14Kc]. So $\mathscr{F}(A\uparrow)$ is Cauchy, i.e. $\langle \sum_{i<n} x_i \rangle_{n \in \mathbf{N}}$ is a Cauchy sequence. So $\langle x_n \rangle_{n \in \mathbf{N}} \to 0$.

***24I Corollary** Let $\boldsymbol{E}$ be an Archimedean Riesz space, and $\mathfrak{T}$ a linear space topology on $\boldsymbol{E}$. Then $\mathfrak{T}$ is Lebesgue iff (i) order-bounded sets are bounded (ii) $\mathfrak{T}$ has a base of neighbourhoods U of 0 such that $\mathscr{S}U \subseteq U$ (iii) whenever $\langle x_n \rangle_{n \in \mathbf{N}}$ is a disjoint sequence in $\boldsymbol{E}^+$ which is bounded above, $\langle x_n \rangle_{n \in \mathbf{N}} \to 0$ for $\mathfrak{T}$.

Proof **(a)** If $\mathfrak{T}$ is Lebesgue, it satisfies conditions (i) and (ii) by 24Bb and 24Db, and condition (iii) by 24Da and 24H(ii) ⇒ (i).

(b) Conversely, suppose that $\mathfrak{T}$ satisfies the conditions. Suppose that $\varnothing \subset A \downarrow 0$ in $\boldsymbol{E}$, and let U be any neighbourhood of 0 for $\mathfrak{T}$. Let

V be a neighbourhood of 0 such that $\mathscr{I}V \subseteq U$. Now by 24H, inverted, $\mathscr{F}(A\downarrow)$ is Cauchy; let $x \in A$ be such that $x-y \in V$ whenever $y \in A$ and $y \leqslant x$. Now as $A\downarrow 0$,

$$\{x-y : y \in A, \, y \leqslant x\}\uparrow x,$$

and $x \in \mathscr{I}V \subseteq U$. As U is arbitrary, $0 \in \bar{A}$; as A is arbitrary, $\mathfrak{T}$ is Lebesgue.

***24J Corollary** Let $\boldsymbol{E}$ be an Archimedean Riesz space and $\mathfrak{T}$ a Fatou topology on $\boldsymbol{E}$. Then $\mathfrak{T}$ is Lebesgue iff whenever $\langle x_n\rangle_{n\in\mathbf{N}}$ is a disjoint sequence in $\boldsymbol{E}^+$ which is bounded above, $\langle x_n\rangle_{n\in\mathbf{N}} \to 0$ for $\mathfrak{T}$.

Proof Immediate from 24I.

***24K Corollary** Let $\boldsymbol{E}$ be an Archimedean Riesz space and $\mathfrak{T}$ a locally convex linear space topology on $\boldsymbol{E}$. Then $\mathfrak{T}$ is Lebesgue iff (i) $\boldsymbol{E}' \subseteq \boldsymbol{E}^\times$, (ii) whenever $\langle x_n\rangle_{n\in\mathbf{N}}$ is a disjoint sequence in $\boldsymbol{E}^+$, bounded above, $\langle x_n\rangle_{n\in\mathbf{N}} \to 0$ for $\mathfrak{T}$.

Proof **(a)** If $\mathfrak{T}$ is Lebesgue, it satisfies the conditions (i) and (ii) by 24F and 24I.

(b) The point is that condition (i) above implies both (i) and (ii) of 24I. If $\boldsymbol{E}' \subseteq \boldsymbol{E}^\times$, then order-bounded sets are $\mathfrak{T}_s(\boldsymbol{E}, \boldsymbol{E}')$-bounded, therefore $\mathfrak{T}$-bounded. At the same time, if $A \subseteq \boldsymbol{E}^\times$ has polar $A^0 \subseteq \boldsymbol{E}$, then $\mathscr{I}A^0 \subseteq A^0$. **P** Suppose that $\varnothing \subset B \subseteq A^0$ and that $B\uparrow x$ in $\boldsymbol{E}$. Then $\{x-y : y \in B\}\downarrow 0$, so for any $f \in A$,

$$\inf_{y\in B}|f(x-y)| \leqslant \inf_{y\in B}|f|\,(x-y) = 0,$$

and $|fx| \leqslant \sup_{y\in B}|fy| \leqslant 1$. As f is arbitrary, $x \in A^0$. **Q** So if U is any closed absolutely convex neighbourhood of 0 for $\mathfrak{T}$, $\mathscr{I}U \subseteq U^{00} = U$.

So $\mathfrak{T}$ satisfies the conditions of 24I, and is Lebesgue.

24L Exercises **(a)** Let $\boldsymbol{E}$ be a Riesz space with a Lebesgue linear space topology. Then any continuous Riesz pseudo-norm on $\boldsymbol{E}$ is a Fatou pseudo-norm.

(b) Let $\boldsymbol{E}$ be a Riesz space with a compatible linear space topology $\mathfrak{T}$ for which order-bounded sets are relatively compact. Then $\mathfrak{T}$ is Lebesgue and $\boldsymbol{E}$ is Dedekind complete. [Use 21Be.]

***(c)** Let $\boldsymbol{E}$ be an Archimedean Riesz space with a metrizable Lebesgue linear space topology. Then $\boldsymbol{E}$ has the countable sup property. [Show that if $A \uparrow x$, there is a sequence in A converging to x; or, use 24I and 18D.]

★(**d**) Let E be an Archimedean Riesz space with a Lebesgue linear space topology $\mathfrak{T}$. Then there is a locally solid linear space topology on E which is Lebesgue and finer than $\mathfrak{T}$. [Hint: 25F, proof.]

★(**e**) Let E be a Riesz space with a complete compatible linear space topology $\mathfrak{T}$ satisfying the conditions of 24H. Then $\mathfrak{T}$ is Lebesgue. [Use 21Bd.]

★(**f**) Let E be a Riesz space with a locally solid linear space topology $\mathfrak{T}$ satisfying the conditions of 24H. Then the topology $\mathfrak{T}^\wedge$ on its completion $E^\wedge$ [22F] is Lebesgue. [Hint: show that $\mathfrak{T}^\wedge$ also satisfies the condition 24H(ii).]

(**g**) Let E be a Dedekind σ-complete Riesz space with a Lebesgue linear space topology $\mathfrak{T}$. Let $\langle x_n \rangle_{n\in\mathbf{N}}$ be an order-bounded sequence in E such that $x = \inf_{n\in\mathbf{N}} \sup_{i\geqslant n} = \sup_{n\in\mathbf{N}} \inf_{i\geqslant n} x_i$. Then $\langle x_n \rangle_{n\in\mathbf{N}} \rightarrow x$ for $\mathfrak{T}$.

Notes and comments Lebesgue topologies are a perpetual source of surprises. The early results of this section show how quite simple arguments can have unexpected consequences; moreover, subject to suitable modifications, many of them can be generalized to much wider contexts. The subtlest appears to be 24Bc; this is the foundation of the most interesting results.

24F and 24G seem to sort out locally solid locally convex Lebesgue topologies, which are of course the most important ones. Note that the condition 'locally solid' in 24G cannot be dispensed with [8XB]. However, the general condition 24I seems to add something even in this case. In § 82 this condition is tied in with duality theory in some general remarks on locally convex Lebesgue topologies.

24H is a very powerful result. Apart from its application to 24I above, it has a striking consequence in 81H. Perhaps it would be profitable to study spaces satisfying the conditions of 24H in their own right. The locally solid ones are of course dealt with by 24Lf.

Of course there are many important examples of Lebesgue topologies. Some are given in 1XD, 1XF, 26B, 2XB, 2XC, 2XF, 2XG, 63K, 6XF, 6XI and 82H.

*The name 'Lebesgue topology' is suggested by 24Lg, which is an abstract version of Lebesgue's Dominated Convergence Theorem [63Md]. It corresponds to 'condition A(ii)' of LUXEMBURG & ZAANEN B.F.S. [note X, § 33]. NAKANO L.T. [§ 6] gives a definition of 'continuous topology' which is equivalent to my 'sequentially Lebesgue topology' [83 Ka], but in a context which ensures that it will be actually Lebesgue [NAKANO L.T., theorem 6.2].

25 Complete metrizable topologies

As in other branches of functional analysis, Banach spaces have a pre-eminent position in the theory of topological Riesz spaces. Banach lattices, that is, Riesz spaces endowed with Riesz norms under which they are complete, were the first topological Riesz spaces to be studied, and remain the most important examples. One of the purposes of this chapter is to exhibit their properties in such a way that it is clear which aspects of their structure are involved at each point. In the present section I give those results which are consequences of the closed graph theorem and therefore apply to all complete metrizable linear space topologies.

We find that a Riesz space structure on a Banach space, if related to the topology at all, is very strongly related; the same is true, of course, of other kinds of algebraic structure. If we restrict our attention to spaces in which the topology is compatible, they seem to divide naturally into three types: (i) not locally solid; (ii) locally solid, not Lebesgue; (iii) locally solid and Lebesgue. The first class will not concern us, though I gave an example in 2XD. The other two are both important. Since we know from 25A that the topology can be regarded as defined by the Riesz space structure, it is natural to ask what Riesz space properties correspond to the classification above. The division between (i) and the others can be expressed directly in terms of a concept, 'uniform completeness', which will be useful elsewhere; so I include this result [25K]. In the locally convex case, the three classes are distinguished very simply by the nature of the dual space [25G, 25M].

25A Proposition Let E be a Riesz space. Then there is at most one compatible complete metrizable linear space topology on E, and for any such topology $(U \cap E^+) - (U \cap E^+)$ is a neighbourhood of 0 whenever U is.

Proof Let $\mathfrak{T}$ and $\mathfrak{T}'$ be two compatible complete metrizable linear space topologies on E. Choose sequences $\langle U_n \rangle_{n \in \mathbf{N}}$ and $\langle V_n \rangle_{n \in \mathbf{N}}$ of balanced sets in E such that

$$\{U_n : n \in \mathbf{N}\} \text{ is a neighbourhood basis at 0 for } \mathfrak{T},$$

$$\{V_n : n \in \mathbf{N}\} \text{ is a neighbourhood basis at } 0 \text{ for } \mathfrak{T}',$$

$$U_{n+1} + U_{n+1} \subseteq U_n \quad \text{and} \quad V_{n+1} + V_{n+1} \subseteq V_n \quad \forall\ n \in \mathbf{N}.$$

Set
$$W_n = (U_n \cap V_n \cap E^+) - (U_n \cap V_n \cap E^+).$$

Then for every $n \in \mathbf{N}$, W_n is balanced, absorbent [because $E^+ - E^+ = E$], $W_{n+1} + W_{n+1} \subseteq W_n$, and $W_{n+1} \subseteq U_n \cap V_n$. So $\{W_n : n \in \mathbf{N}\}$ is a neighbourhood basis at 0 for a metrizable linear space topology $\mathfrak{S}$ finer than either $\mathfrak{T}$ or $\mathfrak{T}'$.

But $\mathfrak{S}$ is complete. **P** Let $\langle x_n \rangle_{n\in\mathbf{N}}$ be a sequence in E such that $x_{n+1} - x_n \in W_n$ for every $n \in \mathbf{N}$. Then we can choose sequences $\langle y_n \rangle_{n\in\mathbf{N}}$ and $\langle z_n \rangle_{n\in\mathbf{N}}$ such that

$$y_n \in U_n \cap V_n \cap E^+, \quad z_n \in U_n \cap V_n \cap E^+, \quad y_n - z_n = x_{n+1} - x_n. \quad \forall\ n \in \mathbf{N}.$$

Now $\sum_{i\in\mathbf{N}} y_i$ exists for both $\mathfrak{T}$ and $\mathfrak{T}'$; let the limits be y and y' respectively. Since $\langle \sum_{i>n} y_i \rangle_{n\in\mathbf{N}} \uparrow$ and $\mathfrak{T}$ is compatible,

$$y = \sup_{n\in\mathbf{N}} \textstyle\sum_{i<n} y_i$$

[21Bd]; but the same applies to y', so $y = y'$. Also $y - \sum_{i<n} y_i$ belongs to the $\mathfrak{T}$-closure of $\sum_{i\geqslant n} U_i$, which must be included in $U_n + U_n + U_n$; so

$$y - \textstyle\sum_{i<n} y_i = y' - \sum_{i<n} y_i \in (U_n + U_n + U_n) \cap (V_n + V_n + V_n) \cap E^+ \subseteq W_{n-2}$$

for every $n \geqslant 2$, and $y = \sum_{i\in\mathbf{N}} y_i$ for $\mathfrak{S}$. Similarly, $z = \sum_{i\in\mathbf{N}} z_i$ exists for $\mathfrak{S}$, and $\langle x_n \rangle_{n\in\mathbf{N}} \to y - z + x_0$ for $\mathfrak{S}$.

As $\mathfrak{S}$ is metrizable, this is enough to show that it is complete. **Q**

But now, by the closed graph theorem, $\mathfrak{S} = \mathfrak{T}' = \mathfrak{T}$; which proves both assertions of the proposition.

25B Corollary Let E be a Riesz space with a compatible complete metrizable linear space topology, and suppose that $\langle x_n \rangle_{n\in\mathbf{N}}$ is a sequence in E converging to 0. Then there exist non-negative sequences $\langle y_n \rangle_{n\in\mathbf{N}}$ and $\langle z_n \rangle_{n\in\mathbf{N}}$, also converging to 0, such that $x_n = y_n - z_n$ for every $n \in \mathbf{N}$.

Proof Let $\langle U_k \rangle_{k\in\mathbf{N}}$ be a decreasing sequence of sets forming a neighbourhood basis at 0 in E, and let $\langle n_k \rangle_{k\in\mathbf{N}}$ be a strictly increasing sequence in $\mathbf{N}$ such that $x_i \in (U_k \cap E^+) - (U_k \cap E^+)$ whenever $i \geqslant n_k$. Now, for $n_k \leqslant i < n_{k+1}$, choose y_i and z_i in $U_k \cap E^+$ such that $x_i = y_i - z_i$. (For $i < n_0$, choose y_i and z_i arbitrarily such that $x_i = y_i - z_i$.) Then $\langle y_i \rangle_{i\in\mathbf{N}}$ and $\langle z_i \rangle_{i\in\mathbf{N}}$ are the required sequences.

Note It is **not** the case that $\langle x_n^+ \rangle_{n\in\mathbf{N}}$ and $\langle x_n^- \rangle_{n\in\mathbf{N}}$ necessarily converge to 0.

25C Corollary Let E be a Riesz space with a compatible complete metrizable linear space topology, and let $\langle x_n \rangle_{n \in \mathbf{N}}$ be a sequence in E converging to 0. Then $\langle x_n \rangle_{n \in \mathbf{N}}$ has a subsequence $\langle y_n \rangle_{n \in \mathbf{N}}$ such that $\{2^n |y_n| : n \in \mathbf{N}\}$ is bounded above.

Proof As E is metrizable, $\langle x_n \rangle_{n \in \mathbf{N}}$ has a subsequence $\langle y_n \rangle_{n \in \mathbf{N}}$ such that $\langle 4^n y_n \rangle_{n \in \mathbf{N}} \to 0$. Now let $\langle w_n \rangle_{n \in \mathbf{N}}$ and $\langle z_n \rangle_{n \in \mathbf{N}}$ be non-negative sequences, convergent to 0, such that $w_n - z_n = 4^n y_n$ for every $n \in \mathbf{N}$. As $\{w_n + z_n : n \in \mathbf{N}\}$ is bounded, and E is complete, $y = \sum_{n \in \mathbf{N}} 2^{-n}(w_n + z_n)$ exists in E; as $\mathfrak{T}$ is compatible, $y \geqslant 2^{-n}(w_n + z_n) \geqslant 2^n |y_n|$ for every $n \in \mathbf{N}$, as required.

25D Proposition Let E be a Riesz space with a compatible complete metrizable linear space topology, and F a Riesz space with a linear space topology such that order-bounded sets are bounded. Then every increasing linear map $T: E \to F$ is continuous.

Proof Let $\langle x_n \rangle_{n \in \mathbf{N}}$ be a sequence converging to 0 in E. Let $\langle y_n \rangle_{n \in \mathbf{N}}$ be a subsequence of $\langle x_n \rangle_{n \in \mathbf{N}}$ such that $\{2^n |y_n| : n \in \mathbf{N}\}$ has an upper bound y say [25C]. Then

$$-y \leqslant 2^n y_n \leqslant y,$$

so

$$-Ty \leqslant 2^n T y_n \leqslant Ty,$$

i.e.

$$Ty_n \in 2^{-n}[-Ty, Ty] \quad \forall \ n \in \mathbf{N}.$$

Since, by hypothesis, $[-Ty, Ty]$ is bounded in F, $\langle Ty_n \rangle_{n \in \mathbf{N}} \to 0$. But as E is metrizable, this is enough to show that T is continuous.

25E Corollary Let E be a Riesz space with a compatible complete metrizable linear space topology. Then $E^{\sim} \subseteq E'$.

25F Proposition Let E be a Riesz space with a compatible complete metrizable linear space topology $\mathfrak{T}$ for which order-bounded sets are bounded. Then $\mathfrak{T}$ is locally solid.

Proof If U is any neighbourhood of 0 for $\mathfrak{T}$, set

$$W_U = \{x : [-|x|, |x|] \subseteq U\}.$$

Then W_U is absorbent [as order-bounded sets are bounded],

$$W_U + W_U \subseteq W_{U+U}$$

[using 14Bn and 14Jb to show that

$$[-|x+y|, |x+y|] \subseteq [-|x|, |x|]+[-|y|, |y|] \quad \text{for every} \quad x, y \in E],$$

W_U is solid, and $W_U \subseteq U$. So

$$\{W_U : U \text{ is a neighbourhood of } 0 \text{ for } \mathfrak{T}\}$$

is a neighbourhood basis at 0 for a locally solid linear space topology $\mathfrak{S} \supseteq \mathfrak{T}$. But now order-bounded sets are still bounded for $\mathfrak{S}$ [22Bd] so by 25D the identity map $(E, \mathfrak{T}) \to (E, \mathfrak{S})$ is continuous, and $\mathfrak{S} = \mathfrak{T}$. So $\mathfrak{T}$ must have been locally solid.

25G Corollary Let E be a Riesz space with a locally convex compatible complete metrizable linear space topology $\mathfrak{T}$. Then $\mathfrak{T}$ is locally solid iff $E' \subseteq E^{\sim}$, and in this case $E' = E^{\sim}$.

Proof (a) If $\mathfrak{T}$ is locally solid, then $E' \subseteq E^{\sim}$ by 22D; but we also know that $E^{\sim} \subseteq E'$ by 25E above, so $E' = E^{\sim}$. (b) Conversely, if $E' \subseteq E^{\sim}$, then order-bounded sets, which are certainly bounded for the weak topology $\mathfrak{T}_s(E, E^{\sim})$, must be bounded for $\mathfrak{T}_s(E, E')$, and therefore for $\mathfrak{T}$; so by 25F $\mathfrak{T}$ is locally solid.

25H Let us now investigate a general condition for the compatible complete metrizable linear space topology (if any) on a given Riesz space to be locally solid.

(a) Definition Let E be a Riesz space with an order unit e [definition: 14I]. Define $\| \ \|_e : E \to \mathbf{R}$ by

$$\|x\|_e = \inf\{\alpha : \alpha \geqslant 0, |x| \leqslant \alpha e\},$$

this being defined because e is an order unit. Then $\| \ \|_e$ is a Fatou seminorm, and if E is Archimedean it is a norm. The topology it defines is Levi; indeed, all bounded sets are order-bounded.

(b) Definition Let E be an Archimedean Riesz space. For each $e \in E^+$, let E_e be the solid linear subspace of E generated by e, i.e. $E_e = \{x : \exists \ n \in \mathrm{N}, |x| \leqslant ne\}$. Then E_e is in itself an Archimedean Riesz space with an order unit, e. Now we say that E is **uniformly complete** if E_e is complete under the norm $\| \ \|_e$ for every $e \in E^+$.

25I I give a couple of sufficient conditions for a space to be uniformly complete.

Lemma Let E be a Riesz space with a compatible complete linear space topology for which order-bounded sets are bounded. Then E is uniformly complete.

Proof Let $e \in E^+$, and let $\langle x_n \rangle_{n\in\mathbf{N}}$ be a sequence in E_e which is Cauchy for $\| \ \|_e$. As the unit ball of E_e is $[-e, e]$, which is bounded in E, the identity map $E_e \to E$ is continuous; so $\langle x_n \rangle_{n\in\mathbf{N}}$ is Cauchy in E, and has a limit x say in E. Next, the unit ball of E_e is closed in E, because the topology on E is compatible; it follows at once that $\langle x_n \rangle_{n\in\mathbf{N}}$ converges to x for $\| \ \|_e$. [For given $\epsilon > 0$, there is an $n \in \mathbf{N}$ such that

$$x_i \in \{z : \|z - x_j\|_e \leqslant \epsilon\} \quad \text{for every} \quad i, j \geqslant n;$$

but this is closed in E, so $x \in \{z : \|z - x_j\|_e \leqslant \epsilon\}$ for every $j \geqslant n$.] As $\langle x_n \rangle_{n\in\mathbf{N}}$ is arbitrary, E_e is complete; as e is arbitrary, E is uniformly complete.

25J Lemma A Dedekind σ-complete Riesz space E is uniformly complete.

Proof Let $e \in E^+$. Then E_e is also Dedekind σ-complete. Let $\langle x_n \rangle_{n\in\mathbf{N}}$ be a sequence in E_e such that $\|x_{n+1} - x_n\|_e \leqslant 2^{-n}$ for every $n \in \mathbf{N}$. Then $\{x_n : n \in \mathbf{N}\}$ is bounded, therefore order-bounded. So $\langle x_n \rangle_{n\in\mathbf{N}}$ converges to $\sup_{m\in\mathbf{N}} \inf_{n\geqslant m} x_n$, by 23D.

25K Proposition Let E be a Riesz space with a compatible complete metrizable linear space topology $\mathfrak{T}$. Then $\mathfrak{T}$ is locally solid iff E is uniformly complete.

Proof (a) If $\mathfrak{T}$ is locally solid, then E must be uniformly complete by 25I.

(b) Suppose that E is uniformly complete. Then each set $[-e, e]$, where $e \in E^+$, is bounded. **P** For consider the identity map $E_e \to E$. Suppose that $\langle x_n \rangle_{n\in\mathbf{N}}$ is a sequence in E_e which converges both for $\| \ \|_e$ and for $\mathfrak{T}$. Let x be its limit for $\| \ \|_e$ and x' its limit for $\mathfrak{T}$. Set

$$\alpha_n = \sup_{i\geqslant n} \|x - x_i\|_e,$$

so that $\langle \alpha_n \rangle_{n\in\mathbf{N}} \to 0$. Now we know that $x_i \in [x - \alpha_n e, x + \alpha_n e]$ for every $i \geqslant n$, and this set is closed for $\mathfrak{T}$ because $\mathfrak{T}$ is compatible; so

$$x' \in [x - \alpha_n e, x + \alpha_n e],$$

i.e. $|x-x'| \leqslant \alpha_n e$ for every $n \in \mathbf{N}$. Because E is certainly Archimedean [21Ba], $x = x'$. This shows that the identity map $(E_e, \| \, \|_e) \to (E, \mathfrak{T})$ has closed graph. By the closed graph theorem (this is where we use the hypothesis that E is both complete and uniformly complete), the identity map $E_e \to E$ is continuous. So the unit ball of E_e, $[-e, e]$, must be bounded in E. **Q**

But this shows that all order-bounded sets in E are bounded for $\mathfrak{T}$; and $\mathfrak{T}$ is locally solid by 25F.

25L I now proceed to examine Lebesgue compatible complete metrizable linear space topologies.

Proposition Let E be a Riesz space with a compatible complete metrizable Lebesgue linear space topology $\mathfrak{T}$. Then

(a) $\mathfrak{T}$ is Fatou;

(b) E is Dedekind complete;

*(c) E has the countable sup property;

(d) if F is any Archimedean Riesz space, $L^{\times}(E; F) = L^{\sim}(E; F)$ [definitions: 16B, 16G].

Proof (a) By 21Ba, E is Archimedean. So by 24Db, order-bounded sets are bounded. By 25F, $\mathfrak{T}$ is locally solid; by 24C, it is Fatou.

(b) follows from 24E.

*(c) is given as an exercise in 24Lc.

(d) Suppose that $T: E \to F$ is an increasing linear map and that $\varnothing \subset A \downarrow 0$ in E. Then there is a sequence $\langle x_n \rangle_{n \in \mathbf{N}}$ in A converging to 0 in E, simply because $0 \in \bar{A}$. This has a subsequence $\langle y_n \rangle_{n \in \mathbf{N}}$ such that $\{2^n y_n : n \in \mathbf{N}\}$ is bounded above by y say [25C]. So

$$\inf_{x \in A} Tx \leqslant \inf_{n \in \mathbf{N}} Ty_n \leqslant \inf_{n \in \mathbf{N}} 2^{-n} Ty = 0$$

as F is Archimedean. As A is arbitrary, this shows that T is order-continuous and belongs to $L^{\times}(E; F)$; as T is arbitrary,

$$L^{\sim}(E; F) \subseteq L^{\times}(E; F).$$

25M Corollary Let E be a Riesz space with a locally convex compatible complete metrizable linear space topology $\mathfrak{T}$. Then $\mathfrak{T}$ is Lebesgue iff $E' \subseteq E^{\times}$, and in this case $E' = E^{\times} = E^{\sim}$.

Proof If $\mathfrak{T}$ is Lebesgue, then it is locally solid [25La], so $E' = E^{\sim}$ [25G]; also, $E^{\sim} = E^{\times}$ by 25Ld. Conversely, if $E' \subseteq E^{\times}$, then $\mathfrak{T}$ is locally solid by 25G again, and therefore Lebesgue by 24G.

25N Exercises (a) If E is a Riesz space with a metrizable locally solid linear space topology for which increasing Cauchy sequences are convergent, it is complete.

*(b) If E is a uniformly complete Riesz space with a Levi metrizable locally solid linear space topology, it is complete.

(c) If E is a Riesz space with a Levi metrizable locally solid linear space topology, and F is a Riesz space with a linear space topology such that order-bounded sets are bounded, then every increasing linear map from E to F is continuous.

(d) Let E be a Riesz space with a barrelled locally solid linear space topology. Then E' is a band in $E^{\sim}$.

(e) Let E and F be Riesz spaces with compatible complete metrizable linear space topologies. Then every increasing linear map from E to F is continuous.

*(f) Let E be a Riesz space with a compatible complete metrizable linear space topology. Then the lattice operations on E have closed graphs.

*(g) Let E be a Riesz space with a locally solid complete metrizable linear space topology. Then a set $A \subseteq E$ is closed iff it is **relatively uniformly closed** in E, that is, if $A \cap E_e$ is $\| \ \|_e$-closed for every $e \in E^+$ [notations as 25H].

*(h) Let E be an Archimedean Riesz space and F a solid linear subspace of E. Then the quotient Riesz space E/F [construction: 14G] is Archimedean iff F is relatively uniformly closed in E.

(i) A solid linear subspace of a uniformly complete Archimedean Riesz space is uniformly complete.

*(j) A relatively uniformly closed Riesz subspace of a uniformly complete Archimedean Riesz space is uniformly complete.

*(k) Let E be a uniformly complete Archimedean Riesz space, and F a solid linear subspace of E such that the quotient space E/F is Archimedean. Then E/F is uniformly complete.

Notes and comments The closed graph theorem is employed in three places above; 25A, 25K (in part (b) of the proof), and 25Ne. 25A is really two separate propositions, run together merely because they can be proved simultaneously. The uniqueness of the topology is a special case of 25Ne; 25Ne is got from the closed graph theorem and 25C by the method used in 25K; 25C is a corollary of 25B; and 25B is the other half of 25A. Of course the use of the closed graph theorem in 25K is just another special case of 25Ne.

At a number of points [25D, 25I, 25Nc] I refer to topologies for which order-bounded sets are bounded. The examples we have seen are locally solid topologies [22Bd], $\mathfrak{T}_s(E, E^\sim)$ [25G, proof], and Lebesgue topologies on Archimedean spaces [24Db]. The first two both have the property that the positive cone is 'normal' [see the end-note to § 22], which is sufficient to make order-bounded sets bounded.

The property of uniform completeness is a half-way house on the way to Dedekind σ-completeness which will turn up again in § 81, where it has an application very similar to that of 25K. Apart from spaces satisfying the criteria of 25I to 25J, observe that every $C(X)$ [A2A] is uniformly complete; more examples follow from 25Ni and 25Nk. In fact it is clear that most important Riesz spaces are uniformly complete; the major exceptions we shall see are the spaces $S(\mathfrak{A})$ [§ 42].

26 *L*-spaces and *M*-spaces

In this section we shall study two special classes of Banach lattices. They are both abstractly defined versions of spaces which are important in measure theory; L-spaces are essentially the same as L^1-spaces [§§ 52, 63], while M-spaces with units are essentially the same as C_b-spaces [A2B], and include L^∞-spaces [§§ 43, 52, 62] as a subclass. In some ways they are opposite ends of the spectrum of spaces which interest use; at the same time they are dual to each other [26C, 26D]. Thus they are leading cases in the development of the general theory. They also have important special qualities; those of M-spaces are fairly straightforward, but L-spaces have many remarkable properties, among which 26E–26G are outstanding.

26A Definitions (a) A **Banach lattice** is a pair $(E, \| \ \|)$ where E is a Riesz space and $\| \ \|$ is a Riesz norm on E under which E is complete.

(b) An **L-space** is a Banach lattice E such that

$$\|x+y\| = \|x\| + \|y\| \quad \forall\ x, y \in E^+.$$

(c) An **M-space** is a Banach lattice E such that

$$\|x \vee y\| = \max(\|x\|, \|y\|) \quad \forall\ x, y \in E^+.$$

(d) An ***M*-space with unit** is an M-space E in which the unit ball has a least upper bound, e say, for which $\|e\| \leqslant 1$. In this case, the unit ball of E must be $[-e, e]$, and $\| \; \| = \| \; \|_e$ in the notation of 25Ha. e is called the **unit** of E. Conversely, of course, if E is a uniformly complete Riesz space with an order unit e, then under $\| \; \|_e$ E is an M-space with unit.

In an M-space with unit, the norm topology is Fatou and Levi [25Ha].

26B Theorem Let E be an L-space and $\mathfrak{T}$ the norm topology on E. Then

(a) $\mathfrak{T}$ is Levi and Lebesgue;

(b) E is Dedekind complete;

*(c) E has the countable sup property;

*(d) any order-dense Riesz subspace of E is super-order-dense.

Proof **(a)** Let A be a non-empty set in E^+ such that $A\uparrow$ and A is bounded. Then $\mathscr{F}(A\uparrow)$ is Cauchy. **P** Let $\alpha = \sup_{x \in A} \|x\| < \infty$. For every $\epsilon > 0$, there is an $x \in A$ such that $\|x\| \geqslant \alpha - \epsilon$. Now if $y \in A$ and $y \geqslant x$, $\|y-x\| = \|y\| - \|x\| \leqslant \alpha - (\alpha - \epsilon) = \epsilon$, by the additive property of $\| \; \|$. So the diameter of

$$\{y : y \in A, y \geqslant x\}$$

is less than or equal to 2ϵ, and this set belongs to $\mathscr{F}(A\uparrow)$. **Q** As E is complete, $\mathscr{F}(A\uparrow)$ has a limit x_0 say; by 21Bd, $x_0 = \sup A$. (Of course $\mathfrak{T}$ is compatible by 22Ea.)

Thus $\mathfrak{T}$ is Levi.

Now suppose that B is a non-empty set in E^+ such that $B\downarrow 0$. Let $x_0 \in B$ and set $A = \{x_0 - x : x \in B, x \leqslant x_0\}$. Then $A \uparrow x_0$ and A is certainly bounded, so by the argument above $\mathscr{F}(A\uparrow) \to x_0$ and $x_0 \in \bar{A}$. Subtracting both sides from x_0 we see that $0 \in \bar{B}$. As B is arbitrary, $\mathfrak{T}$ is Lebesgue.

(b) and **(c)** now follow from 25L; **(d)** follows immediately from (c), or otherwise.

26C Theorem Let E be an L-space. Then $E' = E^\sim = E^\times$ is an M-space with unit; its unit is the functional $\int$ given by

$$\int x = \|x^+\| - \|x^-\| \quad \forall \; x \in E.$$

Proof. The three spaces are equal by 25M and 26Ba above.

The additive property of the norm of $\boldsymbol{E}$ means that there is a linear functional $\int : \boldsymbol{E} \to \mathbf{R}$ as defined above, which is plainly in the positive cone of $\boldsymbol{E}^\sim = \boldsymbol{E}'$. Now suppose that $f \in \boldsymbol{E}'$. Then

$$\begin{aligned} \|f\| \leqslant \alpha &\Rightarrow |fx| \leqslant \alpha\|x\| = \alpha \textstyle\int x \quad \forall\ x \in \boldsymbol{E}^+ \\ &\Rightarrow -\alpha \textstyle\int \leqslant f \leqslant \alpha \textstyle\int \quad \text{in } \boldsymbol{E}^\sim \\ &\Rightarrow |f| \leqslant \alpha \textstyle\int \quad \text{in } \boldsymbol{E}^\sim \\ &\Rightarrow |fx| \leqslant |f|\,(|x|) \leqslant \alpha \textstyle\int |x| = \alpha \|\,|x|\,\| = \alpha\|x\| \quad \forall\ x \in \boldsymbol{E} \\ &\Rightarrow \|f\| \leqslant \alpha. \end{aligned}$$

So $\|f\| = \inf\{\alpha : \alpha \geqslant 0,\ |f| \leqslant \alpha \int\}$, and $\int$ must be an order unit of $\boldsymbol{E}^\sim$ which defines the norm of $\boldsymbol{E}^\sim = \boldsymbol{E}'$. As $\boldsymbol{E}'$ is certainly complete under its norm, it is an M-space with unit.

26D Proposition Let $\boldsymbol{G}$ be any Riesz space with a Riesz norm $\|\ \|$ such that $\|x \vee y\| = \max(\|x\|, \|y\|)$ for all $x,\ y \in \boldsymbol{G}^+$. Then $\boldsymbol{G}'$ is an L-space.

Proof Recall that $\boldsymbol{G}'$ is a solid linear subspace of $\boldsymbol{G}^\sim$ [22D], and the argument of 22D shows also that the norm of $\boldsymbol{G}'$ is a Riesz norm [22Ga], so $\boldsymbol{G}'$ is a Banach lattice. Now suppose that f and g are non-negative elements of $\boldsymbol{G}'$. Of course $\|f+g\| \leqslant \|f\| + \|g\|$. On the other hand, given $\epsilon > 0$, we can find x and y in the unit ball of $\boldsymbol{G}$ such that

$$|fx| \geqslant \|f\| - \epsilon, \quad |gy| \geqslant \|g\| - \epsilon.$$

Set $z = |x| \vee |y|$; then

$$\|z\| = \max(\|\,|x|\,\|, \|\,|y|\,\|) = \max(\|x\|, \|y\|) \leqslant 1.$$

So
$$\begin{aligned} \|f+g\| &\geqslant (f+g)(z) = fz + gz \\ &\geqslant f(|x|) + g(|y|) \geqslant |fx| + |gy| \\ &\geqslant \|f\| + \|g\| - 2\epsilon. \end{aligned}$$

As ϵ is arbitrary, $\|f+g\| = \|f\| + \|g\|$. Thus $\boldsymbol{G}'$ is an L-space.

***26E** We come now to a very special property of L-spaces.

Theorem Let $\boldsymbol{E}$ be an L-space, and $\boldsymbol{F}$ a Dedekind complete Riesz space with a locally convex locally solid Levi linear space topology. Then $L^\sim(\boldsymbol{E}; \boldsymbol{F})$ is precisely the set of continuous linear maps from $\boldsymbol{E}$ to $\boldsymbol{F}$.

Proof (a) We know already that every member of $L^\sim(E; F)$ is continuous, by 25D.

(b) Now suppose that $T: E \to F$ is a continuous linear map. To show that it is in $L^\sim(E; F)$, I shall use the approach of 16F. Let $x \in E^+$ and write

$$A_x = \{\sum_{i<n} |Tx_i| : \langle x_i \rangle_{i<n} \text{ is a finite sequence in } E^+ \text{ such that } \sum_{i<n} x_i = x\}.$$

Then $A_x \uparrow$ by 16F. Moreover, it is bounded. **P** Let V be any convex neighbourhood of 0 in F, and let U be a solid neighbourhood of 0 included in V. Then there is an $\alpha \in \mathbf{R}^+$ such that

$$Ty \in \alpha \|y\| U \quad \forall\ y \in E,$$

so that
$$|Ty| \in \alpha \|y\| U \subseteq \alpha \|y\| V \quad \forall\ y \in E.$$

Now if z is any member of A_x there is a finite sequence $\langle x_i \rangle_{i<n}$ in E^+ such that $\sum_{i<n} x_i = x$ and

$$z = \sum_{i<n} |Tx_i| \in \sum_{i<n} \alpha \|x_i\| V \subseteq (\sum_{i<n} \alpha \|x_i\|) V = \alpha \|x\| V,$$

as V is convex. So V absorbs A_x. **Q**

As $A_x \uparrow$ and the topology on F is Levi, A_x is bounded above; now $T \in L^\sim(E; F)$ by the criterion of 16F.

***26F Proposition** Let E be an L-space and F a Dedekind complete Riesz space with a Fatou norm under which it is complete. Then $L^\sim(E; F)$ is a closed linear subspace of the normed space $L(E; F)$ of all continuous linear maps from E to F. The induced norm on $L^\sim(E; F)$ is a Fatou norm.

Proof (a) We know by 25D that $L^\sim(E; F) \subseteq L(E; F)$. I show first that the norm on $L^\sim(E; F)$ is a Riesz norm. **P** Let $T \in L^\sim(E; F)$ and set $\alpha = \|T\|$. Let $x \in E^+$. As in 16F and 26E above, set

$$A_x = \{\sum_{i<n} |Tx_i| : \langle x_i \rangle_{i<n} \text{ is a finite sequence in } E^+ \text{ such that } \sum_{i<n} x_i = x\}.$$

Now $A_x \uparrow |T|(x)$ [16F], and the norm on F is Fatou, so

$$\| |T|(x) \| = \sup\{\|z\| : z \in A_x\}.$$

On the other hand, if $z \in A_x$, then z can be expressed as $\sum_{i<n} |Tx_i|$, where $\sum_{i<n} x_i = x$ and $x_i \geqslant 0$ for every $i < n$. So

$$\|z\| \leqslant \sum_{i<n} \|Tx_i\| \leqslant \sum_{i<n} \alpha \|x_i\| = \alpha \|x\|.$$

Thus $\qquad \| |T| (x) \| \leqslant \alpha \|x\| \quad \text{for every} \quad x \in E^+.$

Now suppose that $|S| \leqslant |T|$ in $L^\sim(E; F)$. Then

$$\|Sx\| = \| |Sx| \| \leqslant \| |S| (|x|)\| \leqslant \| |T| (|x|)\|$$
$$\leqslant \alpha \| |x| \| = \alpha \|x\|$$

for every $x \in E$. So $\|S\| \leqslant \alpha = \|T\|$. As S and T are arbitrary, the norm of $L^\sim(E; F)$ is a Riesz norm. **Q**

(b) Consequently it is a Fatou norm. **P** Suppose that $\varnothing \subset A \uparrow T_0$ in $L^\sim(E; F)^+$. Then, for any $x \in E$,

$$\{T(|x|) : T \in A\} \uparrow T_0(|x|)$$

by 16Db. So

$$\|T_0 x\| = \| |T_0 x| \| \leqslant \|T_0(|x|)\| = \sup_{T \in A} \|T(|x|)\|$$
$$\leqslant \sup_{T \in A} \|T\| \, \| |x| \| = \sup_{T \in A} \|T\| \, \|x\|.$$

As x is arbitrary, $\|T_0\| \leqslant \sup_{T \in A} \|T\|$. Of course, as the norm of $L^\sim(E; F)$ is a Riesz norm, $\|T_0\| = \sup_{T \in A} \|T\|$. **Q**

(c) Finally, suppose that T belongs to the closure of $L^\sim(E; F)$ in $L(E; F)$. Then there is a sequence $\langle T_n \rangle_{n \in \mathbf{N}}$ in $L^\sim(E; F)$ such that

$$\|T_n - T\| \leqslant 2^{-n} \quad \text{for every} \quad n \in \mathbf{N}.$$

Now $\qquad \| |T_{n+1}| - |T_n| \| = \| \, | |T_{n+1}| - |T_n| | \, \| \leqslant \| \, |T_{n+1} - T_n| \, \|$

$$= \|T_{n+1} - T_n\| \leqslant 2^{-n+1} \quad \forall \; n \in \mathbf{N},$$

so $\langle |T_n| \rangle_{n \in \mathbf{N}}$ has a limit S say in $L(E; F)$ [this is where we use the hypothesis that F is complete]. Now for any $x \in E^+$,

$$Sx = \lim_{n \to \infty} |T_n| (x) \geqslant 0$$

and $\qquad Sx - Tx = \lim_{n \to \infty} (|T_n| (x) - T_n x) \geqslant 0,$

so S and $S - T$ are both increasing linear maps, and $T = S - (S - T)$ belongs to $L^\sim(E; F)$. As T is arbitrary, $L^\sim(E; F)$ is closed.

***26G Corollary** Let E be an L-space and F a Dedekind complete Riesz space with a Fatou norm which generates a Levi topology. Then $L^\sim(E; F) = L^\times(E; F)$ is precisely the space $L(E; F)$ of continuous linear maps from E to F, and the norm on $L(E; F)$ is Fatou; moreover, the topology it generates is Levi.

Proof By 26E, $L(E; F) = L^\sim(E; F)$; by 25Ld, $L^\sim(E; F) = L^\times(E; F)$. F is complete under its norm by Nakano's theorem [23K†] or 25Nb,

† See the remarks on Nakano's theorem in the end-note to §23.

so by 26F the norm on $\boldsymbol{L}(\boldsymbol{E}; \boldsymbol{F}) = \boldsymbol{L}^{\sim}(\boldsymbol{E}; \boldsymbol{F})$ is Fatou. Finally, suppose that A is a non-empty bounded set in $\boldsymbol{L}(\boldsymbol{E}; \boldsymbol{F})$ which is directed upwards. Then for every $x \in \boldsymbol{E}^+$ the set $\{Tx : T \in A\}$ is bounded and directed upwards. As the topology on $\boldsymbol{F}$ is Levi, $\sup_{T \in A} Tx$ always exists, and by the criterion of 16Db A is bounded above in $\boldsymbol{L}^{\sim}(\boldsymbol{E}; \boldsymbol{F})$. As A is arbitrary, this shows that the topology of $\boldsymbol{L}(\boldsymbol{E}; \boldsymbol{F})$ is Levi.

26H Exercises (**a**) Let $\boldsymbol{E}$ be a Riesz space with a Fatou norm $\| \ \|$ such that $\|x+y\| = \|x\| + \|y\|$ for every x, $y \in \boldsymbol{E}^+$. Then the norm topology on $\boldsymbol{E}$ is Lebesgue.

(**b**) Let $\boldsymbol{E}$ be a Riesz space with a linear space topology $\mathfrak{T}$. Suppose that $e \in \text{int}\, \boldsymbol{E}^+$ for $\mathfrak{T}$. Then e is an order unit of $\boldsymbol{E}$. If moreover order-bounded sets of $\boldsymbol{E}$ are bounded for $\mathfrak{T}$, then $\mathfrak{T}$ is precisely the topology generated by the seminorm $\| \ \|_e$.

(**c**) Let $\boldsymbol{G}$ be a Dedekind complete M-space with unit and $\boldsymbol{E}$ any Riesz space with a locally solid linear space topology. Then every continuous linear map $T: \boldsymbol{E} \to \boldsymbol{G}$ belongs to $\boldsymbol{L}^{\sim}(\boldsymbol{E}; \boldsymbol{G})$.

(**d**) Let $\boldsymbol{G}$ be a Dedekind complete M-space with unit and $\boldsymbol{E}$ a Banach lattice. Then $\boldsymbol{L}^{\sim}(\boldsymbol{E}; \boldsymbol{G}) = \boldsymbol{L}(\boldsymbol{E}; \boldsymbol{G})$ and the norm on $\boldsymbol{L}(\boldsymbol{E}; \boldsymbol{G})$ is Fatou and Levi.

(**e**) Let $\boldsymbol{G}$ be an M-space. Let $\langle x_i \rangle_{i<n}$ and $\langle y_i \rangle_{i<n}$ be finite sequences in $\boldsymbol{G}$, where $n > 0$. Then

$$\|\sup_{i<n} x_i - \sup_{i<n} y_i\| \leqslant \max_{i<n} \|x_i - y_i\|.$$

*(**f**) Let $\boldsymbol{G}$ be an M-space. Let $A \subseteq \boldsymbol{G}$ be a non-empty compact set. Then $\sup A$ exists and

$$\{\sup B : \varnothing \subset B \subseteq A\}$$

is compact. [Hints: (i) show that $\{\sup B : \varnothing \subset B \subseteq A, B \text{ finite}\}$ is totally bounded; (ii) show that the map $B \mapsto \sup B$ is continuous for the Hausdorff uniformity on the closed subsets of A.]

*(**g**) Let $\boldsymbol{E}$ be a Riesz space with a Hausdorff linear space topology $\mathfrak{T}$ for which $\boldsymbol{E}^+$ has a 'compact base' i.e. there is an $f \in \boldsymbol{E}^*$ such that $\{x : x \in \boldsymbol{E}^+, fx = 1\}$ is non-empty and compact for $\mathfrak{T}$. Then, writing $\|x\| = f(|x|)$ for every $x \in \boldsymbol{E}$, $(\boldsymbol{E}, \| \ \|)$ is an L-space.

*(**h**) Let $\boldsymbol{G}$ be an M-space with unit in which the norm topology is Lebesgue. Then $\boldsymbol{G}$ is finite-dimensional. [Hint: use 15G to show that if $\boldsymbol{G}$ is an Archimedean Riesz space which is not finite-dimensional there is an infinite disjoint sequence in $\boldsymbol{G}^+$; now use 24I.]

*(**i**) Use 25Na to shorten part (c) of the proof of 26F.

(**j**) (B. Levi's theorem.) Let E be an L-space, and $\langle x_n \rangle_{n \in \mathbf{N}}$ an increasing sequence in E such that $\alpha = \sup_{n \in \mathbf{N}} \int x_n < \infty$. Then $x = \sup_{n \in \mathbf{N}} x_n$ exists in E and $\int x = \alpha$.

(**k**) (Fatou's lemma). Let E be an L-space, and $\langle x_n \rangle_{n \in \mathbf{N}}$ a bounded sequence in E^+. Then $x = \sup_{n \in \mathbf{N}} \inf_{i \geqslant n} x_i$ exists in E, and

$$\|x\| \leqslant \liminf_{n \to \infty} \|x_n\|.$$

(**l**) (Lebesgue's Dominated Convergence Theorem.) Let E be an L-space, and $\langle x_n \rangle_{n \in \mathbf{N}}$ an order-bounded sequence in E such that

$$\sup_{n \in \mathbf{N}} \inf_{i \geqslant n} x_i = \inf_{n \in \mathbf{N}} \sup_{i \geqslant n} x_i = x.$$

Then $\lim_{n \to \infty} \int x_n$ exists and is equal to $\int x$. [See 24Lg.]

Notes and comments For simple examples of L-spaces and M-spaces, see 2XB, 2XA and A2B. More sophisticated examples must await the chapters on measure theory, though the duals of these M-spaces are already non-trivial L-spaces.

Both L-spaces and M-spaces with units are Banach lattices with Fatou Levi topologies. L-spaces are always Dedekind complete; M-spaces with units need not be [A2B], though many are (e.g. the dual of any L-space, and the most important L^∞-spaces [2XA, § 53]). L-spaces are always Lebesgue, M-spaces with units hardly ever are [26Hh].

For both L-spaces and M-spaces with units effective representation theorems are available. Every L-space can be represented as the L^1-space of a suitable measure space (in fact a Radon measure space [§ 73]); see KAKUTANI or SEMADENI [§ 26.3]. Thus the results 26Hj–26Hl, abstract versions of the famous convergence theorems, are in fact equivalent to them. While every M-space with unit is isomorphic to a space $C(X)$, where X is a compact Hausdorff topological space; this result is rather easier, and is given by KELLEY & NAMIOKA [24.6] and by SCHAEFER T.V.S. [V.8.5].

Many proofs of 26E have been given; see, for example, CHACON & KRENGEL. Clearly 26E and 26Hc are connected, for if T is a continuous linear map with domain an L-space, its transpose T' has range an M-space, and vice versa. The conditions of 26E cannot be relaxed at all easily; the counterexample 2XE shows some of the difficulties. In particular, the range space F does have to be locally convex, as is shown by an ingenious example due to PRYCE U.S., unfortunately too sophisticated for inclusion here. And it can also be shown that the result is characteristic of L-spaces, in that any Banach lattice E for which 26E is true has an equivalent norm under which it is an L-space. KRENGEL [Satz 7.1] has shown that, under hypotheses similar to those of 26E, the space of compact linear operators from $\boldsymbol{E}$ to $\boldsymbol{F}$ is also a Riesz space.

2X Examples for Chapter 2

Most of this section will be taken up with descriptions of familiar sequence spaces, showing how their natural Riesz space structures interact with topologies that can be imposed on them. The spaces $\boldsymbol{l}^\infty$, $\boldsymbol{l}^1$ and $\boldsymbol{c}_0$, with their norm topologies, are dealt with in 2XA–2XC. Many of their properties can be deduced from the general theorems in this book; however, I have tried here to set out more elementary proofs, as these show some of the same ideas in simplified forms, and a comparison is often edifying. In 2XF and 2XG I show how we may use the ideas of this chapter to throw light on other topologies which arise naturally in the theory of linear topological spaces.

2XD and 2XE are more in the nature of counterexamples. 2XD is just a simple example of a compatible complete metrizable linear space topology which is not locally solid. 2XE is more important; it marks one of the boundaries of the theorem 26E.

2XA The space $\boldsymbol{l}^\infty(X)$ Let X be any non-empty set. Let $\boldsymbol{l}^\infty(X)$ be the set of all bounded functions from X to $\mathbf{R}$, i.e. the solid linear subspace of $\mathbf{R}^X$ generated by e where $e(t) = 1$ for every $t \in X$. Then $\boldsymbol{l}^\infty(X)$ is a Dedekind complete Riesz space in which e is an order unit.

On $\boldsymbol{l}^\infty(X)$, $\| \; \|_\infty$ or $\| \; \|_e$, given by

$$\|x\|_\infty = \sup\{|x(t)| : t \in X\} = \inf\{\alpha : |x| \leqslant \alpha e\}$$

is a Fatou norm [25Ha]. $\boldsymbol{l}^\infty(X)$ is complete under $\| \; \|_\infty$ [25J, or directly], so is an M-space with unit [26Ab]. The norm topology is Levi [25Ha].

$(\boldsymbol{l}^\infty(X))^\times$ can be identified with $\boldsymbol{l}^1(X)$ [2XB below], so $\boldsymbol{l}^\infty(X)$ is a perfect Riesz space [33A].

*$\boldsymbol{l}^\infty(X)$ has the countable sup property iff X is countable; the norm topology is Lebesgue iff X is finite [cf. 26Hh].

[See also 2XE.]

2XB The space $\boldsymbol{l}^1(X)$ Let X be any non-empty set. Let $\boldsymbol{l}^1(X)$ be the set of all functions $x: X \to \mathbf{R}$ such that

$$\|x\|_1 = \textstyle\sum_{t \in X} |x(t)| < \infty.$$

Then $l^1(X)$ is a solid linear subspace of $\mathbf{R}^X$, so is in its own right a Dedekind complete Riesz space. $\| \ \|_1$ is a Riesz norm on $l^1(X)$ such that $\|x+y\|_1 = \|x\|_1 + \|y\|_1$ for all $x, y \geqslant 0$. $l^1(X)$ is complete under $\| \ \|_1$ [TAYLOR, p. 100; or use 25Na], so it is an L-space.

There is a natural duality between $l^1(X)$ and $l^\infty(X)$ given by

$$\langle x, y\rangle = \sum_{t \in X} x(t)\, y(t),$$

the sum being absolutely convergent for every $x \in l^1(X)$ and $y \in l^\infty(X)$. This identifies $l^\infty(X)$, as normed Riesz space, with

$$l^1(X)' = l^1(X)^\sim = l^1(X)^\times$$

[26C; the identification $l^\infty(X) = l^1(X)'$ is proved by KÖTHE, 14.7.8, and TAYLOR, theorem 4.32–A; or we may use 65B/6XD to prove that $l^\infty(X) = l^1(X)^\times$].

At the same time it identifies $l^1(X)$ with $l^\infty(X)^\times$.

P This is a consequence of 33F. Alternatively, it can be proved directly as follows.

(a) Suppose that $x \geqslant 0$ in $l^1(X)$ and that $\varnothing \subset B \downarrow 0$ in $l^\infty(X)$. Then $0 = \inf B$ in $\mathbf{R}^X$, i.e.

$$\inf\{y(t) : y \in B\} = 0 \quad \forall \ t \in X.$$

Fix $y_0 \in B$ and let $\epsilon > 0$. Then there is a finite $I \subseteq X$ such that

$$\sum_{t \in X \setminus I} |x(t)| \leqslant \epsilon.$$

Since, for each $t \in I$, there is a $y \in B$ such that $y(t) \leqslant \epsilon$, and since $B\downarrow$, there is a $y_1 \in B$ such that $y_1 \leqslant y_0$ and $y_1(t) \leqslant \epsilon$ for every $t \in I$. Now

$$\begin{aligned} \sum_{t \in X} x(t)\, y_1(t) &= \sum_{t \in I} x(t)\, y_1(t) + \sum_{t \in X \setminus I} x(t)\, y_1(t) \\ &\leqslant \sum_{t \in I} \epsilon x(t) + \|y_0\|_\infty \sum_{t \in X \setminus I} x(t) \\ &\leqslant \epsilon \|x\|_1 + \epsilon \|y_0\|_\infty. \end{aligned}$$

As ϵ is arbitrary, $\inf_{y \in B} \langle x, y\rangle = 0$. Thus x acts on $l^\infty(X)$ as a member of $l^\infty(X)^\times$. Since $l^\infty(X)$ certainly separates its points, $l^1(X)$ may be regarded as embedded in $l^\infty(X)^\times$.

(b) Now suppose that $f \geqslant 0$ in $l^\infty(X)^\times$. For each $t \in X$, let e_t be the member of $l^\infty(X)$ given by

$$e_t(t) = 1, \quad e_t(u) = 0 \quad \forall \ u \in X \setminus \{t\}.$$

Define $x: X \to \mathbf{R}^+$ by

$$x(t) = f(e_t) \quad \forall \ t \in X.$$

Then if $I \subseteq X$ is finite, $\sum_{t\in I} e_t$ is defined in $l^\infty(X)$ and $0 \leqslant \sum_{t\in I} e_t \leqslant e$, where e is the unit of $l^\infty(X)$. So

$$\textstyle\sum_{t\in I} x(t) = \sum_{t\in I} f(e_t) = f(\sum_{t\in I} e_t) \leqslant fe.$$

As I is arbitrary $\qquad \sum_{t\in X} x(t) \leqslant fe < \infty,$

and $x \in l^1(X)$. Now consider $g \in l^\infty(X)^\times$ defined by $gy = \langle x, y\rangle$ for every $y \in l^\infty(X)$. We know that $g(e_t) = f(e_t)$ for every $t \in X$. But suppose that y is any member of $l^\infty(X)^+$. Consider

$$B = \{\textstyle\sum_{t\in I} y(t)\, e_t : I \subseteq X,\ I \text{ finite}\}.$$

Then $fz = gz$ for every $z \in B$, and $B \uparrow y$. So, as f and g are both order-continuous,

$$fy = \sup_{z\in B} fz = \sup_{z\in B} gz = gy.$$

As y is arbitrary, $f = g$.

Thus every positive element of $l^\infty(X)^\times$ can be represented by an element of $l^1(X)$; it follows that the duality between $l^\infty(X)$ and $l^1(X)$ induces a linear space isomorphism between $l^\infty(X)^\times$ and $l^1(X)$.

(c) Finally, we see that for $x \in l^1(X)$,

$$\begin{aligned} x \geqslant 0 &\Leftrightarrow x(t) \geqslant 0 \quad \forall\ t \in X \\ &\Leftrightarrow \langle x, e_t\rangle \geqslant 0 \quad \forall\ t \in X \\ &\Leftrightarrow \langle x, y\rangle \geqslant 0 \quad \forall\ y \in l^\infty(X)^+, \end{aligned}$$

so that the ordering on $l^1(X)$ induced by its embedding in $l^\infty(X)^\times$ is correct, and the isomorphism is a Riesz space isomorphism. **Q**

(The argument above is of course a particularly easy special case of that in 65A.)

Hence $l^1(X)$ is perfect [33A, 33F].

2XC The space $c_0(X)$ Let X be any non-empty set. Let $c_0(X)$ be the set of all functions $x: X \to \mathbf{R}$ such that

$$\{t : t \in X,\ |x(t)| \geqslant \epsilon\}$$

is finite for every $\epsilon > 0$. Then $c_0(X)$ is a solid linear subspace of $l^\infty(X)$ [2XA above]. It is easy to see that $c_0(X)$ is closed under $\| \ \|_\infty$, so $(c_0(X), \| \ \|_\infty)$ is a Dedekind complete Riesz space with a Fatou norm under which it is complete. The norm topology is Lebesgue; but it will be Levi only if X is finite. (So $c_0(X)$ does not satisfy the hypotheses of Nakano's theorem, 23K.) Because $l^\infty(X)$ is an M-space, so is $c_0(X)$ [26Ac].

*Note that a subset of $c_0(X)$ is relatively compact for $\| \|_\infty$ iff it is order-bounded; cf. 26Hf.

[See also 2XF, 4XCb.]

2XD Let E be the set of sequences $x \in \mathbf{R}^{\mathbf{N}}$ such that

$$\|x\| = \max\left(|x(0)|, \sup_{i \in \mathbf{N}} |x(i+1) - x(i)|\right) < \infty.$$

Then E is an order-dense Riesz subspace of $\mathbf{R}^{\mathbf{N}}$ (for if $x \in E$ then $|x| \in E$ and $\| |x| \| \leqslant \|x\|$). $\| \|$ is a norm on E under which E is a Banach space isomorphic (as Banach space) with $l^\infty(\mathrm{N})$. The topology $\mathfrak{T}$ derived from $\| \|$ is compatible (since for each $n \in \mathrm{N}$

$$|x(n)| \leqslant (n+1)\|x\| \quad \forall\ x \in E,$$

so the coordinate map $x \mapsto x(n)$ is continuous). E has an order unit e given by

$$e(n) = n+1 \quad \forall\ n \in \mathrm{N}.$$

But the interval $[-e, e]$ is not bounded, so $\mathfrak{T}$ is an example of a compatible complete metrizable linear space topology which is not locally solid. Since $[-e, e]$ includes the unit ball, $\mathfrak{T}$ must be Levi; but of course it is not Lebesgue [24Db].

Exercise Show that the lattice operations on E are not continuous.

***2XE An example for 26E** Recall that the space $l^2 = l^2(\mathrm{N})$, defined by

$$l^2 = \{x : x \in \mathbf{R}^{\mathbf{N}}, \textstyle\sum_{i \in \mathbf{N}} |x(i)|^2 < \infty\}$$

is a Hilbert space under the inner product

$$(x|y) = \textstyle\sum_{i \in \mathbf{N}} x(i)\, y(i).$$

Clearly it is also a solid linear subspace of $\mathbf{R}^{\mathbf{N}}$, and its norm is a Riesz norm. Moreover, it is easy to see (adapting the arguments of 2XB, or using 65A) that the inner product identifies l^2 with $(l^2)^\times$, and therefore that the norm topology is Lebesgue [25M]; alternatively, it is equally easy to show directly that the norm topology is Lebesgue, and therefore that $l^2 \cong (l^2)' = (l^2)^\times$ [25M].

[See also 6XG.]

Now let us examine $H = L^\sim(l^2; l^2)$. We know that $H \subseteq L(l^2; l^2)$, the space of continuous linear endomorphisms of l^2 [25D]. The norm of $L(l^2; l^2)$ induces a metrizable linear space topology on H. The positive cone of H is

$$\{T : T \in L(l^2; l^2),\ Tx \geqslant 0 \quad \forall\ x \geqslant 0\},$$

which must be closed in $L(l^2; l^2)$, and therefore in H, as the positive cone of l^2 itself is closed. Next, if $0 \leqslant S \leqslant T$ in H,

$$\begin{aligned} \|S\| &= \sup\{\|Sx\| : \|x\| \leqslant 1\} \\ &\leqslant \sup\{\|S\,|x|\,\| : \|x\| \leqslant 1\} \\ &\leqslant \sup\{\|T\,|x|\,\| : \|x\| \leqslant 1\} \\ &= \|T\| \end{aligned}$$

because the norm on l^2 is a Riesz norm. So the interval $[0, T]$ is bounded in H; it follows that all order-bounded sets are bounded.

However, given $n \in \mathbf{N}$, there is a $2^n \times 2^n$ matrix $\mathbf{A}_n = \langle \alpha_{nij} \rangle$ such that $|\alpha_{nij}| = 1$ for all $i, j < 2^n$ and

$$\textstyle\sum_k \alpha_{nik}\alpha_{njk} = 2^n \quad \text{if } i = j, \quad 0 \text{ otherwise.}$$

P Set $\mathbf{A}_0 = (1)$ and

$$\mathbf{A}_{n+1} = \begin{pmatrix} \mathbf{A}_n & \mathbf{A}_n \\ -\mathbf{A}_n & \mathbf{A}_n \end{pmatrix}$$

for each $n \in N$. **Q** Consider $\mathbf{A}_n$ as a linear transformation of l^2, given by

$$\begin{aligned} (\mathbf{A}_n x)(i) &= \textstyle\sum_{j<2^n} \alpha_{nij} x(j) \quad \forall\ i < 2^n \\ &= 0 \quad \forall\ i \geqslant 2^n. \end{aligned}$$

Then $2^{-(n/2)}\mathbf{A}_n$ is unitary, so

$$\|\mathbf{A}_n\| = 2^{n/2} \quad \forall\ n \in \mathbf{N}.$$

On the other hand, it is easy to see that $|\mathbf{A}_n|$ is represented by the $2^n \times 2^n$ matrix with all entries 1. So

$$\|\,|\mathbf{A}_n|\,\| = 2^n.$$

Thus $$\{2^{-(n/2)}\mathbf{A}_n : n \in \mathbf{N}\}$$

is a bounded set in H which has a solid hull which is not bounded. So H cannot be complete in its norm [25F, 22Bc]. In particular, $H \neq L(l^2; l^2)$. Thus $L(l^2; l^2) \nsubseteq L^{\sim}(l^2; l^2)$. So we cannot substitute l^2 for E in 26E.

Exercise The norm topology on H is not Lebesgue or Fatou; but the unit ball is order-closed.

***2XF** I shall conclude this section with a description of a couple of less familiar topological Riesz spaces exemplifying the concepts of this chapter.

The topology $\mathfrak{T}_k(l^\infty, l^1)$ For a non-empty set X, consider the Mackey topology on $l^\infty = l^\infty(X)$ derived from its duality with $l^1 = l^1(X)$ [2XB above]. There is a well-known characterization of the $\mathfrak{T}_s(l^1, l^\infty)$-compact sets when $X = \mathbf{N}$ [see KÖTHE 22.4.3], which extends naturally to the general case in the form: $A \subseteq l^1$ is relatively $\mathfrak{T}_s(l^1, l^\infty)$-compact iff (i) it is $\| \ \|_1$-bounded, (ii) for every $\epsilon > 0$ there is a finite $I \subseteq X$ such that

$$\sum_{t\in X\setminus I} |x(t)| \leqslant \epsilon \quad \forall\ x \in A.$$

[This is 83F below.] This makes it obvious that the solid convex hull of A is still relatively weakly compact, from which it follows that $\mathfrak{T}_k(l^\infty, l^1)$ is locally solid. [See also 82H.] So it must be Lebesgue [24G, 82F].

A $\mathfrak{T}_k(l^\infty, l^1)$-bounded set must be $\mathfrak{T}_s(l^\infty, l^1)$-bounded; by the uniform boundedness theorem (l^∞ being $(l^1)'$), it is $\| \ \|_\infty$-bounded, i.e. order-bounded. Thus $\mathfrak{T}_k(l^\infty, l^1)$ must be Levi. [See also 33B.] It is also well known that l^∞ is complete under $\mathfrak{T}_k(l^\infty, l^1)$; this could be proved as a corollary of Nakano's theorem [23K], but the usual proof is from the fact that l^∞ is the dual of the Banach space l^1 [see A1E].

Thus $\mathfrak{T}_k(l^\infty, l^1)$ is a complete Hausdorff Lebesgue Levi Fatou topology.

***2XG The topology $\mathfrak{T}_k(l^1, c_0)$** Now let us consider the Mackey topology on $l^1 = l^1(X)$ derived from the duality between l^1 and $c_0 = c_0(X)$ [2XC]. It is coarser than the Lebesgue topology derived from $\| \ \|_1$ [2XB], so it must be Lebesgue. It has the same bounded sets as $\mathfrak{T}_s(l^1, c_0)$, which (because l^1 is the dual of c_0) has the same bounded sets as $\| \ \|_1$; so, like $\| \ \|_1$, it must be Levi. Finally, it is locally solid.

P First method Let B be the unit ball of $l^\infty = l^\infty(X)$. Then B can be identified with $[-1, 1]^X$. Now it is easy to see that the product topology of B is the topology on B induced by $\mathfrak{T}_s(l^\infty, l^1)$, under which B is compact. Consider the map

$$(x, y) \mapsto xy : B \times B \to B,$$

where we define $(xy)(t) = x(t)\,y(t)$ for every $t \in X$. Then this is continuous for the product topology. So if $A \subseteq B$ is compact for $\mathfrak{T}_s(l^\infty, l^1)$,

$$\{xy : x \in A,\ y \in B\}$$

is also compact. But this is precisely the solid hull of A in l^∞.

Now suppose that $U \subseteq l^1$ is a neighbourhood of 0 for $\mathfrak{T}_k(l^1, c_0)$. Then its polar $U^0 \subseteq c_0$ is compact for $\mathfrak{T}_s(c_0, l^1)$ or $\mathfrak{T}_s(l^\infty, l^1)$. So certainly U^0 is bounded for $\| \ \|_\infty$, and there is an $\alpha > 0$ such that $\alpha U^0 \subseteq B$. Now the solid hull C of αU^0 is compact, and $D = \alpha^{-1}C$, which is the solid hull of U^0 itself, is compact, But D is still included in $\boldsymbol{c}_0$. We must now observe that by Krein's theorem [A1G], the convex hull of D is relatively compact for $\mathfrak{T}_s(\boldsymbol{c}_0, \boldsymbol{l}^1)$. So D^0, the polar of D in $\boldsymbol{l}^1$, is a neighbourhood of 0 for $\mathfrak{T}_k(\boldsymbol{l}^1, \boldsymbol{c}_0)$; but of course it is solid, so it is the required solid neighbourhood of 0 included in U.

Second method It is in fact not difficult to show that a set $A \subseteq \boldsymbol{c}_0$ is relatively $\mathfrak{T}_s(\boldsymbol{c}_0, \boldsymbol{l}^1)$-compact iff:

(i) A is bounded for $\| \; \|_\infty$;

(ii) whenever $\langle t_i \rangle_{i \in \mathbf{N}}$ is a sequence of distinct points in X,

$$\inf_{k \in \mathbf{N}} \sup_{y \in A} \inf_{i \leqslant k} |y(t_i)| = 0.$$

From this it follows at once that the solid hull of a relatively compact set in $\boldsymbol{c}_0$ is relatively compact. The rest of the work proceeds as above.
Q

Now we can say that $\boldsymbol{l}^1$ is complete under $\mathfrak{T}_k(\boldsymbol{l}^1, \boldsymbol{c}_0)$, either by Nakano's theorem [23K], or by A1E.

(The results above have been generalized by MOORE & REBER.)

Further reading for Chapter 2 The principal omission from this chapter has been the concept of 'normal cone'; at the same time, generalizations to the theory of partially ordered topological linear spaces have hardly been considered. Both of these subjects are tackled by SCHAEFER T.V.S. and PERESSINI. For further ideas on these lines, and for results concerning complete metrizable partially ordered linear spaces, see JAMESON. There is an alternative introduction to the elementary theory of L- and M-spaces in KELLY & NAMIOKA, with a few additional results. For more about normed Lebesgue and Fatou topologies, see LUXEMBURG & ZAANEN B.F.S. [notes X–XVI].

3. Dual spaces

In the theory of linear topological spaces, the concept of a dual space is of paramount importance. In this short chapter I supplement this theory with a brief discussion of the dual spaces $E^\sim$ and $E^\times$ which we have already seen, and which are fundamental to the ideas of the rest of the book. The most important new results concern the canonical evaluation map from a Riesz space E to the space of linear functionals on $E^\times$ [32B below].

In the third section I give an equally abbreviated note on 'perfect' Riesz spaces, which include the majority of the spaces which interest us. Further general results concerning spaces in duality will be derived in the course of Chapter 8, while studying weak compactness.

31 The space $E^\sim$

This is the basic dual space associated with a given Riesz space E. Recall that so far we know that $E^\sim$ is the set of linear functionals on E which are bounded on order-bounded sets, and that it is a Dedekind complete Riesz space [16C, 16D]. The importance of the results below is that they apply to the evaluation map from E into G^* for any solid linear subspace G of $E^\sim$; their principal application is of course in the case $G = E^\times$.

31A Let us begin with a list of definitions, repeating some given in Chapter 1.

Definitions (a) Let E be any linear space over $\mathbf{R}$. Then E^* will denote the linear space of all linear maps from E to $\mathbf{R}$.

(b) Let E be a Riesz space. Then $E^\sim$ is the Dedekind complete Riesz space $L^\sim(E; \mathbf{R})$ [16B, 16D].

(c) Let E be a Riesz space. Then $E^\times$ is the band $L^\times(E; \mathbf{R})$ in $E^\sim$ [16G, 16H].

(d) Let E be a linear topological space over $\mathbf{R}$. Then E' is the linear space of continuous linear maps from E to $\mathbf{R}$.

(e) Suppose that E is a linear space over $\mathbf{R}$ and that F is a linear subspace of $\boldsymbol{E}^*$. For any $x \in \boldsymbol{E}$, I shall write $\hat{x}$ for the member of $\boldsymbol{F}^*$ given by

$$\hat{x}(f) = f(x) \quad \forall\ f \in \boldsymbol{F}.$$

31B Lemma Let $\boldsymbol{E}$ be a Riesz space; suppose that $f \in \boldsymbol{E}^{\sim +}$ and that $x \in \boldsymbol{E}^+$. Then there is a $g \in \boldsymbol{E}^{\sim}$ such that $0 \leqslant g \leqslant f$, $gx = fx$, and $gy = 0$ whenever $x \wedge y = 0$ in $\boldsymbol{E}$.

Proof Define g: $\boldsymbol{E}^+ \to \mathbf{R}$ by

$$gz = \sup\{f(z \wedge \alpha x) : \alpha \in \mathbf{R}^+\} \leqslant fz,$$

as f is increasing. Then without difficulty (using nothing stronger than 14Ja), we can see that g is linear on $\boldsymbol{E}^+$ (I mean that it satisfies the conditions of 16A), so extends to become a member of $\boldsymbol{E}^*$. The rest of its properties are immediate [using 14Kb for the last].

31C Proposition Let $\boldsymbol{E}$ be a Riesz space and $\boldsymbol{G}$ a solid linear subspace of $\boldsymbol{E}^{\sim}$. Then the canonical map $x \mapsto \hat{x}$: $\boldsymbol{E} \to \boldsymbol{G}^*$ is a Riesz homomorphism from $\boldsymbol{E}$ into $\boldsymbol{G}^{\times}$.

Proof If $x \in \boldsymbol{E}^+$, $\hat{x} \in \boldsymbol{G}^{\times}$ by 16Ec [for if $\varnothing \subset A \downarrow 0$ in $\boldsymbol{G}$, then $A \downarrow 0$ in $\boldsymbol{E}^{\sim}$, so $\inf_{f \in A} \hat{x}(f) = \inf_{f \in A} fx = 0$]. As $\boldsymbol{G}^{\times}$ is a linear subspace of $\boldsymbol{G}^*$, $\hat{x} \in \boldsymbol{G}^{\times}$ for every $x \in \boldsymbol{E}$.

The map $x \mapsto \hat{x}$ is plainly linear and increasing.

Finally, suppose that $x \wedge y = 0$ in $\boldsymbol{E}$ and that $f \in \boldsymbol{G}^+$. Then there is a $g \in \boldsymbol{E}^{\sim}$ such that $0 \leqslant g \leqslant f$, $gx = fx$, and $gy = 0$ [31B]. But $\boldsymbol{G}$ is solid in $\boldsymbol{E}^{\sim}$, so $g \in \boldsymbol{G}$. Now

$$(\hat{x} \wedge \hat{y})(f) \leqslant \hat{x}(f-g) + \hat{y}(g) = 0$$

by 16Eb. As f is arbitrary, $\hat{x} \wedge \hat{y} = 0$; by 14Eb, this is enough to show that the map $x \mapsto \hat{x}$ is a Riesz homomorphism.

31D Corollary Let $\boldsymbol{E}$ be a Riesz space and $f \in \boldsymbol{E}^{\sim +}$. Then

$$f(|x|) = \sup\{gx : g \in \boldsymbol{E}^{\sim},\ |g| \leqslant f\}$$

$$f(x \vee y) = \sup\{gx + (f-g)y : 0 \leqslant g \leqslant f\}$$

for any $x, y \in \boldsymbol{E}$.

Proof These follow from applying 16Ea, 16Eb to the elements $\hat{x}$, $\hat{y}$ of $\boldsymbol{E}^{\sim\sim}$, and then going back to $\boldsymbol{E}$ with 31C; or, of course, by applying 31B directly.

***31E Exercises** (a) Let E be a Riesz space and G a locally order-dense [definition: 14F] Riesz subspace of $E^\sim$. Then the canonical map $x \mapsto \hat{x}: E \to G^*$ is a Riesz homomorphism from E to $G^\times$.

(b) Let E and F be Riesz spaces and $T: E \to F$ a Riesz homomorphism which is onto. Define $T': F^* \to E^*$ by $(T'g)(x) = g(Tx)$ for every $x \in E$ and $g \in F^*$. Then $T'[F^\sim] \subseteq E^\sim$ and $T': F^\sim \to E^\sim$ is a one-to-one Riesz homomorphism.

Notes and comments Of course 31B is the central idea of this section. In effect, this asserts that $E^\sim$ contains enough functionals to make 31C and 31D true. 31D and 16E together indicate a curious symmetry in the relationship between E and $E^\sim$.

32 The space $E^\times$

This seems to be the most interesting of the Riesz space duals of a given Riesz space E, especially in view of the remarkable topological properties discussed in § 82 below. At the same time it turns out that, in particular cases, the 'natural' dual of E is $E^\times$ as often as it is $E^\sim$ [see, for instance, 2XA, 52E, 6XD]. Of course they are frequently identical [as in 25M], which leads to particularly powerful results. In this section I give the fundamental embedding theorem 32B and an elementary, but useful, characterization of members of $E^\times$ [32D].

32A I begin with a vital technical result.

Lemma Let E be an Archimedean Riesz space. Let $f > 0$ in $E^\times$. Then there is an $x > 0$ in E such that:

(i) $fx > 0$;

(ii) $0 < y \leqslant x \Rightarrow fy > 0$;

(iii) if $g \wedge f = 0$ in $E^\times$, $gx = 0$.

Proof (a) Let

$$F = \{z : z \in E, f(|z|) = 0\}.$$

Then F is a band in E [cf. 23Fa]. But $F \neq E$ because $f \neq 0$. So $F^d \neq \{0\}$ [definition: 15F], because [by 15F] $F = F^{dd}$. Now if $x > 0$ in F^d, then

$$0 < y \leqslant x \Rightarrow x \wedge y = y > 0 \Rightarrow y \notin F \Rightarrow fy > 0,$$

and of course $fx > 0$.

(b) Now suppose that $g \wedge f = 0$ in $E^\times$; of course $g \geqslant 0$. Let $\epsilon > 0$. By 16Eb we may choose, for each $n \in \mathbf{N}$, an element z_n of E such that $0 \leqslant z_n \leqslant x$ and

$$fz_n + g(x - z_n) \leqslant 2^{-n}\epsilon.$$

For each $n \in \mathbf{N}$, let $w_n = \inf_{i \leqslant n} z_i$. Then

$$x - w_n = \sup_{i \leqslant n}(x - z_i) \leqslant \textstyle\sum_{i \leqslant n}(x - z_i),$$

and
$$g(x - w_n) \leqslant \textstyle\sum_{i \leqslant n} g(x - z_i) \leqslant 2\epsilon$$

for every $n \in \mathbf{N}$. At the same time, $\langle w_n \rangle_{n \in \mathbf{N}} \downarrow$, and if $0 \leqslant w \leqslant w_n$ for every $n \in \mathbf{N}$,

$$fw \leqslant fz_n \leqslant 2^{-n}\epsilon \quad \forall\ n \in \mathbf{N},$$

so $fw = 0$; as $0 \leqslant w \leqslant x$, $w = 0$. This shows that $0 = \inf_{n \in \mathbf{N}} w_n$. So $\inf_{n \in \mathbf{N}} g w_n = 0$ because g is order-continuous. So

$$gx \leqslant 2\epsilon + \inf_{n \in \mathbf{N}} g w_n \leqslant 2\epsilon.$$

As ϵ is arbitrary, $gx = 0$, as required.

32B Theorem Let E be an Archimedean Riesz space and G a solid linear subspace of $E^\times$. Then the canonical map $x \mapsto \hat{x}$ is an order-continuous Riesz homomorphism from E onto an order-dense Riesz subspace of $G^\times$.

Proof (a) By 31C, $x \mapsto \hat{x}: E \to G^\times$ is a Riesz homomorphism. It is order-continuous because, if $\varnothing \subset A \downarrow 0$ in E,

$$\inf_{x \in A} \hat{x}(f) = \inf_{x \in A} fx = 0 \quad \forall\ f \in G^+,$$

and $\inf_{x \in A} \hat{x} = 0$ by 16Ec.

So we are left considering the Riesz subspace $\{\hat{x} : x \in E\}$ in $G^\times$. I shall use 15E to show that it is order-dense.

(b) Perhaps I should first remark that G, and similarly $G^\times$, are Archimedean, because they are Riesz subspaces of the Dedekind complete Riesz spaces $E^\sim$ and $G^\sim$.

Suppose that $\phi > 0$ in $G^\times$. Then there is an $f \in G^+$ such that (i) $f > 0$, (ii) if $\psi \wedge \phi = 0$ in $G^\times$, $\psi f = 0$ [32A]. Now there is an $x \in E^+$ such that $fx > 0$. Since $\hat{x}(f) = fx > 0$, $\hat{x} \wedge \phi > 0$.

Because $G^\times$ is Archimedean,

$$\inf\{\alpha\hat{x} : \alpha > 0\} = 0$$

[15D], and there is an $\alpha > 0$ such that

$$\psi = (\hat{x} \wedge \phi - \alpha\hat{x})^+ > 0.$$

Let $g \in G^+$ be such that (i) $\psi g > 0$, (ii) $0 < h \leqslant g \Rightarrow \psi h > 0$, (iii) if $\theta \wedge \psi = 0$ in $G^\times$, $\theta g = 0$ [32A again]. Let $y \in E^+$ be such that (i) $gy > 0$, (ii) $0 < z \leqslant y \Rightarrow gz > 0$, (iii) if $g \wedge h = 0$ in G, $hy = 0$ [32A once more]. Now

$$\hat{y}(g) = gy > 0,$$

so $\hat{y} \wedge \psi > 0$, by the choice of g; as $\psi \leqslant \hat{x}$, $\hat{y} \wedge \hat{x} > 0$; by 14Kb, $\hat{y} \wedge \alpha\hat{x} > 0$. Set $z = y \wedge \alpha x$; then $\hat{z} = \hat{y} \wedge \alpha\hat{x}$ by 31C, so $\hat{z} > 0$.

(**c**) In fact, $\hat{z} \leqslant \phi$. **P?** Otherwise, $(\hat{z} - \phi)^+ > 0$, so there is an $h_0 \in G^+$ such that (i) $(\hat{z} - \phi)^+(h_0) > 0$ (ii) $0 < h \leqslant h_0 \Rightarrow (\hat{z} - \phi)^+(h) > 0$ [32A, for the last time]. Now examine

$$(\hat{z} - \phi)^+(h_0 \wedge g) \leqslant (\alpha\hat{x} - \phi \wedge \hat{x})^+(h_0 \wedge g) \leqslant (\alpha\hat{x} - \phi \wedge \hat{x})^+(g).$$

But $\psi \wedge (\alpha\hat{x} - \phi \wedge \hat{x})^+ = 0$, so $(\alpha\hat{x} - \phi \wedge \hat{x})^+(g) = 0$, by the choice of g. So $h_0 \wedge g = 0$, by the choice of h_0. Now, however, $h_0(y) = 0$, by the choice of y. So

$$(\hat{z} - \phi)^+(h_0) \leqslant \hat{z}(h_0) = h_0(z) \leqslant h_0(y) = 0,$$

which contradicts the orginal requirements on h_0. **XQ**

(**d**) Thus we have $0 < \hat{z} \leqslant \phi$. As ϕ is arbitrary, $\{\hat{w} : w \in E\}$ is order-dense in $G^\times$, by 15E.

32C Corollary Let E be a Dedekind complete Riesz space and G a solid linear subspace of $E^\times$. Then the map $x \mapsto \hat{x}$ takes E onto an order-dense solid linear subspace of $G^\times$.

Proof Immediate from 32B and 17D.

32D The following is useful for proving that functionals belong to $E^\times$. It works only for Archimedean spaces.

Lemma Let E be an Archimedean Riesz space, and let $f \in E^*$ be such that $\inf_{x \in A} |fx| = 0$ whenever $\varnothing \subset A \downarrow 0$ in E. Then $f \in E^\times$.

Proof Let $\mathfrak{T}$ be the linear space topology on E defined by the seminorm $x \mapsto |fx| : E \to \mathbf{R}^+$. The condition on f states precisely that $\mathfrak{T}$ is Lebesgue. So $E' \subseteq E^\times$ [24F]. But of course $f \in E'$.

Notes and comments The argument of 32A–32B is particularly concentrated, not to say opaque. It can perhaps be clarified by following it through in a special case. Let $E = \mathbf{c}_0(X)$ [2XC], and set $G = E^\times$, which is identified with $l^1(X)$; in this case $G^\times$ is identified with $l^\infty(X)$ [2XB]. Now suppose that $\phi > 0$ in $G^\times$, as in part (b) of the proof of 32B. Let A be the set $\{t : t \in X, \phi(t) > 0\}$. Then f must be such that

$$B = \{t : f(t) > 0\} \subseteq A,$$

and x such that

$$C = \{t : x(t) > 0\} \subseteq B.$$

Since $x > 0$, $C \neq \varnothing$, and obviously $\phi \wedge \hat{x} > 0$, as $(\phi \wedge \hat{x})(t) > 0$ for every $t \in C$.

Now if
$$D = \{t : t \in X, \psi(t) > 0\},$$
we see that

$$\{t : z(t) > 0\} \subseteq \{t : y(t) > 0\} \subseteq \{t : g(t) > 0\}$$
$$\subseteq D = \{t : (\phi \wedge \hat{x})(t) > \alpha \hat{x}(t)\}.$$

Since $z(t) \leqslant \alpha x(t)$ for every t, it is obvious that $\hat{z} \leqslant \phi \wedge \hat{x}$.

Of course this sudden conclusion of the argument relies on the successful identification of E as a subspace of $G^\times$. But in fact the essential idea here, which is to identify the relationship in the lemma 32A with properties of the sets A, B, etc., above, is widely applicable. 65A exhibits a more general situation in which it can be used.

Readers familiar with the Radon–Nikodým theorem will recognize, in the construction of ψ, an old device. It will be used again for the Radon–Nikodým theorem itself [52D below].

The ideas of this section are taken from LUXEMBURG & ZAANEN B.F.S. [note VII, §§ 27–8]. These authors are interested in a space intermediate between $E^\times$ and $E^\sim$, the space of functions $f \in E^*$ such that

$$\langle x_n \rangle_{n \in \mathbf{N}} \downarrow 0 \quad \text{in} \quad E \Rightarrow \langle f x_n \rangle_{n \in \mathbf{N}} \to 0$$

[LUXEMBURG & ZAANEN B.F.S., note VI, § 20]. This also is a band in $E^\sim$. In its properties it appears to be a pale shadow of $E^\times$; for this reason I ignore it in most of this book, though it is never far away [see 42P, 83K]. See also LUXEMBURG.

*33 Perfect Riesz spaces

Here we have a class of Riesz spaces which includes a large proportion of those arising naturally in analysis. A Riesz space E is 'perfect' iff the evaluation map from E to $E^{\times\times}$ is an isomorphism between E and $E^{\times\times}$. This section concentrates on methods of showing that Riesz spaces are perfect. The consequences of the property will be dealt with later, especially in § 82.

33A Definition A Riesz space E is **perfect** if the canonical map $x \mapsto \hat{x}: E \to (E^\times)^\times$ [31Ae, 32B] is a Riesz space isomorphism.

Remarks If E is perfect, it is Dedekind complete (for $E^{\times\times}$, being a solid linear subspace of the Dedekind complete space $E^{\times\sim}$, is always Dedekind complete), and $E^\times$ must separate the points of E (as the evaluation map is one-to-one). These conditions are not enough, as they are satisfied by $\boldsymbol{c}_0(\mathbf{N})$, which is not perfect [2XC]. But the concept of 'Levi topology', already used in § 23, provides us with a necessary and sufficient condition.

33B Proposition Let E be a Dedekind complete Riesz space and G a solid linear subspace of $E^\times$. Then the canonical map $x \mapsto \hat{x}: E \to G^\times$ is one-to-one and onto iff the weak topology $\mathfrak{T}_s(E, G)$ is Levi [definition: 23I].

Proof (a) If the map $x \mapsto \hat{x}: E \to G^\times$ is one-to-one and onto, it is a Riesz space isomorphism, by 31C. And we know by 23M that $\mathfrak{T}_s(G^\times, G)$ is Levi; so $\mathfrak{T}_s(E, G)$ must be Levi.

(b) Conversely, suppose that $\mathfrak{T}_s(E, G)$ is Levi. Then the map $x \mapsto \hat{x}$ is one-to-one. **P** If $x \neq 0$ in E, then $|x| > 0$, so $A = \{n|x| : n \in \mathbf{N}\}$ is not bounded above [because E is Dedekind complete and therefore Archimedean]. But A is directed upwards, so it cannot be bounded for $\mathfrak{T}_s(E, G)$; and there is an $f \in G$ such that $f|x| \neq 0$. By 31C,

$$|\hat{x}|(f) = f(|x|),$$

so $\hat{x} \neq 0$. **Q**

At the same time, the map $x \mapsto \hat{x}$ is onto. **P** Suppose that $\phi \geqslant 0$ in $G^\times$, and let

$$A = \{x : x \in E,\ 0 \leqslant \hat{x} \leqslant \phi\}.$$

Then for any $f \in G$ $\quad \sup_{x\in A} |fx| \leqslant \phi(|f|) < \infty,$

so that A is bounded for $\mathfrak{T}_s(E, G)$; by 31C again, $A\uparrow$, so A is bounded above in E; as E is Dedekind complete, $x_0 = \sup A$ exists in E. But now

$$\hat{x}_0 = \sup\{\hat{x} : x \in A\} = \phi,$$

using the full strength of 32B.

Thus $\{\hat{x} : x \in E\}$ includes $G^{\times+}$; but it is a linear subspace, so must be $G^\times$. **Q**

33C Corollary A Dedekind complete Riesz space E is perfect iff $\mathfrak{T}_s(E, E^\times)$ is Levi.

33D Corollary If E is any Riesz space, $E^{\sim}$ and $E^{\times}$ are perfect.

Proof For $\mathfrak{T}_s(E^{\sim}, E^{\sim\times})$ is finer than the Levi topology $\mathfrak{T}_s(E^{\sim}, E)$ [31C, 23M] and is therefore Levi. The same argument applies to $E^{\times}$.

33E Corollary If E is a Dedekind complete Riesz space with a locally convex Lebesgue Levi linear space topology $\mathfrak{T}$, E is perfect.

Proof Since $\mathfrak{T}$ and $\mathfrak{T}_s(E, E')$ have the same bounded sets, $\mathfrak{T}_s(E, E')$ is Levi. But $E' \subseteq E^{\times}$ [24F], so $\mathfrak{T}_s(E, E^{\times})$ is Levi.

33F Corollary Any L-space is perfect.

Proof 26B.

33G Exercises (a) Let E be any Riesz space and F a perfect Riesz space. Then $L^{\sim}(E; F)$ and $L^{\times}(E; F)$ are perfect.

(b) If E is a perfect Riesz space and F is a band in E, then F is perfect.

*(c) Let E be a Dedekind complete Riesz space such that $E^{\times}$ separates the points of E, and suppose that there is a sequence $\langle x_n \rangle_{n \in \mathbf{N}}$ in E^+ such that

$$\forall\ x \in E \quad \exists\ n \in \mathbf{N}, \quad |x| \leqslant x_n.$$

Then E is perfect.

Notes and comments The original 'vollkommene Räume', studied in a series of papers by G. Köthe, were sequence spaces (that is to say, subspaces of $\mathbf{R}^{\mathbf{N}}$); they were defined in a way which we can now recognize [see 6XD below] as the form $E^{\times}$, for an arbitrary solid linear subspace E of $\mathbf{R}^{\mathbf{N}}$; so Köthe's first theorem was that $E^{\times\times\times} = E^{\times}$. His most important results are repeated in KÖTHE, § 30. They were taken up by DIEUDONNÉ, who generalized them to include spaces of equivalence classes of measurable functions, as in § 65 below. The abstract approach above comes from LUXEMBURG & ZAANEN B.F.S. [note VII, § 28]; the two systems are linked by FREMLIN A.K.S. II, in which I gave a representation theorem showing that an arbitrary perfect Riesz space is isomorphic to one of Dieudonné's 'espaces de Köthe'.

Examples for Chapter 3 I give no special examples for this chapter. Whenever we encounter a Riesz space E, one of our first tasks is to seek to identify $E^{\sim}$ and $E^{\times}$. The latter is generally easier, especially when we can use 65A or 65B. Important examples from other chapters include the elementary sequence spaces [2XA–2XC], function spaces [§ 65], and, for some insight into less familiar cases, 8XA–8XC.

Further reading for Chapter 3 Much the most important reference for this work is LUXEMBURG & ZAANEN B.F.S. [notes VI–X].

4. Riesz spaces on Boolean rings

The main purpose of this book is to show how the abstract theory of Riesz spaces may be applied to the study of function spaces arising in measure theory. In Chapters 1–3 I have set out the most important concepts needed for this approach. But in this chapter I propose to open up another line of attack. My eventual aim is to describe the relationship between the measure algebra of a measure space and its function spaces. In order to do this, I demonstrate methods of constructing Riesz spaces from Boolean algebras which, when applied to measure algebras, will produce isomorphic copies of the basic function spaces L^1 and L^∞. At the same time I shall apply the concepts of the first three chapters to describe the properties of these Riesz spaces.

The technical problems encountered along the way are considerable. An intuitive understanding, however, of the basic constructions $\boldsymbol{S}$ and $\boldsymbol{L}^\infty$, is not hard to attain; this is because some relatively easy examples already offer most of the principal aspects of the theory [4XA–4XD]. The construction $\boldsymbol{L}^\#$ is essentially deeper, and requires faith in Chapters 1–3 to be meaningful at all; its real significance will not appear until § 62.

One of the advantages of this method is that all the constructions are functors, and behave reasonably when the right kind of homomorphism is applied. This will enable us to see deeper into one of the more puzzling aspects of measure theory; details are in §§ 45, 54 and 61E.

For readers who have not encountered measure algebras before, I recommend taking § 61 in parallel with this chapter.

41 Boolean rings

The theory of Boolean rings is an abstract framework for the study of rings of sets. If X is any set, then its power set $\mathscr{P}X$, with the operations of symmetric difference and intersection, forms a Boolean ring [4XA]. Conversely, any Boolean ring is isomorphic to a 'ring of sets', i.e. a subring of some $\mathscr{P}X$ [41D].

A ring of sets has a natural ordering $\subseteq$, under which it is a lattice. Since $A \subseteq B$ iff $A \cap B = A$, this lattice structure can be applied to any Boolean ring [41F].

41A Definition A **Boolean ring** is a ring $(\mathfrak{A}, +, .)$ such that $a^2 = a$ for every $a \in \mathfrak{A}$.

[Examples: 4XA, 4XCb, 4XF.]

41B Lemma If $\mathfrak{A}$ is a Boolean ring:

(a) $a+a = 0 \quad \forall\ a \in \mathfrak{A}$;

(b) $ab = ba \quad \forall\ a, b \in \mathfrak{A}$.

Proof

$$\text{(a)}\ a+a = (a+a)^2 = a^2+a^2+a^2+a^2 = a+a+a+a.$$

$$\text{(b)}\ a+b = (a+b)^2 = a^2+ab+ba+b^2 = a+b+ab+ba,$$

so

$$ab = -ba = ba.$$

41C Lemma Let $\mathfrak{A}$ be a Boolean ring, and a a non-zero element of $\mathfrak{A}$. Let I be an ideal of $\mathfrak{A}$ which is maximal subject to $a \notin I$. Then

(a) if $c \in \mathfrak{A}$ and $ac \in I$, then $c \in I$;

(b) the ideal generated by a and I is $\mathfrak{A}$;

(c) I is maximal.

Proof If $c \in \mathfrak{A}$, the set

$$\{b : b \in \mathfrak{A},\ b+bc \in I\}$$

is an ideal (because $\mathfrak{A}$ is commutative), includes I, and contains c (because $c+c^2 = c+c = 0$). So it is the ideal generated by I and c.

(a) If $c \notin I$, then a belongs to the ideal generated by I and c, so $a+ac \in I$; but we are supposing that $a \notin I$, so $ac \notin I$. Conversely, if ac does belong to I, so does c.

(b) In fact, if c is any member of $\mathfrak{A}$, then $a(c+ca) = (a+a^2)c = 0 \in I$. So $c+ca \in I$, and c belongs to the ideal of $\mathfrak{A}$ generated by a and I.

(c) Now any ideal of $\mathfrak{A}$ properly including I must contain a, by hypothesis, so is $\mathfrak{A}$ itself.

41D M. H. Stone's theorem Let $\mathfrak{A}$ be a Boolean ring. Then there is a set X such that $\mathfrak{A}$ is isomorphic to a subring of $\mathscr{P}X$ [definition: 4XA].

Proof Let X be the set of all maximal ideals t of $\mathfrak{A}$. Define $\phi: \mathfrak{A} \to \mathscr{P}X$ by

$$\phi a = \{t : t \in X,\ a \notin t\} \quad \forall\ a \in \mathfrak{A}.$$

Now if $a, b \in \mathfrak{A}$ and $t \in X$, then

(i) $a \in t \Rightarrow ab \in t$;

(ii) $a \notin t, b \notin t \Rightarrow ab \notin t \Rightarrow a+b \in t$ [applying 41Ca first to t and a, and then to t and ab, as $(a+b)ab = 0 \in t$];

(iii) $a \in t, b \notin t \Rightarrow a+b \notin t$;

(iv) $a, b \in t \Rightarrow a+b \in t$.

So $\phi(ab) = \phi a \cap \phi b$ [using (i) and (ii)], and

$$\phi(a+b) = (\phi a \backslash \phi b) \cup (\phi b \backslash \phi a)$$

[using (ii), (iii) and (iv)]. Thus ϕ is a ring homomorphism.

If a is any non-zero element of $\mathfrak{A}$, let $\mathscr{X}$ be the set of all ideals of $\mathfrak{A}$ not containing a. By Zorn's lemma, $\mathscr{X}$ has a maximal element t say; now by 41Cc, t is a maximal ideal of $\mathfrak{A}$, i.e. $t \in X$. And $t \in \phi a$, so $\phi a \neq \varnothing$.

Thus ϕ is one-to-one. So $\mathfrak{A} \simeq \phi[\mathfrak{A}]$, which is a subring of $\mathscr{P}X$.

41E Notation If $\mathfrak{A}$ is a Boolean ring, I shall call the space X of maximal ideals of $\mathfrak{A}$ the **Stone space** of $\mathfrak{A}$, and the map $\phi: \mathfrak{A} \to \mathscr{P}X$ the **Stone representation**. In the next section I shall use ϕ and X in further constructions, so it is important to note now that they are canonical.

41F The ordering of a Boolean ring If $\mathfrak{A}$ is a Boolean ring, we may define a relation $\subseteq$ on $\mathfrak{A}$ by

$$a \subseteq b \Leftrightarrow ab = a.$$

Now we see that if $\phi: \mathfrak{A} \to \mathscr{P}X$ is the Stone representation of $\mathfrak{A}$, then $a \subseteq b \Leftrightarrow \phi a \subseteq \phi b$. From this it follows at once that $\subseteq$ is a partial ordering under which $\mathfrak{A}$ is a lattice. The lattice operations on $\mathfrak{A}$ are given by

$$\inf\{a, b\} = ab,$$
$$\sup\{a, b\} = a+b+ab.$$

Of course, these facts can readily be verified without appeal to Stone's theorem.

$\mathfrak{A}$ always has a least element, which is 0; it has a greatest element iff it has a multiplicative identity, and they are then the same.

41G Notation If $\mathfrak{A}$ is a Boolean ring, I shall use the symbols $\cap$, $\cup$ and $\backslash$ for the operations on $\mathfrak{A}$ corresponding to $\cap$, $\cup$ and $\backslash$, that is,

$$a \cap b = \inf\{a, b\} = ab, \quad a \cup b = \sup\{a, b\} = a+b+ab,$$
$$a \backslash b = a+ab,$$

as well as $\subseteq$ defined above and its inverse $\supseteq$. In a Boolean ring, the words sup and inf will always refer to the ordering $\subseteq$. So, for instance, $\mathfrak{A}$ is Dedekind complete or σ-complete iff it is complete in the sense of 13B with respect to this ordering.

If $\mathfrak{A}$ is a Boolean ring, a subring $\mathfrak{B}$ of $\mathfrak{A}$ (i.e. a subset containing 0 and closed under addition and multiplication) is a Boolean ring in its own right. The ordering on $\mathfrak{B}$ induced by the ordering on $\mathfrak{A}$ coincides with the ordering on $\mathfrak{B}$ derived from its own Boolean ring structure. As $\mathfrak{B}$ is a sublattice of $\mathfrak{A}$, the lattice operations on $\mathfrak{B}$ are also those derived from $\mathfrak{A}$.

It is arbitrary, but convenient, to use the phrase '**Boolean algebra**' to mean a Boolean ring with a 1, a multiplicative identity. A **subalgebra** of a Boolean algebra is now a subring containing 1.

*If $\mathfrak{A}$ is a Boolean ring, a **σ-subring** of $\mathfrak{A}$ is a subring which is a σ-sublattice in the sense of 13F, i.e. one which is closed under countable sups and infs in so far as these exist in $\mathfrak{A}$. A **σ-subalgebra**, or **σ-ideal**, of $\mathfrak{A}$ is a σ-subring which is also a subalgebra or ideal respectively.

***41H A lemma on suprema and infima** Let $\mathfrak{A}$ be a Boolean ring, and A a non-empty subset of $\mathfrak{A}$.

(a) If the set B of upper bounds of A is not empty, then

$$\inf\{b+a : a \in A, b \in B\} = \inf\{b \backslash a : a \in A, b \in B\} = 0.$$

(b) Given $a_0 \in \mathfrak{A}$, $A \uparrow a_0$ iff a_0 is an upper bound for A and

$$\{a+a_0 : a \in A\} = \{a_0 \backslash a : a \in A\} \downarrow 0.$$

(c) Given $a_0 \in \mathfrak{A}$, $A \downarrow a_0$ iff a_0 is a lower bound for A and

$$\{a+a_0 : a \in A\} = \{a \backslash a_0 : a \in A\} \downarrow 0.$$

Proof I shall appeal to the correspondence between $\cup$, $\cap$, $\backslash$ and $\cup$, $\cap$, $\backslash$ to justify intuitively the following arguments. They can, of course, be validated formally from the definitions.

We observe first that, for $a, b \in \mathfrak{A}$,

$$a \subseteq b \Leftrightarrow a+b = b \backslash a.$$

(a) If c is a lower bound for $\{b \backslash a : b \in B, a \in A\}$, then fix $b_0 \in B$. As $c \subseteq b_0 \backslash a \quad \forall\, a \in A$, $c \cap a = 0 \quad \forall\, a \in A$. So $b_0 \backslash c$ is an upper bound for A. As $A \neq \varnothing$, $c \subseteq b_0 \backslash c$ and $c = 0$.

(b) If a_0 is an upper bound for A, then the map $b \mapsto a_0 \backslash b$ is an order-reversing involution on $\{b : b \subseteq a_0\}$. So $A \uparrow a_0$ iff $\{a_0 \backslash a : a \in A\} \downarrow 0$.

(c) If a_0 is a lower bound for A, then $A\downarrow$ iff $B = \{a\backslash a_0 : a \in A\}\downarrow$. Now c is a lower bound for B iff $c \cup a_0$ is a lower bound for A and $c \cap a_0 = 0$. So if $a_0 = \inf A$, then for any lower bound c of B, $c \subseteq a_0$ and $c \cap a_0 = 0$, so $c = 0$. Conversely, if $0 = \inf B$, then for any lower bound d of A, $d\backslash a_0 = 0$ and $d \subseteq a_0$.

***41I Corollary** Let $\mathfrak{A}$ be a Boolean ring. Then $\mathfrak{A}$ is Dedekind σ-complete iff every countable non-empty set in $\mathfrak{A}$ has a greatest lower bound.

Proof For now if $A \subseteq \mathfrak{A}$ is a non-empty countable set with an upper bound a_0,

$$\sup A = a_0\backslash\inf\{a_0\backslash a : a \in A\}.$$

41J Ideals and quotients of Boolean rings Observe first that if $\mathfrak{A}$ is a Boolean ring, then a set $I \subseteq \mathfrak{A}$ is an ideal (written $I \lhd \mathfrak{A}$) iff

(i) $0 \in I$;

(ii) $a, b \in I \Rightarrow a \cup b \in I$;

(iii) $a \subseteq b \in I \Rightarrow a \in I$.

For (iii) is precisely the condition that $I\mathfrak{A} \subseteq I$, and now (ii) is equivalent to $I + I \subseteq I$.

Obviously, if $I \lhd \mathfrak{A}$, then $\mathfrak{A}/I$ is a Boolean ring. Examine the identities

$$(ab)^{\bullet} = a^{\bullet}b^{\bullet}$$

$$(a+b+ab)^{\bullet} = a^{\bullet}+b^{\bullet}+a^{\bullet}b^{\bullet}$$

where $a \mapsto a^{\bullet} : \mathfrak{A} \to \mathfrak{A}/I$ is the canonical homomorphism. These show that the canonical homomorphism is a lattice homomorphism [13E], when $\mathfrak{A}$ and $\mathfrak{A}/I$ are given their lattice structures. [See also 45A below.]

Observe that

$$a^{\bullet} \subseteq b^{\bullet} \Leftrightarrow a^{\bullet}b^{\bullet} = a^{\bullet} \Leftrightarrow ab + a \in I \Leftrightarrow a\backslash b \in I.$$

***41K Products** If $\langle\mathfrak{A}_\iota\rangle_{\iota\in I}$ is any family of Boolean rings, the product ring $\prod\langle\mathfrak{A}_\iota\rangle_{\iota\in I}$ is Boolean. Clearly, the lattice structure of $\prod\langle\mathfrak{A}_\iota\rangle_{\iota\in I}$ is precisely the product of the lattice structures of the family $\langle\mathfrak{A}_\iota\rangle_{\iota\in I}$ in the sense of 11G/13H.

41L Exercises (a) Let $\mathfrak{A}$ be a Boolean ring. Then its lattice structure is distributive in the strong sense of 14D; that is, if $A \subseteq \mathfrak{A}$ is such that $\sup A$ exists, and if $b \in \mathfrak{A}$, then

$$\sup\{b \cap a : a \in A\} \text{ exists } = b \cap \sup A.$$

Similarly, if $\inf A$ exists, then

$$\inf\{b \cup a : a \in A\} \text{ exists} = b \cup \inf A.$$

***(b)** If $\mathfrak{A}$ is a Boolean ring and $A \subseteq \mathfrak{A}$, set

$$A^d = \{b : b \in \mathfrak{A}, \quad b \cap a = 0 \quad \forall\ a \in A\}.$$

Then A^d is an order-closed ideal of $\mathfrak{A}$. If A is an order-closed ideal of $\mathfrak{A}$, then $A = A^{dd}$ [cf. 15F].

***(c)** Suppose that $\mathfrak{A}$ is a Boolean ring and that $\mathfrak{B}$ is a subring of $\mathfrak{A}$ such that for every non-zero $a \in \mathfrak{A}$ there is a non-zero $b \in \mathfrak{B}$ such that $b \subseteq a$. Then

$$a = \sup\{b : b \in \mathfrak{B}, b \subseteq a\} \quad \forall\ a \in \mathfrak{A}$$

[cf. 15E].

***(d)** If $\mathfrak{A}$ is a Boolean ring and $I \lhd \mathfrak{A}$, the canonical map from $\mathfrak{A}$ to $\mathfrak{A}/I$ is order-continuous iff I is order-closed, and sequentially order-continuous iff I is a σ-ideal [cf. 14Lb].

***(e) The topology of the Stone space** Let $\mathfrak{A}$ be a Boolean ring and $\phi: \mathfrak{A} \to \mathscr{P}X$ its Stone representation. Let $\mathfrak{T}$ be the topology on X generated by $\phi[\mathfrak{A}]$. Then $\mathfrak{T}$ is Hausdorff, locally compact and totally disconnected, and $\phi[\mathfrak{A}]$ is precisely the ring of compact open subsets of X.

Notes and comments The reason for the abstract study of Boolean rings is our need to cope with quotient rings of rings of sets; just as abstract Riesz spaces are important because of the quotients of function spaces which we wish to study. Stone's representation theorem means that up to a point Boolean rings can safely be thought of as rings of sets. The point at which this conception breaks down is when we begin to discuss the suprema and infima of infinite sets. The Stone representation is hardly ever order-continuous, or even sequentially order-continuous. So, in general, if we see a supremum or an infimum, we are safer if we forget about the representation. 41H is an example of the kind of technique we must employ; 4XF is an example of a ring in which care is necessary.

In 41La–d is a little group of results showing analogies between Boolean rings and Archimedean Riesz spaces. Another is 41Ha, corresponding to 15C. It appears that no satisfying account has yet been given of these.

42 The space $S(\mathfrak{A})$

This is the fundamental Riesz space associated with a given Boolean ring $\mathfrak{A}$. When $\mathfrak{A}$ is a ring of sets, $S(\mathfrak{A})$ can be regarded as the linear space of 'simple functions' generated by the characteristic functions of members of $\mathfrak{A}$ [4XB]. Its most important property is the universal mapping theorem 42C, which establishes a one-to-one correspondence between 'additive' functions on $\mathfrak{A}$ and linear functions on $S(\mathfrak{A})$. Simple universal mapping theorems of this kind are cheap; what makes the construction important is the fact that $S(\mathfrak{A})$ has a canonical normed Riesz space structure [42E]. From this we deduce that $S(\mathfrak{A})$ is universal for many other classes of function. Particularly important are 'completely additive' and 'countably additive' real-valued functionals [42K *et seq.*].

42A Definition Let $\mathfrak{A}$ be a Boolean ring, and E a linear space over $\mathbf{R}$. A function $\nu: \mathfrak{A} \to E$ is **additive** if

$$\nu(a+b) = \nu a + \nu b \quad \text{whenever} \quad ab = 0 \text{ in } \mathfrak{A}.$$

In this case $\nu(0+0) = \nu(0) = 0$.

Remark This definition is familiar when $\mathfrak{A}$ is a ring of sets; of course, if $a \cap b = ab = 0$, then $a+b = a \cup b$. It remains a curious concept, because the $+$ signs on the two sides of the equation refer to very different operations.

42B Construction Let $\mathfrak{A}$ be any Boolean ring, and $\phi: \mathfrak{A} \to \mathscr{P}X$ its Stone representation [41D–41E]. For each $a \in \mathfrak{A}$, let χa be the characteristic function of ϕa, that is,

$$(\chi a)(t) = 1 \quad \text{if} \quad t \in \phi a, \quad (\chi a)(t) = 0 \quad \text{if} \quad t \in X \backslash \phi a.$$

Let $S(\mathfrak{A})$ be the linear subspace of $\mathbf{R}^X$ generated by $\{\chi a : a \in \mathfrak{A}\}$. Then it is easy to see that $\chi: \mathfrak{A} \to S(\mathfrak{A})$ is additive in the sense of 42A.

42C Theorem Let $\mathfrak{A}$ be a Boolean ring, and suppose that $S = S(\mathfrak{A})$ is constructed as above. Let E be any linear space over $\mathbf{R}$. Then there is a one-to-one correspondence between additive maps $\nu: \mathfrak{A} \to E$ and linear maps $T: S \to E$ given by $\nu = T\chi$.

Proof Let $\phi: \mathfrak{A} \to \mathscr{P}X$ be the Stone representation of $\mathfrak{A}$.

(a) Let us observe first that if $a, b \in \mathfrak{A}$, then $\phi(a \cap b) = \phi a \cap \phi b$. So

$$\chi(a \cap b) = \chi a \times \chi b$$

where $\times$ is ordinary pointwise multiplication on $\mathbf{R}^X$, i.e.

$$(x \times y)(t) = x(t)\,y(t) \quad \forall\ t \in X \quad \forall\ x, y \in \mathbf{R}^X.$$

(**b**) The core of the proof is the following observation. Let $\nu: \mathfrak{A} \to \boldsymbol{E}$ be additive. If $\langle \alpha_i \rangle_{i<n}$ and $\langle a_i \rangle_{i<n}$ are finite sequences in $\mathbf{R}$ and $\mathfrak{A}$ respectively such that $\sum_{i<n} \alpha_i \chi a_i = 0$ in $\boldsymbol{S}$, then $\sum_{i<n} \alpha_i \nu a_i = 0$.

P Induce on n. For $n = 0$, the result is trivial. For $n = 1$, we have $\alpha_0 \chi a_0 = 0$; so either $\alpha_0 = 0$ or $\phi a_0 = \varnothing$ and $a_0 = 0$; in either case $\alpha_0 \nu a_0 = 0$. Now the inductive step to $n+1$, where $n \geqslant 1$, is as follows. Given that $x = \sum_{i<n+1} \alpha_i \chi a_i = 0$, we consider

$$x_0 = \textstyle\sum_{i\leqslant n} \alpha_i \chi(a_i \cap a_{n-1} \cap a_n) = x \times \chi(a_{n-1} \cap a_n) = 0,$$

$$x_1 = \textstyle\sum_{i\leqslant n} \alpha_i \chi(a_i \cap a_{n-1} \backslash a_n) = x \times \chi(a_{n-1}) - x_0 = 0,$$

$$x_2 = \textstyle\sum_{i\leqslant n} \alpha_i \chi(a_i \backslash a_{n-1}) = x - x \times \chi(a_{n-1}) = 0.$$

Now these can be re-expressed as

$$x_0 = \textstyle\sum_{i<n-1} \alpha_i \chi(a_i \cap a_{n-1} \cap a_n) + (\alpha_{n-1} + \alpha_n)\,\chi(a_{n-1} \cap a_n),$$

$$x_1 = \textstyle\sum_{i<n-1} \alpha_i \chi(a_i \cap a_{n-1} \backslash a_n) + \alpha_{n-1} \chi(a_{n-1} \backslash a_n),$$

$$x_2 = \textstyle\sum_{i<n-1} \alpha_i \chi(a_i \backslash a_{n-1}) + \alpha_n \chi(a_n \backslash a_{n-1}).$$

By the inductive hypothesis,

$$\begin{aligned} 0 &= \textstyle\sum_{i<n-1} \alpha_i \nu(a_i \cap a_{n-1} \cap a_n) + (\alpha_{n-1} + \alpha_n)\,\nu(a_{n-1} \cap a_n) \\ &= \textstyle\sum_{i<n-1} \alpha_i \nu(a_i \cap a_{n-1} \backslash a_n) + \alpha_{n-1} \nu(a_{n-1} \backslash a_n) \\ &= \textstyle\sum_{i<n-1} \alpha_i \nu(a_i \backslash a_{n-1}) + \alpha_n \nu(a_n \backslash a_{n-1}). \end{aligned}$$

Adding these three equations,

$$0 = \textstyle\sum_{i\leqslant n} \alpha_i \nu a_i,$$

and the induction proceeds. **Q**

(**c**) But this is precisely the condition for the existence of a function $T: \boldsymbol{S} \to \boldsymbol{E}$ given by

$$T(\textstyle\sum_{i<n} \alpha_i \chi a_i) = \sum_{i<n} \alpha_i \nu a_i.$$

For if $\sum_{i<n} \alpha_i \chi a_i = \sum_{j<m} \beta_j \chi b_j$, then by applying (b) to the form

$$\textstyle\sum_{i<n} \alpha_i \chi a_i - \sum_{j<m} \beta_j \chi b_j = 0,$$

we see that $\sum_{i<n} \alpha_i \nu a_i = \sum_{j<m} \beta_j \nu b_j$. And, given that such a function T exists, it is clearly the unique linear map from $\boldsymbol{S}$ to $\boldsymbol{E}$ such that $T\chi = \nu$.

(**d**) Conversely, if $T: \boldsymbol{S} \to \boldsymbol{E}$ is linear, then $T\chi: \mathfrak{A} \to \boldsymbol{E}$ is additive, because χ is additive.

42D Notation If $\mathfrak{A}$ is a Boolean ring and $\boldsymbol{E}$ a linear space, then for each additive function $\nu: \mathfrak{A} \to \boldsymbol{E}$ I shall write $\hat{\nu}$ for the corresponding linear map from $\boldsymbol{S}(\mathfrak{A})$ to $\boldsymbol{E}$. Thus $\nu = \hat{\nu}\chi$.

42E The Riesz space structure of $S(\mathfrak{A})$ Let $\mathfrak{A}$ be a Boolean ring, and let $\boldsymbol{S} = \boldsymbol{S}(\mathfrak{A})$ be defined as in 42B.

(a) $\boldsymbol{S}$ is a Riesz subspace of $\boldsymbol{l}^\infty(X)$ [2XA], where X is the Stone space of $\mathfrak{A}$.

(b) Every non-zero member x of $\boldsymbol{S}$ can be expressed in the form

$$x = \textstyle\sum_{i<n} \alpha_i \chi a_i$$

where $\langle a_i \rangle_{i<n}$ is a disjoint sequence in $\mathfrak{A}$ [i.e. $a_i \cap a_j = 0$ if $i \neq j$; see 13G], and $a_i \neq 0$ and $\alpha_i \neq 0$ for each $i < n$. In this case,

$$x > 0 \Leftrightarrow \alpha_i > 0 \quad \forall\ i < n,$$

$$x^+ = \textstyle\sum_{i<n} \alpha_i^+ \chi a_i,$$

$$\|x\|_\infty = \max_{i<n} |\alpha_i|.$$

(c) Every non-zero member x of $\boldsymbol{S}^+$ can be expressed in the form

$$x = \textstyle\sum_{j<n} \beta_j \chi b_j,$$

where $\langle b_j \rangle_{j<n}$ is a decreasing sequence in $\mathfrak{A}$ [i.e. $b_i \supseteq b_j$ if $i \leqslant j$; see 11D], and $b_j \neq 0$ and $\beta_j > 0$ for each $j < n$. In this case,

$$\|x\|_\infty = \textstyle\sum_{j<n} \beta_j \quad \text{and} \quad x \leqslant \|x\|_\infty \chi b_0.$$

*(d) Suppose that α, $\beta \in \mathbf{R}$, a, $b \in \mathfrak{A}$, and $0 < \alpha\chi a \leqslant \beta\chi b$. Then $0 < \alpha \leqslant \beta$ and $0 \neq a \subseteq b$.

Proof Let $\phi: \mathfrak{A} \to \mathscr{P}X$ be the Stone representation of $\mathfrak{A}$. Let $\boldsymbol{S}_0$ be the set of those functions $x: X \to \mathbf{R}$ such that (i) $x[X]$ is finite, (ii) for each $\alpha \neq 0$, $\{t : x(t) = \alpha\} \in \phi[\mathfrak{A}]$. Then it is clear that $\boldsymbol{S}_0$ is a linear subspace of $\mathbf{R}^X$ (because $\phi[\mathfrak{A}]$ is a subring of $\mathscr{P}X$), that $\chi a \in \boldsymbol{S}_0$ for each $a \in \mathfrak{A}$, and that each non-zero member x of $\boldsymbol{S}_0$ can be expressed in the form of (b) (the α_i being just the non-zero values taken by the function x). Consequently $\boldsymbol{S}_0 = \boldsymbol{S}$.

But $\boldsymbol{S}_0$ is clearly a Riesz subspace of $\boldsymbol{l}^\infty(X)$, which proves (a). The identities in (b) are now elementary. For (c), express $x = \sum_{i<n} \alpha_i \chi a_i$ as in (b), but with the sum so arranged that $0 < \alpha_0 < \ldots < \alpha_{n-1}$. Set $\beta_0 = \alpha_0$ and $\beta_i = \alpha_i - \alpha_{i-1}$ for $1 \leqslant i < n$. Set $b_j = \sup_{j \leqslant i < n} a_i$ for $j < n$, so that $\chi b_j = \sum_{j \leqslant i < n} \chi a_i$. Then $x = \sum_{j<n} \beta_j \chi b_j$ is in the required form. The other properties are immediate.

Finally, (d) is obvious, remembering that $\phi a \subseteq \phi b \Leftrightarrow a \subseteq b$.

42F Increasing additive functions If $\mathfrak{A}$ is a Boolean ring and $\boldsymbol{E}$ a partially ordered linear space [§ 12], then a function $\nu: \mathfrak{A} \to \boldsymbol{E}$ is **increasing** if $\nu a \leqslant \nu b$ whenever $a \subseteq b$ [11D].

Lemma (a) An additive function $\nu: \mathfrak{A} \to \boldsymbol{E}$ is increasing iff $\nu a \geqslant 0$ for every $a \in \mathfrak{A}$. (b) In particular, $\chi: \mathfrak{A} \to \boldsymbol{S}(\mathfrak{A})$ is increasing. (c) If $\boldsymbol{E}$ is a partially ordered linear space and $\nu: \mathfrak{A} \to \boldsymbol{E}$ is an additive function, then ν is increasing iff $\hat{\nu}: \boldsymbol{S}(\mathfrak{A}) \to \boldsymbol{E}$ is increasing.

Proof (**a**) If ν is increasing, then $0 = \nu(0) \leqslant \nu(a)$ for every $a \in \mathfrak{A}$. Conversely, if $\nu a \geqslant 0$ for every $a \in \mathfrak{A}$, then

$$\nu b = \nu c + \nu(c+b) \geqslant \nu c$$

whenever $c \subseteq b$ in $\mathfrak{A}$, so ν is increasing.

(**b**) is trivial, and (**c**) follows at once from either 42Eb or 42Ec.

42G Locally bounded additive functions If $\mathfrak{A}$ is a Boolean ring, let us call an additive functional $\nu: \mathfrak{A} \to \mathbf{R}$ **locally bounded** if $\{\nu b : b \subseteq a\}$ is bounded above for each $a \in \mathfrak{A}$.

42H Proposition Let $\mathfrak{A}$ be a Boolean ring and $\nu: \mathfrak{A} \to \mathbf{R}$ an additive functional. Then ν is locally bounded iff $\hat{\nu} \in \boldsymbol{S}(\mathfrak{A})^{\sim}$, and in this case

$$\hat{\nu}^{+} = (\nu^{+})^{\wedge}$$

where $\nu^{+}: \mathfrak{A} \to \mathbf{R}$ is given by

$$\nu^{+}(a) = \sup\{\nu b : b \subseteq a\} \quad \forall\ a \in \mathfrak{A},$$

and is an increasing additive functional.

Proof (**a**) Suppose first that $\hat{\nu} \in \boldsymbol{S}^{\sim}$. In this case, for any $a \in \mathfrak{A}$,

$$\{\nu b : b \subseteq a\} = \{\hat{\nu}(\chi b) : b \subseteq a\}$$

is bounded above by $\hat{\nu}^{+}(\chi a)$, since $b \subseteq a \Rightarrow \chi b \leqslant \chi a$. Thus ν is locally bounded and $\nu^{+}: \mathfrak{A} \to \mathbf{R}$ is defined, with

$$\nu^{+}(a) \leqslant \hat{\nu}^{+}(\chi a) \quad \forall\ a \in \mathfrak{A}.$$

(**b**) Conversely, suppose that ν is locally bounded. Suppose that $x > 0$ in $\boldsymbol{S}$, and consider

$$A = \{\hat{\nu} y : 0 \leqslant y \leqslant x\}.$$

We know that there is an $a_0 \in \mathfrak{A}$ such that $x \leqslant \|x\|_\infty \chi a_0$ [42Ec]. Now if $0 < y \leqslant x, y$ can be expressed as $\sum_{i<n} \beta_i \chi b_i$, where each $\beta_i > 0$ and $\sum_{i<n} \beta_i = \|y\|_\infty \leqslant \|x\|_\infty$ [42Ec again]. But each $b_i \subseteq a_0$, since

$$\beta_i \chi b_i \leqslant y \leqslant \|x\|_\infty \chi a_0$$

[42Ed]. So

$$\hat{\nu} y = \sum_{i<n} \beta_i \nu b_i \leqslant \sum_{i<n} \beta_i \nu^+(a_0) \leqslant \|x\|_\infty \nu^+(a_0).$$

Thus A is bounded above by $\|x\|_\infty \nu^+(a_0)$. As x is arbitrary, $\hat{\nu} \in S^\sim$ [16C]. And this argument shows at the same time that

$$\hat{\nu}^+(\chi a) \leqslant \nu^+(a) \quad \forall \; a \in \mathfrak{A} \quad \text{[using 16D]}.$$

(c) Putting (a) and (b) together, we see that if ν is locally bounded, $\nu^+(a) = \hat{\nu}^+(\chi a)$ for every $a \in \mathfrak{A}$. So ν^+ is additive and increasing and $(\nu^+)^\wedge = \hat{\nu}^+$.

***42I Lemma** Let $\mathfrak{A}$ be a Boolean ring, and $\nu: \mathfrak{A} \to \mathbf{R}$ an additive functional which is *not* locally bounded. Then there is a decreasing sequence $\langle a_n \rangle_{n \in \mathbf{N}}$ in $\mathfrak{A}$ such that $|\nu a_n| \geqslant n$ and $|\nu(a_n \backslash a_{n+1})| \geqslant n+1$ for every $n \in \mathbf{N}$.

Proof We know that there is an $a_0 \in \mathfrak{A}$ such that

$$\sup \{\nu b : b \subseteq a_0\} = \infty.$$

Now we can choose $\langle a_n \rangle_{n \in \mathbf{N}}$ inductively such that, for each n,

$$a_{n+1} \subseteq a_n, \quad |\nu a_n| \geqslant n, \quad |\nu(a_n \backslash a_{n+1})| \geqslant n+1,$$

and

$$\sup \{\nu b : b \subseteq a_n\} = \infty.$$

P We already have a_0. Given a_n, choose a $c \subseteq a_n$ such that

$$\nu c \geqslant |\nu a_n| + n + 1.$$

Then for $b_0 \subseteq a_n$,

$$\begin{aligned} \nu b_0 &= \nu(b_0 \cap c) + \nu(b_0 \backslash c) \\ &\leqslant \sup \{\nu b : b \subseteq c\} + \sup \{\nu b : b \subseteq a_n \backslash c\}. \end{aligned}$$

So one of these sups must be infinite. Let a_{n+1} be either c or $a_n \backslash c$, such that $\sup \{\nu b : b \subseteq a_{n+1}\} = \infty$. Now νa_{n+1} is either νc or $\nu a_n - \nu c$; in either case, $|\nu a_{n+1}| \geqslant n+1$; at the same time, $|\nu(a_n \backslash a_{n+1})| \geqslant n+1$, as required. **Q**

42J Order-continuous increasing additive functions: proposition Let $\mathfrak{A}$ be a Boolean ring.

(a) For any Riesz space E, an increasing additive function $\nu: \mathfrak{A} \to E$ is order-continuous iff

$$\inf_{a \in A} \nu a = 0 \text{ in } E \quad \text{whenever} \quad \varnothing \subset A \downarrow 0 \quad \text{in } \mathfrak{A}.$$

(b) In particular, $\chi: \mathfrak{A} \to S(\mathfrak{A})$ is order-continuous.

(c) If $\nu: \mathfrak{A} \to \mathbf{R}$ is an increasing additive function, then it is order-continuous iff $\hat{\nu}: S(\mathfrak{A}) \to \mathbf{R}$ is order-continuous.

Proof (a) [cf. 14Ec]. This follows at once from 41Hb and 41Hc and the corresponding results 12Ca and 12Cb.

(b) **?** Otherwise, there is a non-empty set $A \downarrow 0$ in $\mathfrak{A}$ and an $x > 0$ in $S = S(\mathfrak{A})$ such that $x \leqslant \chi a$ for every $a \in A$. Now by either 42Eb or 42Ec, there is a $\beta > 0$ and a $b \neq 0$ such that $\beta \chi b \leqslant x$. But now, by 42Ed, $b \subseteq a$ for every $a \in A$, which is impossible if $\inf A = 0$. **X**

(c) From (b) we see that if $\hat{\nu}: S \to \mathbf{R}$ is order-continuous, then $\nu = \hat{\nu}\chi: \mathfrak{A} \to \mathbf{R}$ is order-continuous.

Conversely, suppose that ν is order-continuous. To show that $\hat{\nu}$ is order-continuous, we must examine an arbitrary non-empty set $A \downarrow 0$ in S. Let $\epsilon > 0$. Fix $x_0 \in A$. By 42Ec, there is an $\alpha_0 \in \mathbf{R}^+$ and an $a_0 \in \mathfrak{A}$ such that $x_0 \leqslant \alpha_0 \chi a_0$. Let $\delta > 0$ be such that $\delta \nu a_0 \leqslant \epsilon$.

For each $x \in S^+$, define $b_x \in \mathfrak{A}$ as follows. Let $\phi: \mathfrak{A} \to \mathscr{P}X$ be the Stone representation of $\mathfrak{A}$. Then $x \in \mathbf{R}^X$. Let

$$E_x = \{t : x(t) > \delta\}.$$

Then $E_x \subseteq X$ and in fact $E_x \in \phi[\mathfrak{A}]$ (see the characterization of S given in the proof of 42E). So there is a unique $b_x \in \mathfrak{A}$ such that $\phi b_x = E_x$. Now

$$\delta \chi b_x \leqslant x \leqslant \alpha_0 \chi b_x + \delta \chi a_0$$

whenever $0 \leqslant x \leqslant x_0$. Observe also that

$$0 \leqslant x \leqslant y \Rightarrow E_x \subseteq E_y \Rightarrow b_x \subseteq b_y.$$

so that $B = \{b_x : x \in A,\, x \leqslant x_0\}\downarrow$.

If b is a lower bound for B in $\mathfrak{A}$, then

$$\delta \chi b \leqslant \delta \chi b_x \leqslant x \quad \text{whenever} \quad x \in A,\, x \leqslant x_0.$$

As $A \downarrow 0$, $\delta \chi b = 0$; as $\delta > 0$, $b = 0$ [42Ed]. Thus $B \downarrow 0$ in $\mathfrak{A}$. As ν is order-continuous, there is an $x \in A$ such that $x \leqslant x_0$ and $\alpha_0 \nu b_x \leqslant \epsilon$. Now

$$\hat{\nu} x \leqslant \hat{\nu}(\alpha_0 \chi b_x + \delta \chi a_0) = \alpha_0 \nu b_x + \delta \nu a_0 \leqslant 2\epsilon.$$

As ϵ is arbitrary, $\inf_{x \in A} \hat{\nu} x = 0$; as A is arbitrary, $\hat{\nu}$ is order-continuous.

42K Completely additive functionals Let $\mathfrak{A}$ be a Boolean ring. An additive functional $\nu: \mathfrak{A} \to \mathbf{R}$ is called **completely additive** if

$$\inf_{a\in A}|\nu a| = 0 \quad \text{whenever} \quad \varnothing \subset A \downarrow 0 \text{ in } \mathfrak{A}.$$

(The reason for the phrase 'completely additive' lies in 42Rj.)

42L Theorem Let $\mathfrak{A}$ be a Boolean ring. Then an additive functional $\nu: \mathfrak{A} \to \mathbf{R}$ is completely additive iff $\hat{\nu} \in S(\mathfrak{A})^\times$.

Proof (a) If $\nu: \mathfrak{A} \to \mathbf{R}$ is completely additive, it is locally bounded. **P** Use 42I. **?** Otherwise, there is a decreasing sequence $\langle a_n\rangle_{n\in\mathbf{N}}$ in $\mathfrak{A}$ such that $|\nu a_n| \geqslant n$ for each $n\in\mathbf{N}$. Let B be the set of lower bounds of $\{a_n : n\in\mathbf{N}\}$; then $B\uparrow$. By the argument of 41Ha,

$$C = \{a_n \backslash b : n\in\mathbf{N}, b\in B\}\downarrow 0 \quad \text{in } \mathfrak{A}.$$

Let
$$D = \{a_n\backslash b : b\in B,\ n \geqslant |\nu b|+1\} \subseteq C.$$

Then D is cofinal with C [i.e. $\forall\ c\in C\ \ \exists\ d\in D,\ d\subseteq c$] so $D\downarrow 0$. But $|\nu d| \geqslant 1$ for every $d\in D$. **X Q**

(b) If $\nu: \mathfrak{A}\to\mathbf{R}$ is completely additive, $\nu^+: \mathfrak{A}\to\mathbf{R}$ is order-continuous. **P** The argument is similar to that of 24Bc. Suppose that $\varnothing \subset A\downarrow 0$ in $\mathfrak{A}$. **?** Suppose, if possible, that there is an $\epsilon > 0$ such that $\nu^+ a > \epsilon$ for every $a\in A$. Then for every $a\in A$ there is a $b\subseteq a$ such that $\nu b > \epsilon$. But

$$\{c\backslash b : c\in A,\ c\subseteq a\}\downarrow 0.$$

So there is a $c\in A$ such that $c\subseteq a$ and $|\nu(c\backslash b)| \leqslant \nu b - \epsilon$, so that $\nu(b\cup c) = \nu b + \nu(c\backslash b) \geqslant \epsilon$.

Now let

$$B = \{b : b\in\mathfrak{A},\quad \nu b\geqslant\epsilon,\quad \exists\ a, c\in A \quad \text{such that} \quad c\subseteq b\subseteq a\}.$$

Since for every $a\in A$ there is a $b\in B$ such that $b\subseteq a$, $B\downarrow 0$. But $\inf_{b\in B}|\nu b| \geqslant \epsilon > 0$, which is impossible. **X**

Thus $\inf_{a\in A}\nu^+ a$ must be 0; as A is arbitrary, ν^+ is order-continuous [42Ja]. **Q**

(c) Now if $\nu: \mathfrak{A}\to\mathbf{R}$ is completely additive, $\hat{\nu}\in S(\mathfrak{A})^\sim$ by (a) and 42H, and $\hat{\nu}^+ = (\nu^+)^\wedge$. But ν^+ is order-continuous by (b), so $(\nu^+)^\wedge$ is order-continuous [42Jc], and $\hat{\nu}^+\in S(\mathfrak{A})^\times$. Similarly, $\hat{\nu}^- = (-\hat{\nu})^+ = (-\nu)^{\wedge +}\in S^\times$, and $\hat{\nu}\in S^\times$.

(d) Conversely, if $\hat{\nu}\in S^\times$, and $\varnothing\subset A\downarrow 0$ in $\mathfrak{A}$,

$$\inf_{a\in A}|\nu a| = \inf_{a\in A}|\hat{\nu}(\chi a)| \leqslant \inf_{a\in A}|\hat{\nu}|(\chi a) = 0$$

as $|\hat{\nu}|$ and χ are both order-continuous. So ν is completely additive.

***42M Countably additive functionals** A slightly weaker condition than 'completely additive' is 'countably additive'. If $\mathfrak{A}$ is a Boolean ring, an additive functional $\nu: \mathfrak{A} \to \mathbf{R}$ is **countably additive** if $\langle \nu a_n \rangle_{n\in\mathbf{N}} \to 0$ whenever $\langle a_n \rangle_{n\in\mathbf{N}} \downarrow 0$ in $\mathfrak{A}$. [See also 42Rj below.] Clearly, a completely additive functional is countably additive, and an increasing additive functional is countably additive iff it is sequentially order-continuous.

***42N Proposition** Let $\mathfrak{A}$ be a Boolean ring and $\nu: \mathfrak{A} \to \mathbf{R}$ a locally bounded countably additive functional. Then ν is the difference of two increasing countably additive functionals.

Proof We know that $\nu^+: \mathfrak{A} \to \mathbf{R}$ is an additive functional [42H]. Suppose that $\langle a_n \rangle_{n\in\mathbf{N}} \downarrow 0$ in $\mathfrak{A}$. **?** Suppose, if possible, that

$$\langle \nu^+ a_n \rangle_{n\in\mathbf{N}} \not\to 0.$$

Since certainly $\langle \nu^+ a_n \rangle_{n\in\mathbf{N}} \downarrow$, there must be an $\epsilon > 0$ such that $\nu^+ a_n > \epsilon$ for every $n \in \mathbf{N}$. Now for each $n \in \mathbf{N}$ there is a $b \subseteq a_n$ such that $\nu b > \epsilon$. Since $\langle \nu(b \cap a_m) \rangle_{m\in\mathbf{N}} \to 0$, there must be an $m \in \mathbf{N}$ such that

$$|\nu(b \cap a_m)| \leqslant \nu b - \epsilon,$$

and now $\nu(b \backslash a_m) \geqslant \epsilon$. So $\nu^+(a_n \backslash a_m) \geqslant \epsilon$.

Consequently, we can find a strictly increasing sequence $\langle n(i) \rangle_{i\in\mathbf{N}}$ such that

$$\nu^+(a_{n(i)} \backslash a_{n(i+1)}) \geqslant \epsilon, \quad \text{i.e. } \nu^+ a_{n(i)} \geqslant \epsilon + \nu^+ a_{n(i+1)}$$

for every $i \in \mathbf{N}$. But now

$$\nu^+ a_{n(0)} \geqslant k\epsilon + \nu^+ a_{n(k)} \geqslant k\epsilon \quad \forall\ k \in \mathbf{N},$$

which is impossible. **X**

As $\langle a_n \rangle_{n\in\mathbf{N}}$ is arbitrary, this shows that ν^+ is countably additive. But similarly $\nu^- = (-\nu)^+$ is countably additive, and $\nu = \nu^+ - \nu^-$ [using 42H, or otherwise].

***42O Proposition** Let $\mathfrak{A}$ be a Boolean ring and $\nu: \mathfrak{A} \to \mathbf{R}$ an increasing additive functional. Then ν is countably additive iff $\hat{\nu}: S(\mathfrak{A}) \to \mathbf{R}$ is sequentially order-continuous.

Proof The argument is exactly the same as that of 42Jc; the only difference is that arbitrary directed sets are replaced throughout by monotonic sequences.

***42P** These arguments can be elaborated along the lines of 42L to set up a one-to-one correspondence between locally bounded countably additive functionals $\nu: \mathfrak{A} \to \mathbf{R}$ and linear functionals $f: S(\mathfrak{A}) \to \mathbf{R}$ such that $\langle fx_n \rangle_{n\in\mathbf{N}} \to 0$ whenever $\langle x_n \rangle_{n\in\mathbf{N}} \downarrow 0$ in $S(\mathfrak{A})$ [see the remark at the end of § 32, and 83K]. The principal difference is that, unlike completely additive functionals, countably additive functionals need not be locally bounded [42L, proof, part (a), and 4XG]. However, the most important situations are covered by the following lemma.

***42Q Lemma** Let $\mathfrak{A}$ be a Dedekind σ-complete Boolean ring, and $\nu: \mathfrak{A} \to \mathbf{R}$ a countably additive functional. Then ν is locally bounded.

Proof *?* Otherwise, there is a decreasing sequence $\langle a_n \rangle_{n\in\mathbf{N}}$ in $\mathfrak{A}$ such that $\langle |\nu a_n| \rangle_{n\in\mathbf{N}} \to \infty$ [42I]. Let $a = \inf_{n\in\mathbf{N}} a_n$. Then $\{a_n \backslash a : n \in \mathbf{N}\} \downarrow 0$ [41Hc]. But now

$$\langle \nu(a_n \backslash a) \rangle_{n\in\mathbf{N}} \to 0, \quad \text{i.e. } \langle \nu a_n \rangle_{n\in\mathbf{N}} \to \nu a,$$

which is impossible. **X**

42R Exercises (a) Let $\mathfrak{A}$ be a Boolean ring, E a Riesz space, and $\nu: \mathfrak{A} \to E$ an additive function. Then $\hat{\nu}: S(\mathfrak{A}) \to E$ is a Riesz homomorphism iff ν is a lattice homomorphism.

(b) Let $\mathfrak{A}$ be a Boolean ring, E an Archimedean Riesz space, and $\nu: \mathfrak{A} \to E$ an increasing additive function. Then $\hat{\nu}: S(\mathfrak{A}) \to E$ is order-continuous iff ν is order-continuous.

*(c) Let $\mathfrak{A}$ be a Boolean ring, E a normed linear space over $\mathbf{R}$, and $\nu: \mathfrak{A} \to E$ an additive function. Then $\hat{\nu}: S(\mathfrak{A}) \to E$ is continuous for the norm $\| \|_\infty$ on $S(\mathfrak{A})$ iff ν is bounded.

(d) Let $\mathfrak{A}$ be a Boolean ring, E a Dedekind complete Riesz space, and $\nu: \mathfrak{A} \to E$ an additive function. (i) Then $\hat{\nu}: S(\mathfrak{A}) \to E$ belongs to $L^\sim(S; E)$ iff $\nu^+ a = \sup\{\nu b : b \subseteq a\}$ exists in E for every $a \in \mathfrak{A}$, and in this case $\hat{\nu}^+ = (\nu^+)^\wedge$. *(ii) And now $\hat{\nu} \in L^\times(S; E)$ iff $\inf_{a\in A} |\nu a| = 0$ in E whenever $\varnothing \subset A \downarrow 0$ in $\mathfrak{A}$.

(e) For any Boolean ring $\mathfrak{A}$, $\| \|_\infty$ is a Fatou norm on $S(\mathfrak{A})$.

(f) Let $\mathfrak{A}$ be a Boolean ring. Then these are equivalent:

(i) $S(\mathfrak{A})$ is Dedekind complete;

(ii) $S(\mathfrak{A})$ is uniformly complete;

(iii) for every $a \in \mathfrak{A}$, $\{b : b \subseteq a\}$ is finite;

(iv) the Stone representation of $\mathfrak{A}$ expresses it as the ring of finite subsets of its Stone space;

(v) Every real-valued additive functional on $\mathfrak{A}$ is locally bounded.

(**g**) Let $\mathfrak{A}$ be a Boolean ring, and μ and ν locally bounded real-valued functionals on $\mathfrak{A}$. Then

$$\theta a = \sup\{\mu b + \nu(a \backslash b) : b \subseteq a\}$$

exists in $\mathbf{R}$ for every $a \in \mathfrak{A}$. θ: $\mathfrak{A} \to \mathbf{R}$ is locally bounded and additive, and $\hat{\theta} = \hat{\mu} \vee \hat{\nu}$ in $S(\mathfrak{A})^{\sim}$.

(**h**) Let $\mathfrak{A}$ be a Boolean algebra (i.e. a Boolean ring with a 1), and μ and ν locally bounded real-valued additive functionals on $\mathfrak{A}$ of which μ is increasing. Then $\hat{\nu}$ is in the band of $S(\mathfrak{A})^{\sim}$ generated by $\hat{\mu}$ iff

$$\forall\ \epsilon > 0 \quad \exists\ \delta > 0 \quad \text{such that} \quad \mu a \leqslant \delta \Rightarrow |\nu a| \leqslant \epsilon.$$

[Hint: show that the set of $\hat{\nu}$ satisfying the condition is a band in $S^{\sim}$ containing $\hat{\mu}$. Now use 15F and a lemma analogous to (g) above, with $\wedge$ replacing $\vee$.]

Under these conditions, we say that ν is **absolutely continuous** with respect to μ

*(**i**) Let $\mathfrak{A}$ be a Dedekind σ-complete Boolean ring, and μ and ν locally bounded real-valued additive functionals on $\mathfrak{A}$ of which μ is countably additive and increasing. Then $\hat{\nu}$ is in the band of $S(\mathfrak{A})^{\sim}$ generated by $\hat{\mu}$ iff it is countably additive and $\mu a = 0 \Rightarrow \nu a = 0$.

*(**j**) Let $\mathfrak{A}$ be a Boolean ring and ν: $\mathfrak{A} \to \mathbf{R}$ an additive function. Then (i) ν is countably additive if $\sum_{n \in \mathbf{N}} \nu a_n$ exists and is equal to $\nu(\sup_{n \in \mathbf{N}} a_n)$ whenever $\langle a_n \rangle_{n \in \mathbf{N}}$ is a disjoint sequence in $\mathfrak{A}$ such that $\sup_{n \in \mathbf{N}} a_n$ exists (ii) ν is completely additive iff $\sum_{a \in A} \nu a$ exists and is equal to $\nu(\sup A)$ whenever A is a disjoint set in $\mathfrak{A}$ such that $\sup A$ exists. [Hint for (ii): show first that if ν satisfies the condition, it is locally bounded, and ν^+ also satisfies the condition.]

Notes and comments The basic property of $S(\mathfrak{A})$ is the first universal mapping theorem, 42C. It is clear that this defines $S(\mathfrak{A})$, and the function χ, up to linear space isomorphism. What is remarkable is that S is endowed with a natural partial order for which the next universal mapping theorem, 42Fc, is true; that this natural order is a lattice order [42Ea]; and that the remaining universal mapping theorems [42Jc/42Rb, 42O, 42Ra, 42Rc] are all true. In fact we have six distinct theorems based on a single construction.

In 16C we saw how functions which were the difference of increasing linear maps could be characterized; this result corresponds exactly to 42H. Similarly, functionals which are the difference of order-continuous increasing additive functionals are associated with the completely additive functionals [42L, corresponding to 32D]. In the analysis here of completely and countably additive functionals we are beginning the Radon–Nikodým theorem. The reason for studying both is that in the most important applications we are presented with countably additive functionals on rings of sets; but that (as I shall endeavour to show) their most striking properties are a result of the fact that they can be regarded as completely additive functionals on quotient rings.

*Of the arguments above, two are familiar from elementary measure theory. 42C is just the theorem that an additive measure on a ring of sets defines an integral on the space of 'simple functions' [cf. 4XB]. 42O is the theorem that a countably additive measure defines a sequentially order-continuous integral; the same methods yield the more thoroughgoing 42Jc.

*I ought to remark there that the method of constructing $S(\mathfrak{A})$ given in 42B is far from the only one. Indeed, an essentially simpler algebraic construction, not using Stone's theorem, quickly sets up a space for which 42C is true. From this point, the positive cone of $S(\mathfrak{A})$ can be defined as the cone generated by $\{\chi a : a \in \mathfrak{A}\}$, and the other results will follow. However, such an abstract approach introduces substantial additional complications into the proofs; and the temporary avoidance of the axiom of choice seemed an inadequate compensation.

43 The space $L^\infty(\mathfrak{A})$

If we complete $S(\mathfrak{A})$ with respect to the norm $\| \|_\infty$, we obtain the space $L^\infty(\mathfrak{A})$. This is a Banach lattice, in fact an M-space. It is a model for the L^∞ spaces of ordinary measure theory, and shares many of their properties. The main question to which we shall address ourselves here is the Dedekind completeness of $L^\infty(\mathfrak{A})$ [43D].

43A Definition Let $\mathfrak{A}$ be a Boolean ring and X its Stone space. Then $S(\mathfrak{A})$ is a Riesz subspace of $l^\infty(X)$ [42Ea]. Let $L^\infty(\mathfrak{A})$ be the closure of $S(\mathfrak{A})$ in $l^\infty(X)$ for $\| \|_\infty$. Then $L^\infty(\mathfrak{A})$ is a closed Riesz subspace of $l^\infty(X)$, because the lattice operations on $l^\infty(X)$ are continuous. Under $\| \|_\infty$, $L^\infty(\mathfrak{A})$ is an M-space, because $l^\infty(X)$ is.

For examples, see 4XB–4XF and 43Ec below.

43B Lemma Let $\mathfrak{A}$ be a Boolean ring and X its Stone space.

(a) If $\mathbf{1}: X \to \mathbf{R}$ is the function with constant value 1, then

$$y \wedge \alpha \mathbf{1} \in \boldsymbol{L}^\infty(\mathfrak{A})$$

for every $y \in \boldsymbol{L}^\infty$ and $\alpha \geqslant 0$.

(b) If $x \in \boldsymbol{L}^{\infty+}$, there is a sequence $\langle x_n \rangle_{n\in\mathbf{N}}$ in $\boldsymbol{S}(\mathfrak{A})^+$ such that $\langle x_n \rangle_{n\in\mathbf{N}} \uparrow x$ and $\langle x_n \rangle_{n\in\mathbf{N}} \to x$ for $\| \ \|_\infty$.

(c) $\boldsymbol{S}(\mathfrak{A})$ is super-order-dense in $\boldsymbol{L}^\infty(\mathfrak{A})$.

*(d) If $x \in \boldsymbol{L}^{\infty+}$ and $\delta > 0$, then

$$\{a : a \in \mathfrak{A},\ \delta\chi a \leqslant x\}$$

is bounded above in $\mathfrak{A}$.

Proof (a) Observe first that if $y \in \boldsymbol{S} = \boldsymbol{S}(\mathfrak{A})$ and $\alpha \geqslant 0$, $y \wedge \alpha\mathbf{1} \in \boldsymbol{S}$. **P** For if $y = \sum_{i<n} \alpha_i \chi a_i$, where $\langle a_i \rangle_{i<n}$ is disjoint in $\mathfrak{A}$ [42Eb],

$$y \wedge \alpha \mathbf{1} = \textstyle\sum_{i<n} (\alpha_i \wedge \alpha)\, \chi a_i. \quad \mathbf{Q}$$

Consequently $y \wedge \alpha\mathbf{1} \in \boldsymbol{L}^\infty$ for every $y \in \boldsymbol{L}^\infty$, because the map $y \mapsto y \wedge \alpha\mathbf{1} : \boldsymbol{l}^\infty \to \boldsymbol{l}^\infty$ is continuous.

(b) Given $x \in \boldsymbol{L}^{\infty+}$, there must be a sequence $\langle y_n \rangle_{n\in\mathbf{N}}$ in $\boldsymbol{S}^+$ such that $\|x - y_n\|_\infty \leqslant 2^{-n}$ for every $n \in \mathbf{N}$. Set

$$x_n = (y_n - 3.2^{-n}\mathbf{1})^+ = y_n - y_n \wedge 3.2^{-n}\mathbf{1} \in \boldsymbol{S}$$

for each $n \in \mathbf{N}$. Since

$$\|y_n - y_{n+1}\|_\infty \leqslant 3.2^{-(n+1)} \quad \forall\ n \in \mathbf{N},$$

$\langle x_n \rangle_{n\in\mathbf{N}} \uparrow$; also, $\|x_n - x\|_\infty \leqslant 4.2^{-n}$ for each n, so $\langle x_n \rangle_{n\in\mathbf{N}} \to x$. By 21Bd, $\langle x_n \rangle_{n\in\mathbf{N}} \uparrow x$.

(c) By definition, $\boldsymbol{S}$ is super-order-dense in $\boldsymbol{L}^\infty$.

*(d) We know that there is a $y \in \boldsymbol{S}^+$ such that $\|x - y\|_\infty < \delta$. Express y as $\sum_{i<n} \beta_j \chi b_j$ where $\langle b_j \rangle_{j<n}$ is decreasing and $\beta_j \geqslant 0$ for each $j < n$ [42Ec]. Then

$$\{t : y(t) > 0\} = \phi b_0$$

where ϕ is the Stone representation of $\mathfrak{A}$. Now

$$\begin{aligned} \delta\chi a \leqslant x &\Rightarrow x(t) \geqslant \delta \quad \forall\ t \in \phi a \\ &\Rightarrow y(t) > 0 \quad \forall\ t \in \phi a \\ &\Rightarrow \phi b_0 \supseteq \phi a \Rightarrow b_0 \supseteq a. \end{aligned}$$

So $\{a : \delta\chi a \leqslant x\}$ is bounded above by b_0.

43C Corollary For any Boolean ring $\mathfrak{A}$, the norm $\| \ \|_\infty$ is a Fatou norm on $L^\infty(\mathfrak{A})$.

Proof We know that $\| \ \|_\infty$ is a Riesz norm. So suppose that $\varnothing \subset A \uparrow x$ in $L^{\infty+}$. Let $\alpha = \sup_{y \in A} \|y\|_\infty$. Then $x \wedge \alpha \mathbf{1} \in L^\infty$ [43Ba], and clearly $x \wedge \alpha \mathbf{1}$ is an upper bound for A. So $x = x \wedge \alpha \mathbf{1}$ and $\|x\|_\infty \leqslant \alpha$. As A is arbitrary, this shows that $\| \ \|_\infty$ is a Fatou norm.

43D Theorem Let $\mathfrak{A}$ be a Boolean ring. Then

(a) $L^\infty(\mathfrak{A})$ is Dedekind complete iff $\mathfrak{A}$ is;

(b) $L^\infty(\mathfrak{A})$ is Dedekind σ-complete iff $\mathfrak{A}$ is.

Proof The proofs of the two halves of the theorem are formally independent, but almost identical; so I shall give a proof of (a), with occasional words in brackets which may be inserted to give a proof of (b).

(i) Suppose first that $L^\infty = L^\infty(\mathfrak{A})$ is Dedekind (σ-) complete and that $A \subseteq \mathfrak{A}$ is a non-empty (countable) set. Then $\{\chi a : a \in A\}$ has an infimum x in L^∞. Let $y \in S = S(\mathfrak{A})$ be such that $0 \leqslant y \leqslant x$ and $\|x - y\|_\infty < 1$ [using 43Bb]. Then there is a $\beta > 0$ and a $b_0 \in \mathfrak{A}$ such that

$$\beta \chi b_0 \leqslant y \leqslant \|y\|_\infty \chi b_0 \leqslant \chi b_0$$

[using 42Ec; if $y = 0$, set $b_0 = 0$].

Now, if $a \in A$, $\beta \chi b_0 \leqslant y \leqslant x \leqslant \chi a$; by 42Ed, $b_0 \subseteq a$; thus b_0 is a lower bound for A. Conversely, if b is a lower bound for A, χb is a lower bound for $\{\chi a : a \in A\}$, so $\chi b \leqslant x$; now

$$\|\chi(b \backslash b_0)\|_\infty = \|(\chi b - \chi b_0)^+\|_\infty \leqslant \|(x - y)^+\|_\infty < 1.$$

So $b \backslash b_0 = 0$, i.e. $b \subseteq b_0$. As b is arbitrary, $b_0 = \inf A$; as A is arbitrary, $\mathfrak{A}$ is Dedekind (σ-) complete [13Ca or 41I].

(ii) Now suppose that $\mathfrak{A}$ is Dedekind (σ-) complete, and that $A \subseteq L^{\infty+}$ is a non-empty (countable) set with an upper bound $w \in L^\infty$ say. For each $x \in A$, we can choose a non-empty (countable) set $B_x \subseteq S^+$ such that $x = \sup B_x$ [43Bb]. Now $B = \bigcup_{x \in A} B_x$ is a non-empty (countable) set bounded above by w, and A and B have the same upper bounds in L^∞.

To save space, let us write $\alpha_n = 2^{-n} \|w\|_\infty$ for each $n \in \mathbf{N}$, and let C be the set of upper bounds of B in L^∞.

We can construct inductively a sequence $\langle y_n \rangle_{n \in \mathbf{N}}$ in S^+ with the following properties:

(α) $y_{n+1} \geqslant y_n$;

(β) $\|y_{n+1} - y_n\|_\infty \leqslant \alpha_{n+1}$;

(γ) $y_n \leqslant v \quad \forall\ v \in C$;

(δ) $\|(z - y_n)^+\|_\infty \leqslant \alpha_n \quad \forall\ z \in B$,

for every $n \in \mathbf{N}$.

P Set $y_0 = 0$. Having found y_n, consider for each $z \in B$ the set

$$E_z = \{t : (z - y_n)(t) > \alpha_{n+1}\}.$$

Since $z - y_n \in \boldsymbol{S}$, $E_z \in \phi[\mathfrak{A}]$, where ϕ is the Stone representation of $\mathfrak{A}$; let a_z be such that $\phi a_z = E_z$. Then

$$w \geqslant (z - y_n)^+ \geqslant \alpha_{n+1} \chi a_z,$$

and
$$\|(z - y_n - \alpha_{n+1} \chi a_z)^+\|_\infty \leqslant \alpha_{n+1},$$

because we are supposing that $\|(z - y_n)^+\| \leqslant \alpha_n = 2\alpha_{n+1}$. Now

$$D = \{a_z : z \in B\}$$

is a non-empty (countable) set in $\mathfrak{A}$, and

$$D \subseteq \{a : a \in \mathfrak{A},\ \alpha_{n+1} \chi a \leqslant w\},$$

which is bounded above in $\mathfrak{A}$ by 43Bd. So $d = \sup D$ exists in $\mathfrak{A}$. And $\chi d = \sup\{\chi a : a \in D\}$ in $\boldsymbol{S}$ by 42Jb; because $\boldsymbol{S}$ is order-dense in $\boldsymbol{L}^\infty$ [43Bc], $\chi d = \sup\{\chi a : a \in D\}$ in $\boldsymbol{L}^\infty$ [17A].

Set
$$y_{n+1} = y_n + \alpha_{n+1} \chi d.$$

Then of course $y_{n+1} \geqslant y_n$ and $\|y_{n+1} - y_n\|_\infty \leqslant \alpha_{n+1}$. Next,

$$\begin{aligned} y_{n+1} &= y_n + \alpha_{n+1} \sup\{\chi a_z : z \in B\} \\ &= \sup\{y_n + \alpha_{n+1} \chi a_z : z \in B\}. \end{aligned}$$

But if $v \in C$ and $z \in B$,

$$y_n + \alpha_{n+1} \chi a_z \leqslant y_n + (z - y_n)^+ = y_n \vee z \leqslant v;$$

as z is arbitrary, $y_{n+1} \leqslant v$. Finally,

$$\|(z - y_{n+1})^+\|_\infty \leqslant \|(z - y_n - \alpha_{n+1} \chi a_z)^+\|_\infty \leqslant \alpha_{n+1}$$

for every $z \in B$. Thus the induction continues. **Q**

Now $\langle y_n \rangle_{n \in \mathbf{N}}$ is an increasing Cauchy sequence in $\boldsymbol{S}$, so has a limit $y \in \boldsymbol{L}^\infty$; by 21Bd, $y = \sup_{n \in \mathbf{N}} y_n$. So, for any $z \in B$,

$$\|(z - y)^+\|_\infty = \lim_{n \to \infty} \|(z - y_n)^+\|_\infty = 0,$$

i.e. $z \leqslant y$; thus y is an upper bound for B. On the other hand, if v is any upper bound for B, $y_n \leqslant v$ for every $n \in \mathrm{N}$, so $y \leqslant v$. Thus

$$y = \sup B = \sup A.$$

As A is arbitrary, L^∞ is Dedekind (σ-) complete.

43E Exercises (a) If $\mathfrak{A}$ is a Boolean ring, the following are equivalent: (i) $\mathfrak{A}$ has a 1; (ii) $S(\mathfrak{A})$ has a weak order unit; (iii) $L^\infty(\mathfrak{A})$ has an order unit.

*(b) If $\mathfrak{A}$ is a Boolean ring, there is a natural one-to-one correspondence between the elements of $\mathfrak{A}$ and those bands in $L^\infty(\mathfrak{A})$ which are generated by a single element.

*(c) Let $\mathfrak{A}$ be a Boolean ring, and $\phi: \mathfrak{A} \to \mathscr{P}X$ its Stone representation. Let $\mathfrak{T}$ be the topology on X generated by $\phi[\mathfrak{A}]$, as in 41Le. Then $L^\infty(\mathfrak{A}) = C_0(X)$ [definition: A2C]. [Hint: use the Stone–Weierstrass theorem.]

Notes and comments The result 43Da will be basic to the discussion of Maharam algebras [§ 53] and measure spaces [§ 64], after the relationship between measure algebras and ordinary L^∞ spaces has been established [62H].

*An alternative proof of 43D can be got from the ideas of 41Le/43Ec. The Stone space X of $\mathfrak{A}$ has a canonical topology $\mathfrak{T}$ under which $\mathfrak{A}$ is represented as the ring of compact open sets in X [41Le]. Now it can be shown that $\mathfrak{A}$ is Dedekind complete iff $\mathfrak{T}$ is extremally disconnected (i.e. the closure of every open set is open), and that this is so iff $C_0(X)$ is Dedekind complete; but $C_0(X) = L^\infty(\mathfrak{A})$ [43Ec]. A similar characterization of Dedekind σ-complete Boolean rings is not hard to find. When $\mathfrak{A}$ has a 1, it is Dedekind complete iff $\phi[\mathfrak{A}]$ is the algebra of regular open sets of X [see 4XF].

*43Eb shows that the Boolean ring $\mathfrak{A}$ is uniquely determined by the Riesz space $L^\infty(\mathfrak{A})$.

44 The space $L^\#$

Continuing our search for abstract versions of the spaces arising in measure theory, we find that approximate equivalents of the L^1 spaces can be set up as duals of the L^∞ spaces. The equivalence is not in general perfect, so I shall use a new symbol, and write $L^\#(\mathfrak{A})$ for $L^\infty(\mathfrak{A})^\times$. For the moment we shall examine $L^\#$ spaces alone. They are related to L^1 spaces by the Radon–Nikodým theorem, as will appear in §§ 52 and 63.

44A Definition Let $\mathfrak{A}$ be a Boolean ring. Then I shall write $L^\#(\mathfrak{A})$ for $L^\infty(\mathfrak{A})^\times$.

44B Theorem Let $\mathfrak{A}$ be a Boolean ring.

(a) $L^\#(\mathfrak{A})$ is a band in $L^\infty(\mathfrak{A})' = L^\infty(\mathfrak{A})^\sim$; with the induced norm, it is an L-space.

(b) There is a one-to-one correspondence between members f of $L^\#(\mathfrak{A})$ and bounded completely additive functionals $\nu: \mathfrak{A} \to \mathbf{R}$, given by $\nu = f\chi$. $f \geqslant 0$ iff ν is increasing, and in this case

$$\|f\| = \sup\{\nu a : a \in \mathfrak{A}\}.$$

Proof (a) We know that $L^{\infty\prime} = L^{\infty\sim}$ is an L-space because L^∞ is an M-space [43A, 25G, 26D]. Now $L^\# = L^{\infty\times}$ is a band in $L^{\infty\sim}$ so is a closed Riesz subspace [22Ec], and therefore is in its own right an L-space.

(b) (i) Because $L^\# \subseteq L^{\infty\prime}$, and $S = S(\mathfrak{A})$ is dense in L^∞, members of $L^\#$ are determined by their values on S; so the map $f \mapsto f\chi$ is one-to-one [42C].

(ii) Suppose that $f \in L^\#$. Then $f\chi$ is an additive function on $\mathfrak{A}$. For any $a \in \mathfrak{A}$,

$$|f(\chi a)| \leqslant \|f\| \, \|\chi a\|_\infty \leqslant \|f\|,$$

so $f\chi$ is bounded. Now suppose that $\varnothing \subset A \downarrow 0$ in $\mathfrak{A}$. Then $\{\chi a : a \in A\} \downarrow 0$ in S[42Jb], so (because S is order-dense in L^∞, 43Bc), $\{\chi a : a \in A\} \downarrow 0$ in L^∞ [17A]. It follows that

$$\inf_{a \in A} |f(\chi a)| \leqslant \inf_{a \in A} |f|(\chi a) = 0,$$

because $|f|: L^\infty \to \mathbf{R}$ is order-continuous. So $f\chi$ is completely additive.

(iii) Conversely, suppose that $\nu: \mathfrak{A} \to \mathbf{R}$ is bounded and completely additive. In this case, of course, ν is locally bounded and $\lambda = \nu^+$ is completely additive [42L proof, part (b)]; since

$$\lambda a = \sup\{\nu b : b \subseteq a\} \leqslant \sup\{\nu b : b \in \mathfrak{A}\} < \infty$$

for every $a \in \mathfrak{A}$, λ is also bounded; let $\alpha = \sup\{\lambda a : a \in \mathfrak{A}\}$.

Consider $\hat{\lambda}: S \to \mathbf{R}$, the linear functional derived from λ. Then $\hat{\lambda}$ is continuous for $\| \, \|_\infty$ and $\|\hat{\lambda}\| = \alpha$. **P** Suppose that $x \in S^+$. Then x can be expressed as $\sum_{i<n} \beta_i \chi b_i$, where $\|x\|_\infty = \sum_{i<n} |\beta_i|$ [42Ec]. So

$$|\hat{\lambda}x| = |\sum_{i<n} \beta_i \lambda(b_i)| \leqslant \alpha \sum_{i<n} |\beta_i| = \alpha \|x\|_\infty.$$

In general, for any $x \in \boldsymbol{S}$,

$$|\hat{\lambda}x| = |\hat{\lambda}(x^+) - \hat{\lambda}(x^-)| \leqslant \max(\hat{\lambda}(x^+), \hat{\lambda}(x^-))$$
$$\leqslant \alpha \max(\|x^+\|_\infty, \|x^-\|_\infty) = \alpha \| \, |x| \, \|_\infty = \alpha \|x\|_\infty.$$

Thus $\hat{\lambda}$ is continuous, and $\|\hat{\lambda}\| \leqslant \alpha$; now, of course, $\|\hat{\lambda}\| = \alpha$. **Q**

Now it follows that, for any $x \in L^{\infty +}$,

$$\{\hat{\lambda}y : y \in \boldsymbol{S}, 0 \leqslant y \leqslant x\}$$

is bounded above by $\alpha \|x\|_\infty$. Since $\hat{\lambda}: \boldsymbol{S} \to \mathbf{R}$ is order-continuous [42Jc], and $\boldsymbol{S}$ is order-dense in L^∞ [43Bc], there is an order-continuous increasing linear functional $f: L^\infty \to \mathbf{R}$ extending $\hat{\lambda}$, by 17B.

Thus $\nu^+ = f\chi$, where $f \in L^\#$. Similarly, there is a $g \in L^\#$ such that $g\chi = (-\nu)^+$; and now $\nu = (f-g)\chi$. This completes the proof.

44C Exercises (a) Let $\mathfrak{A}$ be a Boolean ring. Then there is a one-to-one correspondence between bounded additive functionals ν on $\mathfrak{A}$ and members f of $L^\infty(\mathfrak{A})'$, given by $\nu = f\chi$.

***(b)** In (a) above, ν is countably additive iff f has the property that $\langle f x_n \rangle_{n \in \mathbf{N}} \to 0$ whenever $\langle x_n \rangle_{n \in \mathbf{N}} \downarrow 0$ in L^∞.

(c) Let $\mathfrak{A}$ be a Boolean ring. Then $L^\infty(\mathfrak{A})^\sim$ can be identified, as normed Riesz space, with $\boldsymbol{S}(\mathfrak{A})'$, which is a solid linear subspace of $\boldsymbol{S}(\mathfrak{A})^\sim$.

(d) Let $\mathfrak{A}$ be a Boolean ring and $f \in L^\infty(\mathfrak{A})^\sim$. Then $f^+\chi = (f\chi)^+$, where f^+ is taken in $L^\infty(\mathfrak{A})^\sim$ and $(f\chi)^+$ is defined as in 42H.

(e) Given a Boolean ring $\mathfrak{A}$, define $\int : L^\#(\mathfrak{A}) \to \mathbf{R}$ by $\int f = \|f^+\| - \|f^-\|$ for every $f \in L^\#$. Then

$$\int f = \lim_{a \uparrow} f(\chi a)$$

in the sense that

$$\forall \, \epsilon > 0 \quad \exists \, b \in \mathfrak{A} \quad \text{such that} \quad |f(\chi a) - \textstyle\int f| \leqslant \epsilon \quad \forall \, a \supseteq b.$$

Notes and comments The fundamental idea here is part (iii) of the proof of 44Bb; if $\lambda : \mathfrak{A} \to \mathbf{R}$ is a bounded order-continuous increasing additive functional, then there is a corresponding order-continuous increasing linear functional on $L^\infty(\mathfrak{A})$. This is a fairly straightforward extension of 42Jc, in which arbitrary order-continuous increasing additive functionals were associated with order-continuous increasing linear functionals on $\boldsymbol{S}(\mathfrak{A})$. Now the bounded completely additive functionals are identified with members of $L^{\infty\times}$, just as general completely additive functionals are identified with members of $\boldsymbol{S}^\times$ [42L].

The value of this refinement lies in the fact that, because L^∞ is an M-space, $L^\# = L^{\infty\times}$ is an L-space. Thus we shall be able to apply some of our most powerful results about Riesz spaces to the space of bounded completely additive functionals.

These expressions represent $L^\#$ as a subspace of $S^\times \subseteq S^\sim$. It is important to note that $L^\#$ is actually a solid linear subspace of $S^\sim$ (though not in general a band). For $L^\#$ is a solid linear subspace of $L^{\infty\sim} = L^{\infty\prime}$, which can be identified with S', which is a solid linear subspace of $S^\sim$ by 22D. The point of this is that the lattice operations on $L^\#$ can be calculated in the same way as those on $S^\sim$, as in 42Rg or 44Cd.

When the ring $\mathfrak{A}$ has a 1, then obviously every locally bounded additive functional on $\mathfrak{A}$ is actually bounded; so $S^\sim = S' \cong L^{\infty\sim}$ and $S^\times \cong L^\#$. There are further results concerning $L^\#$ in § 83.

It is difficult to give convincing examples of $L^\#$ spaces, because simple Boolean rings give rise to trivial cases [see 4XC; also 4XFj]. In fact, as Kakutani's theorem shows, all $L^\#$ spaces can be based on measure algebras [see the end-note to § 26]. So further examples must await Chapters 5 and 6.

45 Ring homomorphisms

An important aspect of the constructions of the last three sections is their behaviour relative to homomorphisms between the underlying rings. This is particularly significant because the rings which are most important to us are defined as quotient rings, that is, as homomorphic images. The fundamental results are that a homomorphism from $\mathfrak{A}$ to $\mathfrak{B}$ gives rise to maps from $S(\mathfrak{A})$ to $S(\mathfrak{B})$, from $L^\infty(\mathfrak{A})$ to $L^\infty(\mathfrak{B})$, and from $L^\#(\mathfrak{B})$ to $L^\#(\mathfrak{A})$. In the language of category theory, S and L^∞ are covariant functors, while $L^\#$ is contravariant. If $\mathfrak{B}$ is a quotient of $\mathfrak{A}$, $S(\mathfrak{B})$ and $L^\infty(\mathfrak{B})$ appear as quotients of $S(\mathfrak{A})$ and $L^\infty(\mathfrak{A})$ respectively; this is the result we need to identify the L^∞ space of an ordinary measure algebra with the usual function space.

45A Definition Let $\mathfrak{A}$ and $\mathfrak{B}$ be Boolean rings. A map $\pi: \mathfrak{A} \to \mathfrak{B}$ is a **ring homomorphism** if $\pi(ab) = \pi a . \pi b$ and $\pi(a+b) = \pi a + \pi b$ for all $a, b \in \mathfrak{A}$. In this case, the formulae of 41G show that

$$\pi(a \cup b) = \pi(a+b+ab) = \pi a + \pi b + \pi a . \pi b = \pi a \cup \pi b,$$

$$\pi(a \cap b) = \pi(ab) = \pi a . \pi b = \pi a \cap \pi b,$$

$$\pi(a \backslash b) = \pi(a+ab) = \pi a + \pi a . \pi b = \pi a \backslash \pi b$$

for all $a, b \in \mathfrak{A}$, so that π is a lattice homomorphism; also, of course, $\pi(0) = 0$.

45B Proposition Let $\mathfrak{A}$ and $\mathfrak{B}$ be Boolean rings, and $\pi: \mathfrak{A} \to \mathfrak{B}$ a ring homomorphism. Then there is a unique Riesz homomorphism $\hat{\pi}: \boldsymbol{S}(\mathfrak{A}) \to \boldsymbol{S}(\mathfrak{B})$ defined by $\hat{\pi}(\chi a) = \chi(\pi a)$ for every $a \in \mathfrak{A}$, and now $\|\hat{\pi}x\|_\infty \leqslant \|x\|_\infty$ for every $x \in \boldsymbol{S}(\mathfrak{A})$. So $\hat{\pi}$ has a unique extension to a norm-decreasing Riesz homomorphism $\hat{\pi}: \boldsymbol{L}^\infty(\mathfrak{A}) \to \boldsymbol{L}^\infty(\mathfrak{B})$.

Proof The map $a \mapsto \chi(\pi a): \mathfrak{A} \to \boldsymbol{S}(\mathfrak{B})$ is easily seen to be additive; so there is a unique linear map $\hat{\pi}: \boldsymbol{S}(\mathfrak{A}) \to \boldsymbol{S}(\mathfrak{B})$ such that $\hat{\pi}(\chi a) = \chi(\pi a)$ for every $a \in \mathfrak{A}$ [42C]. Now if $\langle a_i \rangle_{i<n}$ is a disjoint sequence in $\mathfrak{A}$, $\langle \pi a_i \rangle_{i<n}$ will be a disjoint sequence in $\mathfrak{B}$. So if x is any non-zero member of $\boldsymbol{S}(\mathfrak{A})$, express x as $\sum_{i<n} \alpha_i \chi a_i$ where $\langle a_i \rangle_{i<n}$ is a disjoint sequence of non-zero elements of $\mathfrak{A}$ [42Eb]; then

$$\hat{\pi}x = \textstyle\sum_{i<n} \alpha_i \chi(\pi a_i),$$

so

$$\hat{\pi}(x^+) = \textstyle\sum_{i<n} \alpha_i^+ \chi(\pi a_i) = (\hat{\pi}x)^+$$

and

$$\|\hat{\pi}x\|_\infty = \max\{|\alpha_i| : i < n, \pi a_i \neq 0\} \leqslant \|x\|_\infty.$$

Thus $\hat{\pi}$ is a norm-decreasing Riesz homomorphism from $\boldsymbol{S}(\mathfrak{A})$ to $\boldsymbol{S}(\mathfrak{B})$. Now $\hat{\pi}$ has a unique continuous extension to a map from $\boldsymbol{L}^\infty(\mathfrak{A})$ to $\boldsymbol{L}^\infty(\mathfrak{B})$, since these are the completions of $\boldsymbol{S}(\mathfrak{A})$ and $\boldsymbol{S}(\mathfrak{B})$ respectively. Since the lattice operations on both $\boldsymbol{L}^\infty(\mathfrak{A})$ and $\boldsymbol{L}^\infty(\mathfrak{B})$ are continuous, $\hat{\pi}: \boldsymbol{L}^\infty(\mathfrak{A}) \to \boldsymbol{L}^\infty(\mathfrak{B})$ is a Riesz homomorphism; and of course it is still norm-decreasing.

45C Proposition Let $\mathfrak{A}$ and $\mathfrak{B}$ be Boolean rings, and $\pi: \mathfrak{A} \to \mathfrak{B}$ a ring homomorphism. Let $\hat{\pi}: \boldsymbol{L}^\infty(\mathfrak{A}) \to \boldsymbol{L}^\infty(\mathfrak{B})$ be the associated Riesz homomorphism. Then

(a) if π is one-to-one, then $\hat{\pi}$ is norm-preserving;

(b) if π is onto, then $\hat{\pi}[\boldsymbol{S}(\mathfrak{A})] = \boldsymbol{S}(\mathfrak{B})$ and $\hat{\pi}[\boldsymbol{L}^\infty(\mathfrak{A})] = \boldsymbol{L}^\infty(\mathfrak{B})$; moreover, for each $y \in \boldsymbol{S}(\mathfrak{B})$, $\|y\|_\infty = \inf\{\|x\|_\infty : x \in \boldsymbol{S}(\mathfrak{A}), \hat{\pi}x = y\}$, and for each $y \in \boldsymbol{L}^\infty(\mathfrak{B})$, $\|y\|_\infty = \inf\{\|x\|_\infty : x \in \boldsymbol{L}^\infty(\mathfrak{A}), \hat{\pi}x = y\}$.

Proof (a) On examination of the formula for $\|\hat{\pi}x\|_\infty$ given in the proof of 45B above, it is clear that if π is one-to-one then $\|\hat{\pi}x\|_\infty = \|x\|_\infty$ for every $x \in \boldsymbol{S}(\mathfrak{A})$. Now by continuity $\|\hat{\pi}x\|_\infty$ must be equal to $\|x\|_\infty$ for every $x \in \boldsymbol{L}^\infty(\mathfrak{A})$.

(b) (i) If $\pi: \mathfrak{A} \to \mathfrak{B}$ is onto, then for every disjoint sequence $\langle b_i \rangle_{i<n}$ in $\mathfrak{B}$ there must be a disjoint sequence $\langle a_i \rangle_{i<n}$ in $\mathfrak{A}$ such that $\pi a_i = b_i$

for each $i < n$. **P** Certainly there is a sequence $\langle c_i \rangle_{i<n}$ in $\mathfrak{A}$ such that $\pi c_i = b_i$ for each $i < n$. Now set

$$a_i = c_i \setminus \sup_{j<i} c_j \quad \forall\ i < n.$$

Then $\pi a_i = b_i \setminus \sup_{j<i} b_j = b_i$ for each $i < n$, and $\langle a_i \rangle_{i<n}$ is disjoint. **Q**

(ii) Now suppose that y is a non-zero member of $S(\mathfrak{B})$. Then there is an expression of y as $\sum_{i<n} \alpha_i \chi b_i$ where $\langle b_i \rangle_{i<n}$ is a disjoint sequence in $\mathfrak{B}$ and $\alpha_i \chi b_i \neq 0$ for each $i < n$ [42Eb again]. Let $\langle a_i \rangle_{i<n}$ be a disjoint sequence in $\mathfrak{A}$ such that $\pi a_i = b_i$ for each $i < n$. Set $x = \sum_{i<n} \alpha_i \chi a_i$. Then $\hat{\pi} x = y$ and

$$\|x\|_\infty = \max_{i<n} |\alpha_i| = \|y\|_\infty.$$

So $\hat{\pi}: S(\mathfrak{A}) \to S(\mathfrak{B})$ is onto, and $\|y\|_\infty \geqslant \inf\{\|x\|_\infty : x \in S(\mathfrak{A}),\ \hat{\pi} x = y\}$ for each $y \in S(\mathfrak{B})$. The reverse inequality is trivial, because $\hat{\pi}$ is norm-decreasing.

(iii) The corresponding result for $\hat{\pi}: L^\infty(\mathfrak{A}) \to L^\infty(\mathfrak{B})$ follows at once from the general theory of normed linear spaces. If $y \in L^\infty(\mathfrak{B})$ and $\epsilon > 0$, then there is a sequence $\langle y_n \rangle_{n \in \mathbf{N}}$ in $S(\mathfrak{B})$ such that

$$\|y - y_n\|_\infty \leqslant 2^{-n}\epsilon$$

for each $n \in \mathbf{N}$. Now we can choose a $z_0 \in S(\mathfrak{A})$ such that $\hat{\pi} z_0 = y_0$ and $\|z_0\|_\infty = \|y_0\|_\infty$ and, for each $n \geqslant 1$, a $z_n \in S(\mathfrak{A})$ such that $\hat{\pi} z_n = y_n - y_{n-1}$ and $\|z_n\|_\infty = \|y_n - y_{n-1}\|_\infty \leqslant 3.2^{-n}\epsilon$. But $z = \sum_{n \in \mathbf{N}} z_n$ exists in $L^\infty(\mathfrak{A})$ and $\|z\|_\infty \leqslant \sum_{n \in \mathbf{N}} \|z_n\|_\infty \leqslant \|y\|_\infty + 4\epsilon$, while $\hat{\pi} z = \sum_{n \in \mathbf{N}} \hat{\pi} z_n = y$. As ϵ is arbitrary, this shows that $\|y\|_\infty \geqslant \inf\{\|x\|_\infty : \hat{\pi} x = y\}$; again, the reverse inequality is trivial.

45D Corollary **(a)** If $\mathfrak{B}$ is a Boolean ring and $\mathfrak{A}$ is a subring of $\mathfrak{B}$, we may regard $S(\mathfrak{A})$ and $L^\infty(\mathfrak{A})$ as Riesz subspaces of $S(\mathfrak{B})$ and $L^\infty(\mathfrak{B})$ respectively.

(b) Now if $\mathfrak{A}$ is an ideal of $\mathfrak{B}$, $S(\mathfrak{A})$ and $L^\infty(\mathfrak{A})$ are solid linear subspaces of $S(\mathfrak{B})$ and $L^\infty(\mathfrak{B})$ respectively.

(c) If $\mathfrak{A}$ is a Boolean ring, I an ideal of $\mathfrak{A}$, and $\mathfrak{B}$ the quotient ring $\mathfrak{A}/I$, then $S(\mathfrak{B})$ and $L^\infty(\mathfrak{B})$ can be identified, as normed Riesz spaces, with $S(\mathfrak{A})/S(I)$ and $L^\infty(\mathfrak{A})/L^\infty(I)$ respectively.

Proof (a) Let $\pi: \mathfrak{A} \to \mathfrak{B}$ be the identity map. Then

$$\hat{\pi}: L^\infty(\mathfrak{A}) \to L^\infty(\mathfrak{B})$$

is a norm-preserving Riesz homomorphism, by 45Ca, so embeds the normed Riesz space $L^\infty(\mathfrak{A})$ as a closed Riesz subspace of $L^\infty(\mathfrak{B})$. And of course $\hat{\pi}[S(\mathfrak{A})] \subseteq S(\mathfrak{B})$.

(**b**) Consider first $S(\mathfrak{A})$ as a Riesz subspace of $S(\mathfrak{B})$. Suppose that $x \in S(\mathfrak{A})$, $y \in S(\mathfrak{B})$ and $0 < |y| \leqslant |x|$. Then there is an $a_0 \in \mathfrak{A}$ such that $|x| \leqslant \|x\|_\infty \chi a_0$ [42Ec]. Express y as $\sum_{i<n} \beta_i \chi b_i$, where $\langle b_i \rangle_{i<n}$ is a disjoint sequence in $\mathfrak{B}$ and $\beta_i \chi b_i \neq 0$ for each $i < n$ [42Eb]. Then

$$0 < |\beta_i| \chi b_i \leqslant |y| \leqslant \|x\|_\infty \chi a_0 \quad \forall \ i < n,$$

so $b_i \subseteq a_0$ [42Ed] and $b_i \in \mathfrak{A}$ for each $i < n$. Thus $y \in S(\mathfrak{A})$; as x and y are arbitrary, $S(\mathfrak{A})$ is a solid linear subspace of $S(\mathfrak{B})$.

So $|x| \wedge |y| \in S(\mathfrak{A})$ for every $x \in S(\mathfrak{A})$ and $y \in S(\mathfrak{B})$. By continuity, $|x| \wedge |y|$ belongs to the closure of $S(\mathfrak{A})$, which is $L^\infty(\mathfrak{A})$, for every $x \in L^\infty(\mathfrak{A})$ and $y \in L^\infty(\mathfrak{B})$; so $L^\infty(\mathfrak{A})$ is solid in $L^\infty(\mathfrak{B})$.

(**c**) Let $\pi: \mathfrak{A} \to \mathfrak{B}$ be the canonical map. Since π is onto,

$$\hat{\pi}: S(\mathfrak{A}) \to S(\mathfrak{B}) \quad \text{and} \quad \hat{\pi}: L^\infty(\mathfrak{A}) \to L^\infty(\mathfrak{B})$$

are onto [45Cb]; as $\hat{\pi}$ is a Riesz homomorphism,

$$S(\mathfrak{B}) \cong S(\mathfrak{A})/E, \quad L^\infty(\mathfrak{B}) \cong L^\infty(\mathfrak{A})/F$$

as Riesz spaces, where

$$E = \{x : x \in S(\mathfrak{A}), \quad \hat{\pi}x = 0\}, \quad F = \{x : x \in L^\infty(\mathfrak{A}), \, \hat{\pi}x = 0\}.$$

[See 14G for the construction of quotient Riesz spaces.] 45Cb shows also that the norm on $S(\mathfrak{B})$ is the same as the quotient norm on $S(\mathfrak{A})/E$, and similarly the norm on $L^\infty(\mathfrak{B})$ is the quotient norm of $L^\infty(\mathfrak{A})/F$. So we have only to determine E and F.

Clearly $S(I) \subseteq E$. On the other hand, if $x \in E$, express x as $\sum_{i<n} \alpha_i \chi a_i$ where $\langle a_i \rangle_{i<n}$ is disjoint in $\mathfrak{A}$. Then $\hat{\pi}x = \sum_{i<n} \alpha_i \chi(\pi a_i)$ and $\langle \pi a_i \rangle_{i<n}$ is disjoint in $\mathfrak{B}$. So

$$0 = |\hat{\pi}x| = \textstyle\sum_{i<n} |\alpha_i| \chi(\pi a_i)$$

and $|\alpha_i| \chi(\pi a_i) = 0$ for each $i < n$. Thus, for each $i < n$, either $\alpha_i = 0$ or $\pi a_i = 0$, i.e. $a_i \in I$; so $x \in S(I)$.

Now $L^\infty(I)$ is the closure of $S(I)$; so $\hat{\pi}x = 0$ for every $x \in L^\infty(I)$, i.e. $L^\infty(I) \subseteq F$. Conversely, suppose that $x \in F$. Then $|x| \in F$, since $\hat{\pi}$ is a Riesz homomorphism. Now there is a sequence $\langle x_n \rangle_{n \in \mathbf{N}}$ in $S(\mathfrak{A})^+$ such that $\langle x_n \rangle_{n \in \mathbf{N}} \uparrow |x|$ and $\langle x_n \rangle_{n \in \mathbf{N}} \to |x|$ for $\| \ \|_\infty$ [43Bb]. So

$$0 \leqslant \hat{\pi}x_n \leqslant \hat{\pi}|x| = 0,$$

and $x_n \in S(I)$ for each $n \in \mathbf{N}$, by the argument just above. Hence $|x|$ belongs to the closure of $S(I)$ which is $L^\infty(I)$; since $L^\infty(I)$ is solid [(b) above], $x \in L^\infty(I)$. As x is arbitrary, $F = L^\infty(I)$, as required.

45E Since $\hat{\pi}: \boldsymbol{L}^\infty(\mathfrak{A}) \to \boldsymbol{L}^\infty(\mathfrak{B})$ is norm-decreasing, it always has a norm-decreasing transpose $\hat{\pi}: \boldsymbol{L}^\infty(\mathfrak{B})' \to \boldsymbol{L}^\infty(\mathfrak{A})'$. But in order to ensure that $\hat{\pi}'[\boldsymbol{L}^\#(\mathfrak{B})] \subseteq \boldsymbol{L}^\#(\mathfrak{A})$, we must impose an order-continuity condition on π.

Proposition Let $\mathfrak{A}$ and $\mathfrak{B}$ be Boolean rings, and $\pi: \mathfrak{A} \to \mathfrak{B}$ an order-continuous ring homomorphism. Then there is an order-continuous increasing linear map $\hat{\pi}': \boldsymbol{L}^\#(\mathfrak{B}) \to \boldsymbol{L}^\#(\mathfrak{A})$ given by

$$(\hat{\pi}'g)(x) = g(\hat{\pi}x) \quad \forall \; x \in \boldsymbol{L}^\infty(\mathfrak{A}), \quad g \in \boldsymbol{L}^\#(\mathfrak{B}),$$

or by

$$(\hat{\pi}'g)(\chi a) = g\chi(\pi a) \quad \forall \; a \in \mathfrak{A}, \quad g \in \boldsymbol{L}^\#(\mathfrak{A}).$$

Moreover, $\hat{\pi}'$ is norm-decreasing.

Proof If $g \in \boldsymbol{L}^\#(\mathfrak{B})$, then $g\chi\pi: \mathfrak{A} \to \mathbf{R}$ is completely additive, because $\pi: \mathfrak{A} \to \mathfrak{B}$ is order-continuous and $g\chi: \mathfrak{B} \to \mathbf{R}$ is completely additive; and $g\chi\pi$ is bounded because $g\chi$ is bounded. So there is a unique $f \in \boldsymbol{L}^\#(\mathfrak{A})$ such that

$$g(\hat{\pi}(\chi a)) = g\chi(\pi a) = f(\chi a) \quad \forall \; a \in \mathfrak{A}.$$

Now $g\hat{\pi} = f$ on $\boldsymbol{S}(\mathfrak{A})$, because $g\hat{\pi}$ and f are both linear, and $g\hat{\pi} = f$ on $\boldsymbol{L}^\infty(\mathfrak{A})$ because $g\hat{\pi}$ and f are both continuous. Thus $f = \hat{\pi}'g$. Of course $\|\hat{\pi}'\| \leqslant \|\hat{\pi}\| \leqslant 1$. If $g \geqslant 0$, then $f(\chi a) = g\chi\pi a \geqslant 0$ for every $a \in \mathfrak{A}$, so $f \geqslant 0$; so $\hat{\pi}'$ is increasing. Becauxe $\boldsymbol{L}^\#(\mathfrak{B})$ is an L-space, it follows from 25Ld, or otherwise, that $\hat{\pi}'$ is order-continuous.

45F Proposition Let $\mathfrak{A}$, $\mathfrak{B}$ and $\mathfrak{C}$ be Boolean rings, and let $\pi: \mathfrak{A} \to \mathfrak{B}$ and $\theta: \mathfrak{B} \to \mathfrak{C}$ be ring homomorphisms. Then $\theta\pi: \mathfrak{A} \to \mathfrak{C}$ is a ring homomorphism, and

$$\hat{\theta}\hat{\pi} = (\theta\pi)^\wedge: \boldsymbol{L}^\infty(\mathfrak{A}) \to \boldsymbol{L}^\infty(\mathfrak{C}).$$

If π and θ are order-continuous, so is $\theta\pi$, and in this case

$$\hat{\pi}'\hat{\theta}' = (\theta\pi)^{\wedge\prime}: \boldsymbol{L}^\#(\mathfrak{C}) \to \boldsymbol{L}^\#(\mathfrak{A}).$$

Proof It is easy to see that $\hat{\theta}\hat{\pi} = (\theta\pi)^\wedge$ on $\boldsymbol{S}(\mathfrak{A})$. The rest follows at once.

45G Exercises (a) Let $\mathfrak{A}$ and $\mathfrak{B}$ be Boolean rings, and $\pi: \mathfrak{A} \to \mathfrak{B}$ a lattice homomorphism such that $\pi(0) = 0$. Then π is a ring homomorphism.

(**b**) Let $\mathfrak{A}$ and $\mathfrak{B}$ be Boolean rings and $\pi: \mathfrak{A} \to \mathfrak{B}$ a ring homomorphism, with $\hat{\pi}: L^\infty(\mathfrak{A}) \to L^\infty(\mathfrak{B})$ the associated Riesz homomorphism. Let $I = \{a : a \in \mathfrak{A}, \pi a = 0\} \lhd \mathfrak{A}$. Then

$$\hat{\pi}[S(\mathfrak{A})] \cong S(\pi[\mathfrak{A}]) \cong S(\mathfrak{A}/I) \cong S(\mathfrak{A})/S(I),$$

$$\hat{\pi}[L^\infty(\mathfrak{A})] \cong L^\infty(\pi[\mathfrak{A}]) \cong L^\infty(\mathfrak{A}/I) \cong L^\infty(\mathfrak{A})/L^\infty(I)$$

and $$\hat{\pi}[S(\mathfrak{A})] = S(\mathfrak{B}) \cap \hat{\pi}[L^\infty(\mathfrak{A})].$$

*(**c**) Let $\mathfrak{A}$ and $\mathfrak{B}$ be Boolean rings and $\pi: \mathfrak{A} \to \mathfrak{B}$ a ring homomorphism, with $\hat{\pi}: L^\infty(\mathfrak{A}) \to L^\infty(\mathfrak{B})$ the associated Riesz homomorphism. Then π is order-continuous iff $\hat{\pi}$ is order-continuous. [Hint: use 42Rb and 17B.]

*(**d**) Let $\mathfrak{A}$ be a Boolean ring and I an order-closed ideal of $\mathfrak{A}$. Then the canonical map $\pi: \mathfrak{A} \to \mathfrak{A}/I$ is order-continuous [41Ld], and $\hat{\pi}': L^\#(\mathfrak{A}/I) \to L^\#(\mathfrak{A})$ is a norm-preserving Riesz homomorphism.

(**e**) Let $\langle \mathfrak{A}_\iota \rangle_{\iota \in I}$ be a family of Boolean rings with product $\mathfrak{A}$. Then $\mathfrak{A}$ is Boolean [41K]. For each $\iota \in I$, let $\pi_\iota: \mathfrak{A} \to \mathfrak{A}_\iota$ be the canonical order-continuous ring homomorphism. Then the associated Riesz homomorphisms $\hat{\pi}_\iota: L^\infty(\mathfrak{A}) \to L^\infty(\mathfrak{A}_\iota)$ induce a normed-Riesz-space isomorphism between $L^\infty(\mathfrak{A})$ and the solid linear subspace

$$\{x : \|x\| = \sup_{\iota \in I} \|x(\iota)\|_\infty < \infty\}$$

of the Riesz space product $\prod_{\iota \in I} L^\infty(\mathfrak{A}_\iota)$. *At the same time, the norm-preserving Riesz homomorphisms $\hat{\pi}'_\iota: L^\#(\mathfrak{A}_\iota) \to L^\#(\mathfrak{A})$ induce an isomorphism between $L^\#(\mathfrak{A})$ and the solid linear subspace

$$\{x : \|x\| = \textstyle\sum_{\iota \in I} \|x(\iota)\| < \infty\}$$

of $\prod_{\iota \in I} L^\#(\mathfrak{A}_\iota)$.

Notes and comments The results of this section are essentially algebraic. They rely on a careful analysis of functions on S-spaces which is then extended to functions on L^∞-spaces. So they all seem natural, despite the complexity of some of the proofs.

4X Examples for Chapter 4

Since Boolean rings usually appear as rings of sets or their quotients, these are the important cases to analyse. When $\mathfrak{A}$ is a ring of sets, we have particularly accessible representations of $S(\mathfrak{A})$ and $L^\infty(\mathfrak{A})$ [4XB]. From these we can derive representations of $S(\mathfrak{A}/I)$ and $L^\infty(\mathfrak{A}/I)$, where I is an ideal of $\mathfrak{A}$, as quotients of $S(\mathfrak{A})$ and $L^\infty(\mathfrak{A})$ respectively, as explained in 45Dc. These again are especially simple when $\mathfrak{A}$ is a σ-algebra of sets [4XE].

An interesting special case is the algebra of regular open sets in the real line [4XF]. 4XG is an instructive counterexample.

4XA The algebra $\mathscr{P}X$. Let X be any set, and $\mathscr{P}X$ its power set. On $\mathscr{P}X$ define $\triangle$ by

$$A \triangle B = (A \backslash B) \cup (B \backslash A) \quad \forall\ A, B \subseteq X.$$

Then $(\mathscr{P}X, \triangle, \cap)$ is a Boolean ring [41A]; its zero is the empty set. (The direct verification of the ring postulates is lengthy but elementary. Alternatively, we may identify $\mathscr{P}X$ with $\{0, 1\}^X$, saying that a set $A \subseteq X$ corresponds to the function with value 1 on A, 0 on $X \backslash A$. In this case, $(\mathscr{P}X, \triangle, \cap)$ is isomorphic to $(\mathbf{Z}_2)^X$, where $\mathbf{Z}_2$ is the field with two elements 0 and 1. The extra Boolean postulate, that $A \cap A = A$ for every $A \subseteq X$, is trivial.)

In fact $\mathscr{P}X$ is a Boolean algebra [41G]; its multiplicative identity is X. The lattice operations on $\mathscr{P}X$ are just $\cup$ and $\cap$. $\mathscr{P}X$ is Dedekind complete; if $\mathcal{A} \subseteq \mathscr{P}X$, $\sup \mathcal{A} = \bigcup \mathcal{A}$.

*For each $t \in X$, the set $\mathscr{I}_t = \{A : A \subseteq X, t \notin A\}$ is a maximal ideal of $\mathscr{P}X$. So the map $t \mapsto \mathscr{I}_t$ is an embedding of X in the Stone space of $\mathscr{P}X$. Note however that (unless X is finite) $\mathscr{P}X$ has many more maximal ideals than these.

4XB Rings of sets If Y is a set, a subring of $\mathscr{P}Y$ is a family $\mathfrak{A}$ of subsets of Y containing $\varnothing$ and closed under $\triangle$ and $\cap$; clearly, this is the same as being closed under $\cup$ and $\backslash$.

Suppose that $\mathfrak{A}$ is a subring of $\mathscr{P}Y$. Then $\boldsymbol{S} = \boldsymbol{S}(\mathfrak{A})$ is isomorphic, as normed Riesz space, to the linear subspace $\boldsymbol{E}$ of $\boldsymbol{l}^\infty(Y)$ generated by the characteristic functions of members of $\mathfrak{A}$.

P $\boldsymbol{S}$ is defined as the linear subspace of $\boldsymbol{l}^\infty(X)$ generated by the characteristic functions of members of $\phi[\mathfrak{A}]$, where $\phi: \mathfrak{A} \to \mathscr{P}X$ is the Stone representation of $\mathfrak{A}$. In the general case, the relation between X and Y can be complex. Let us write $1_a \in \mathbf{R}^Y$ for the characteristic function of a set $a \in \mathfrak{A}$, while $\chi a \in \boldsymbol{S} \subseteq \mathbf{R}^X$ is the characteristic function of $\phi a \subseteq X$. Now the map $a \mapsto 1_a : \mathfrak{A} \to \boldsymbol{l}^\infty(Y)$ is clearly additive; so, by 42C, there is a linear map $T: \boldsymbol{S} \to \boldsymbol{l}^\infty(Y)$ such that $T(\chi a) = 1_a$ for every $a \in \mathfrak{A}$. If we examine the description of members of $\boldsymbol{S}$ given in 42Eb, it is clear that T is a norm-preserving Riesz homomorphism, so $\boldsymbol{S} \cong T[\boldsymbol{S}]$ as normed Riesz space. Also, since $\boldsymbol{S}$ is spanned by $\{\chi a : a \in \mathfrak{A}\}$, $T[\boldsymbol{S}]$ must be the linear span of $\{1_a : a \in \mathfrak{A}\}$, which is $\boldsymbol{E}$. **Q**

This isomorphism means that we can identify $\boldsymbol{S}$ with $\boldsymbol{E}$; since the set Y is already to hand, while the Stone space X of $\mathfrak{A}$ must be constructed with the axiom of choice, this is a much more convenient representation of $\boldsymbol{S}$.

It follows also that $\boldsymbol{L}^\infty(\mathfrak{A})$ [43A] can be identified with the $\| \|_\infty$-completion of $\boldsymbol{E}$, which is [isomorphic to] the closure of $\boldsymbol{E}$ in $\boldsymbol{l}^\infty(Y)$. Note that this is indeed a Riesz space isomorphism as well as a normed space isomorphism, because the lattice operations are continuous.

*As a rule, $\boldsymbol{L}^{\#}_{-}(\mathfrak{A})$ is not of much interest in this case. If $\mathfrak{A}$ contains all finite subsets of X, then $\boldsymbol{L}^{\#}(\mathfrak{A})$ can be identified with $\boldsymbol{l}^1(X)$. One proof follows the argument of 2XB; another observes that $\boldsymbol{L}^\infty(\mathfrak{A})$ is order-dense in $\boldsymbol{l}^\infty(X)$, and applies 17B and the result of 2XB; another uses 65B.

4XC $\mathscr{P}X$ and $\mathscr{P}_f X$ (a) If $\mathfrak{A} = \mathscr{P}X$, then 4XB gives us the identifications

$$\begin{aligned} \boldsymbol{S}(\mathscr{P}X) &\cong \{x : x \in \mathbf{R}^X,\ x \text{ takes finitely many values}\}, \\ \boldsymbol{L}^\infty(\mathscr{P}X) &\cong \boldsymbol{l}^\infty(X), \\ \boldsymbol{L}^{\#}(\mathscr{P}X) &\cong \boldsymbol{l}^1(X) \end{aligned}$$

[2XB]. Thus the Dedekind completeness of $\boldsymbol{l}^\infty(X)$ [2XA] is related to the Dedekind completeness of $\mathscr{P}X$ by 43Da.

(b) For any set X, let $\mathscr{P}_f X$ be the family of finite subsets of X. Then $\mathscr{P}_f X$ is an ideal of $\mathscr{P}X$; in its own right, it is a Dedekind complete Boolean ring. Now

$$\begin{aligned} \boldsymbol{S}(\mathscr{P}_f X) &\cong \boldsymbol{s}_0(X) = \{x : x \in \mathbf{R}^X,\ \{t : x(t) \neq 0\} \text{ is finite}\}, \\ \boldsymbol{L}^\infty(\mathscr{P}_f X) &\cong \boldsymbol{c}_0(X), \\ \boldsymbol{L}^{\#}(\mathscr{P}_f X) &\cong \boldsymbol{l}^1(X) \end{aligned}$$

[as in 2XC].

4XD σ-algebras of sets Suppose, in 4XB, that $\mathfrak{A}$ is a σ-subalgebra of $\mathscr{P}X$ [i.e. $X \in \mathfrak{A}$ and $\mathfrak{A}$ is closed under countable unions and intersections, as well as complementation]. In this case, the representation of $\boldsymbol{L}^\infty(\mathfrak{A})$ as a subspace of $\mathbf{R}^X$ expresses $\boldsymbol{L}^\infty(\mathfrak{A})$ as

$$\boldsymbol{E} = \{x : x \in \boldsymbol{l}^\infty(X),\ \{t : x(t) > \alpha\} \in \mathfrak{A} \quad \forall\ \alpha \in \mathbf{R}\}.$$

P (i) Just as in the proof of 42E, $\boldsymbol{S}(\mathfrak{A})$, which we have identified as the linear subspace of $\mathbf{R}^X$ generated by the characteristic functions of members of $\mathfrak{A}$, is

$$\{x : x \in \boldsymbol{E},\ x \text{ takes finitely many values}\}.$$

(ii) If $x \in L^\infty(\mathfrak{A})$, then there is a sequence $\langle x_n \rangle_{n \in \mathbf{N}}$ in $S(\mathfrak{A})$ such that $\langle \|x - x_n\|_\infty \rangle_{n \in \mathbf{N}} \to 0$, so that $x(t) = \lim_{n \to \infty} x_n(t)$ for every $t \in X$. Now for any $\alpha \in \mathbf{R}$,

$$\{t : x(t) > \alpha\} = \bigcup_{m \in \mathbf{N}} \bigcap_{n \geq m} \{t : x_n(t) > \alpha + 2^{-m}\}$$

which must belong to $\mathfrak{A}$. So $x \in E$.

(iii) Conversely, if $x \in E$, then for each $n \in \mathbf{N}$ define x_n by

$$x_n(t) = 2^{-n}[2^n x(t)] \quad \forall\ t \in X$$

where $[\alpha] = \sup\{m : m \in \mathbf{Z},\ m \leq \alpha\}$ for each $\alpha \in \mathbf{R}$. Then

$$\{t : x_n(t) > \alpha\} = \bigcap_{m \in \mathbf{N}} \{t : x(t) > 2^{-n}([2^n \alpha] + 1) - 2^{-m}\} \in \mathfrak{A}$$

for any $\alpha \in \mathbf{R}$, so $x_n \in S$. Since $\|x - x_n\|_\infty \leq 2^{-n}$ for each $n \in \mathbf{N}$, $\langle x_n \rangle_{n \in \mathbf{N}} \to x$ and $x \in L^\infty(\mathfrak{A})$. **Q**

4XE Quotient rings For any Boolean ring $\mathfrak{A}$ and ideal I of $\mathfrak{A}$, $L^\infty(\mathfrak{A}/I)$ can be identified with $L^\infty(\mathfrak{A})/L^\infty(I)$ [45Dc]. When $\mathfrak{A}$ is a subring of $\mathscr{P}X$, then $L^\infty(\mathfrak{A})$ and $L^\infty(I)$ are subspaces of $l^\infty(X)$. In the special case when $\mathfrak{A}$ is a σ-subalgebra of $\mathscr{P}X$, as in 4XD, and I is a σ-ideal of $\mathfrak{A}$ [i.e. I is an ideal closed under countable unions], then $L^\infty(I)$ becomes

$$\{x : x \in L^\infty(\mathfrak{A}),\ \{t : x(t) \neq 0\} \in I\}.$$

(For this is the solid linear subspace of $L^\infty(\mathfrak{A})$ generated by the characteristic functions of members of I, and it is closed for $\| \ \|_\infty$. Remember that by 45Db $L^\infty(I)$ is solid in $L^\infty(\mathfrak{A})$.)

Thus $L^\infty(\mathfrak{A}/I)$ becomes the space of equivalence classes in $L^\infty(\mathfrak{A})$ under the relation

$$x \sim y \quad \text{if} \quad \{t : x(t) \neq y(t)\} \in I.$$

***4XF Algebras of regular open sets** In any topological space, an open set G is **regular** if $G = \operatorname{int} \overline{G}$. The set $\mathscr{G}$ of all regular open sets is partially ordered by $\subseteq$. This ordering is induced by a Boolean ring structure on $\mathscr{G}$ under which $\mathscr{G}$ is a Dedekind complete Boolean algebra. When the underlying topological space is $\mathbf{R}$, the algebra thus obtained has many facets; I shall present it as an example of an algebra on which there are no non-trivial countably additive functionals.

Proofs of these remarks now follow.

(a) Let X be a topological space. Suppose that E and F are closed subsets of X such that $\operatorname{int} E = \operatorname{int} F = \varnothing$. Then $\operatorname{int}(E \cup F) = \varnothing$. [For $X \backslash E$ and $X \backslash F$ are dense open sets, so their intersection is dense.]

(b) Let X be a topological space. Let $\mathfrak{A}$ be the set

$$\{A : A \subseteq X, \operatorname{int}(\partial A) = \varnothing\} \subseteq \mathscr{P}X,$$

where $\partial A = \bar{A} \backslash \operatorname{int} A$ is the boundary of A. Then $\mathfrak{A}$ is a subalgebra of $\mathscr{P}X$. **P** If $A, B \subseteq X$, then $\partial(A \cup B) \subseteq \partial A \cup \partial B$. So by (a) above, $\mathfrak{A}$ is closed under $\cup$. As $\partial(X \backslash A) = \partial A$ for any $A \subseteq X$, $\mathfrak{A}$ is closed under $\backslash$. Finally, $\varnothing \in \mathfrak{A}$, so $\mathfrak{A}$ is a subalgebra of $\mathscr{P}X$. **Q**

Observe that open and closed sets all belong to $\mathfrak{A}$.

(c) Continuing from (b) above, let I be

$$\{A : A \subseteq X, \operatorname{int} \bar{A} = \varnothing\} \subseteq \mathfrak{A}.$$

Using (a) again and 41J, I is an ideal of $\mathfrak{A}$. So we may form the quotient $\mathfrak{A}/I$. For $A \in \mathfrak{A}$, let $A^{\bullet}$ be the image of A in $\mathfrak{A}/I$; let $\subseteq$ be the canonical ordering in $\mathfrak{A}/I$ [41F].

(d) Now $A^{\bullet} \subseteq B^{\bullet} \Leftrightarrow \operatorname{int}\bar{A} \subseteq \operatorname{int} \bar{B}$. **P** If $A \in \mathfrak{A}$, then $\partial A \in I$. So $A^{\bullet} = \bar{A}^{\bullet} = (\operatorname{int} A)^{\bullet}$. (Of course $\bar{A}$ and $\operatorname{int} A$ belong to $\mathfrak{A}$.) Now

$$\begin{aligned} A^{\bullet} \subseteq B^{\bullet} &\Rightarrow \bar{A}^{\bullet} \subseteq \bar{B}^{\bullet} \Rightarrow (\operatorname{int} \bar{A})^{\bullet} \subseteq \bar{B}^{\bullet} \\ &\Rightarrow (\operatorname{int} \bar{A}) \backslash \bar{B} \in I \Rightarrow (\operatorname{int} \bar{A}) \backslash \bar{B} = \varnothing \Rightarrow \operatorname{int} \bar{A} \subseteq \bar{B} \\ &\Rightarrow \operatorname{int} \bar{A} \subseteq \operatorname{int} \bar{B} \\ &\Rightarrow (\operatorname{int} \bar{A})^{\bullet} \subseteq (\operatorname{int} \bar{B})^{\bullet} \Rightarrow \bar{A}^{\bullet} \subseteq \bar{B}^{\bullet} \Rightarrow A^{\bullet} \subseteq B^{\bullet}. \quad \textbf{Q} \end{aligned}$$

(e) Thus $\mathfrak{A}/I$ may be identified, as partially ordered set, with $\mathscr{G} = \{\operatorname{int} \bar{A} : A \in \mathfrak{A}\}$. Now an open set belongs to $\mathscr{G}$ iff it is regular. **P** If $G = \operatorname{int} \bar{G}$, of course $G \in \mathscr{G}$, since $G \in \mathfrak{A}$. Conversely, if $G = \operatorname{int} \bar{A}$, where A is any set, $G \subseteq \bar{G} \subseteq \bar{A}$, so $G \subseteq \operatorname{int} \bar{G} \subseteq \operatorname{int} \bar{A} = G$; thus G is regular. **Q**

(f) Now $\mathscr{G}$ is a Dedekind complete lattice. **P** (i) If $G, H \in \mathscr{G}$, then $G \cap H \in \mathscr{G}$, because

$$G \cap H \subseteq \operatorname{int} \overline{(G \cap H)} \subseteq \operatorname{int}(\bar{G} \cap \bar{H}) \subseteq \operatorname{int} \bar{G} \cap \operatorname{int} \bar{H} = G \cap H.$$

So $G \cap H = \inf\{G, H\}$ in $\mathscr{G}$. (ii) On the other hand, if $\mathscr{E}$ is any subset of $\mathscr{G}$, consider $G_0 = \operatorname{int} \operatorname{cl}(\bigcup \mathscr{E})$. For any $E \in \mathscr{E}$, $E \subseteq \bigcup \mathscr{E} \subseteq \operatorname{cl}(\bigcup \mathscr{E})$, so $E \subseteq G_0$; conversely, if G is any upper bound of $\mathscr{E}$ in $\mathscr{G}$, $\bigcup \mathscr{E} \subseteq G$, so $G_0 \subseteq \operatorname{int} \bar{G} = G$. Thus $\sup \mathscr{E} = G_0$ exists in $\mathscr{G}$. **Q**

(g) Let us give $\mathscr{G}$ the Boolean ring structure induced by its isomorphism with $\mathfrak{A}/I$. Then

$$G \cap H = G \cap H, \quad G \cup H = \operatorname{int} \overline{(G \cup H)} = \operatorname{int}(\bar{G} \cup \bar{H}),$$

$$X | G = X + G = X \backslash \bar{G} = \operatorname{int}(X \backslash G)$$

[for $G \cap (X \backslash \bar{G}) = \varnothing$, $G \cup (X \backslash \bar{G}) = X$]. (It is clear that there can be only one Boolean ring structure on $\mathscr{G}$ associated with the ordering $\subseteq$, for this defines $\cup$ and $\cap$ and therefore $\backslash$ and $+$.)

Now $\mathscr{G}$ is a Dedekind complete Boolean algebra; its maximal element is X.

(h) Let us now consider the case in which X is Hausdorff, separable and without isolated points (e.g. $X = \mathbf{R}$). In this case, any increasing countably additive function $\nu: \mathscr{G} \to \mathbf{R}$ is identically zero. **P** Let Q be a countable dense subset of X; then Q is infinite. Enumerate Q as $\langle q_i \rangle_{i \in \mathbf{N}}$. For each $m, n \in \mathbf{N}$ let G_{mn} be an open set in X such that

$$q_i \notin \bar{G}_{mn} \quad \forall\ i < m, \quad q_i \in G_{mn} \quad \forall\ m \leqslant i < m+n.$$

Set $H_{mn} = \operatorname{int}(\bigcup_{i \leqslant n} \bar{G}_{mi})$; then, for each $m \in \mathbf{N}$, $\langle H_{mn} \rangle_{n \in \mathbf{N}}$ is an increasing sequence in $\mathscr{G}$, and $q_i \notin \bar{H}_{mn}$ if $i < m$.

Let $\epsilon > 0$. Clearly $\bigcup_{n \in \mathbf{N}} H_{mn}$ is dense for each $m \in \mathbf{N}$, that is, $\langle H_{mn} \rangle_{n \in \mathbf{N}} \uparrow X$ in $\mathscr{G}$. So there is, for each m, an $n(m)$ such that

$$\nu H_{m, n(m)} \geqslant \nu X - 2^{-m}\epsilon.$$

Set $E_m = X \backslash \bar{H}_{m, n(m)}$. Then $\nu E_m \leqslant 2^{-m}\epsilon$ and $q_i \in E_m$ for every $i < m$. So $\bigcup_{m \in \mathbf{N}} E_m$ is dense, and $\langle \sup_{r \leqslant m} E_r \rangle_{m \in \mathbf{N}} \uparrow X$ in $\mathscr{G}$. So

$$\nu X = \lim_{m \to \infty} \nu(\sup_{r \leqslant m} E_r) \leqslant \lim_{m \to \infty} \sum_{r \leqslant m} \nu E_r$$
$$= \sum_{r \in \mathbf{N}} \nu E_r \leqslant 2\epsilon.$$

As ϵ is arbitrary, $\nu X = 0$. Because ν is increasing, $\nu = 0$. **Q**

(i) It follows at once that any countably additive functional on $\mathscr{G}$ is zero (for by 42Q and 42N, any countably additive functional on $\mathscr{G}$ is the difference of increasing countably additive functionals). At the same time we see that there can be no finite-valued measure on $\mathscr{G}$ [see 51E]; indeed, it is easy to see that the only measure on $\mathscr{G}$, according to the definition 51A, is the one which takes the value ∞ on every non-zero element.

(j) Consequently, $L^\infty(\mathscr{G})$ is now a Dedekind complete Riesz space – actually, an M-space with unit – such that $L^\infty(\mathscr{G})^\times = \{0\}$ [44Bb]. Of course, $L^\infty(\mathscr{G})^\sim = L^\infty(\mathscr{G})'$ [25G] is large.

(k) Exercise Show that (i) there is a sequence $\langle G_n \rangle_{n \in \mathbf{N}}$ in the algebra $\mathscr{G}(\mathbf{R})$ described above such that, for every non-empty $G \in \mathscr{G}$, there is an $n \in \mathbf{N}$ such that $\varnothing \subset G_n \subset G$ (ii) any Dedekind complete Boolean algebra with this property is isomorphic to $\mathscr{G}$.

(l) Exercise Again supposing that $X = \mathbf{R}$, show that $L^\infty(\mathfrak{A})$ can be identified with the set of those bounded functions $x: \mathbf{R} \to \mathbf{R}$ such that $\{t : x \text{ is continuous at } t\}$ is a dense G_δ set. Consequently, $L^\infty(\mathscr{G})$ is the Riesz space quotient of this by the solid linear subspace

$$\{x : x \text{ is zero on a dense } G_\delta \text{ set}\}.$$

Also, $C_b(X)$ may be regarded as an order-dense Riesz subspace of the Dedekind complete Riesz space $L^\infty(\mathscr{G})$. Hence, or otherwise, show that $C_b(X)^\times = \{0\}$.

***4XG A counterexample for 42Q** Let X be an uncountable set, and let $\mathfrak{A}$ be

$$\{A : A \subseteq X, \textit{ either } A \text{ is finite } \textit{or } X \backslash A \text{ is finite}\}.$$

Then $\mathfrak{A}$ is a subring of $\mathscr{P}X$. Define $\nu : \mathfrak{A} \to \mathbf{R}$ by

$$\begin{aligned} \nu A &= n \quad \text{if } A \text{ has } n \text{ members, where } n \in \mathbf{N}; \\ &= -n \quad \text{if } X \backslash A \text{ has } n \text{ members.} \end{aligned}$$

Now if $\langle A_n \rangle_{n \in \mathbf{N}} \downarrow \varnothing$ in $\mathfrak{A}$, one of the A_n must be finite; so in fact one of them is empty, and $\inf_{n \in \mathbf{N}} |\nu A_n| = 0$. Thus ν is countably additive. But ν is certainly not locally bounded.

So we see that the condition 'Dedekind σ-complete' in 42Q is indeed necessary.

4XH Ring homomorphisms induced by functions Let X and Y be arbitrary sets, and $f : X \to Y$ any function. Define $\pi : \mathscr{P}Y \to \mathscr{P}X$ by $\pi(A) = f^{-1}[A]$ for every $A \subseteq Y$. Then π is an order-continuous ring homomorphism. We see that, for $A \subseteq Y$,

$$\hat{\pi}(\chi A) = \chi(\pi A) = \chi(f^{-1}[A]) = \chi A \circ f,$$

where here we identify χA with the characteristic function of A, as in 4XB. It follows that $\hat{\pi} : L^\infty(\mathscr{P}Y) \to L^\infty(\mathscr{P}X)$ is given by

$$\hat{\pi} y = y \circ f$$

for every $y \in l^\infty \cong L^\infty(\mathscr{P}Y)$ [see 4XCa]. Equally, identifying $L^\#(\mathscr{P}X)$ with $l^1(X)$, $\hat{\pi}' : L^\#(\mathscr{P}X) \to L^\#(\mathscr{P}Y)$ is given by

$$(\hat{\pi}' x)(u) = \sum \{x(t) : f(t) = u\}$$

for every $u \in Y$ and $x \in l^1(X)$.

5. Measure algebras

In this chapter I shall give versions of those results in elementary measure theory which refer to measure algebras or to L^1 and L^∞ spaces. The first two sections apply the concepts of §§ 42–4 to 'measure rings', that is, Boolean rings on which a strictly positive countably additive measure is defined. In this case, a true analogy of L^1 spaces can be found, and the correspondence between L^1 spaces and $L^\#$ spaces is discussed. All measure rings of any significance are 'semi-finite', and consequently their L^1 and $L^\#$ spaces can be identified; this is the basic idea of § 52. The next section deals briefly with Dedekind complete measure algebras, which seem to be central to ordinary measure theory. Finally, in § 54, the ideas of § 45 concerning homomorphisms are reviewed in the new context.

51 Measure rings

In this section, we shall have only definitions and basic properties. A measure ring is a Boolean ring together with a strictly positive measure; this is an extended-real-valued functional which is additive and sequentially order-continuous on the left. The definition of measure ring which I have chosen [51A] follows the ordinary definition of measure space [61A] in allowing 'purely infinite' elements, that is, non-zero elements of infinite measure such that every smaller element is either zero or also of infinite measure. It is an accident of the theory that these need not cause any inconvenience in the early stages; but fairly soon they must be outlawed, and accordingly I shall immediately introduce 'semi-finite' measure rings, that is, measure rings which have no purely infinite elements.

In any measure ring, semi-finite or otherwise, the ideal of elements of finite measure is of great importance, and many of the results here are based on consideration of this ideal as a measure ring in its own right.

All the most important examples of measure rings are derived from measure spaces, as in 61D; so at this point it will be helpful to have in mind Lebesgue measure, as the simplest non-trivial example of a measure space.

51A Definition A **measure ring** is a Boolean ring $\mathfrak{A}$ together with a **measure** $\mu: \mathfrak{A} \to [0, \infty]$ such that:

(i) $\mu(a \cup b) = \mu a + \mu b$ if $a \cap b = 0$ in $\mathfrak{A}$;

(ii) $\mu a = 0 \Leftrightarrow a = 0$;

(iii) if $\langle a_n \rangle_{n \in \mathbf{N}} \uparrow a$ in $\mathfrak{A}$, then $\mu a = \sup_{n \in \mathbf{N}} \mu a_n$.

Note '∞' here, and later, is regarded as an actual point adjoined to $\mathbf{R}^+$. The set $[0, \infty]$ has natural additive-semigroup and total-ordering structures, writing

$$\alpha + \infty = \infty + \alpha = \infty \quad \forall\ \alpha \in [0, \infty],$$

$$\alpha \leqslant \infty \quad \forall\ \alpha \in [0, \infty];$$

and also a multiplicative-semigroup structure, writing

$$0 . \infty = \infty . 0 = 0,$$

$$\alpha . \infty = \infty . \alpha = \infty \quad \forall\ \alpha > 0,$$

which we shall have occasion to use. Subtraction, however, is not fully defined, so we must be wary.

51B Definitions (a) For any measure ring $(\mathfrak{A}, \mu)$, $\mathfrak{A}^f$ will be

$$\{a : a \in \mathfrak{A},\ \mu a < \infty\}.$$

(b) A measure ring $(\mathfrak{A}, \mu)$ is **semi-finite** if, whenever $a \in \mathfrak{A}$ and $\mu a = \infty$, there is a $b \subseteq a$ such that $0 < \mu b < \infty$.

51C Proposition Let $(\mathfrak{A}, \mu)$ be a measure ring.

(a) If $a \subseteq b$ in $\mathfrak{A}$, $\mu a \leqslant \mu b$.

(b) Suppose that $\varnothing \subset B \subseteq \mathfrak{A}^f$ and that $B \downarrow 0$ in $\mathfrak{A}$. Then

(i) there is a sequence $\langle b_n \rangle_{n \in \mathbf{N}}$ in B such that $\langle b_n \rangle_{n \in \mathbf{N}} \downarrow 0$;

(ii) $\inf_{b \in B} \mu b = 0$.

Proof (a) For $\mu b = \mu a + \mu(b \setminus a) \geqslant \mu a$.

(b) Let $\alpha = \inf_{b \in B} \mu b$. Choose a sequence $\langle c_n \rangle_{n \in \mathbf{N}}$ in B such that $\mu c_n \leqslant \alpha + 2^{-n}$ for each $n \in \mathbf{N}$, and now choose a sequence $\langle b_n \rangle_{n \in \mathbf{N}}$ in B such that $b_{n+1} \subseteq b_n \cap c_n$ for each $n \in \mathbf{N}$. Then $\mu b_{n+1} \leqslant \mu c_n \leqslant \alpha + 2^{-n}$ for each $n \in \mathbf{N}$.

Suppose that a is any lower bound of $\{b_n : n \in \mathbf{N}\}$ and that $b \in B$

Then for each $n \in \mathbf{N}$ there is a $c \in B$ such that $c \subseteq b \cap b_{n+1}$, so that $\alpha \leqslant \mu c \leqslant \mu(b \cap b_{n+1})$ for each $n \in \mathbf{N}$, and

$$\mu(b_{n+1} \backslash b) = \mu b_{n+1} - \mu(b_{n+1} \cap b) \leqslant \alpha + 2^{-n} - \alpha = 2^{-n}$$

for each $n \in \mathbf{N}$. (Note that the subtraction here is legitimate because $\mu b_{n+1} < \infty$.) Now $a \backslash b \subseteq b_{n+1} \backslash b$ for each $n \in \mathbf{N}$, so $\mu(a \backslash b) = 0$ and $a \backslash b = 0$, i.e. $a \subseteq b$. As b is arbitrary and $\inf B = 0$, $a = 0$; as a is arbitrary, $\langle b_n \rangle_{n \in \mathbf{N}} \downarrow 0$.

Now however, $\langle b_0 \backslash b_n \rangle_{n \in \mathbf{N}} \uparrow b_0$ [41Hb]. So

$$\mu b_0 = \sup_{n \in \mathbf{N}} \mu(b_0 \backslash b_n) = \lim_{n \to \infty} \mu(b_0 \backslash b_n)$$

$$= \mu b_0 - \lim_{n \to \infty} \mu b_n.$$

Thus $\lim_{n \to \infty} \mu b_n = 0$ and $\alpha \leqslant \inf_{n \in \mathbf{N}} \mu b_n = 0$, as required.

51D Corollary Let $(\mathfrak{A}, \mu)$ be a measure ring. Let $\nu\colon \mathfrak{A}^f \to \mathbf{R}$ be a countably additive functional. Then ν is completely additive.

Proof Suppose that $\varnothing \subset B \downarrow 0$ in $\mathfrak{A}^f$. Of course any lower bound for B in $\mathfrak{A}$ belongs to $\mathfrak{A}^f$, by 51Ca; so $B \downarrow 0$ in $\mathfrak{A}$. By 51Cb there is a sequence $\langle b_n \rangle_{n \in \mathbf{N}}$ in B such that $\langle b_n \rangle_{n \in \mathbf{N}} \downarrow 0$ in $\mathfrak{A}$ and therefore $\langle b_n \rangle_{n \in \mathbf{N}} \downarrow 0$ in $\mathfrak{A}^f$. So

$$\inf_{a \in B} |\nu a| \leqslant \inf_{n \in \mathbf{N}} |\nu b_n| = 0.$$

As B is arbitrary, ν is completely additive.

51E Proposition If $(\mathfrak{A}, \mu)$ is a measure ring, then $\mathfrak{A}^f$ is an ideal of $\mathfrak{A}$. With the appropriate restriction of μ, $\mathfrak{A}^f$ is itself a measure ring. $\mathfrak{A}^f$ is semi-finite, and $\mu\colon \mathfrak{A}^f \to \mathbf{R}$ is completely additive.

Proof If $a, b \in \mathfrak{A}$, then

$$\mu(a \cup b) = \mu a + \mu(b \backslash a) \leqslant \mu a + \mu b$$

[using 51Ca]. So $\mathfrak{A}^f$ is closed under $\cup$; by 51Ca again,

$$a \subseteq b \in \mathfrak{A}^f \Rightarrow a \in \mathfrak{A}^f;$$

and certainly $0 \in \mathfrak{A}^f$, as $\mu(0) = 0$. So $\mathfrak{A}^f$ is an ideal of $\mathfrak{A}$. It follows that if $\langle a_n \rangle_{n \in \mathbf{N}} \uparrow a$ in $\mathfrak{A}^f$, then $\langle a_n \rangle_{n \in \mathbf{N}} \uparrow a$ in $\mathfrak{A}$; so $\mu a = \sup_{n \in \mathbf{N}} \mu a_n$, and $\mathfrak{A}^f$ is a measure ring. Of course it is semi-finite. Now μ is additive on $\mathfrak{A}^f$ by 51A(i), and it is completely additive by 51Cb.

***51F Products** If $\langle(\mathfrak{A}_\iota, \mu_\iota)\rangle_{\iota\in I}$ is a family of measure rings, $\mathfrak{A} = \prod_{\iota\in I}\mathfrak{A}_\iota$ is a Boolean ring [41K]. Define $\mu: \mathfrak{A} \to [0, \infty]$ by

$$\mu\langle a_\iota\rangle_{\iota\in I} = \sum_{\iota\in I}\mu_\iota a_\iota.$$

Then $(\mathfrak{A}, \mu)$ is a measure ring; I will call it the **product** $\prod_{\iota\in I}(\mathfrak{A}_\iota, \mu_\iota)$ even though it is not a product in any obvious category.

If for any $\kappa\in I$ and $a\in\mathfrak{A}_\kappa$, we define $\bar{a} = \langle a_\iota\rangle_{\iota\in I}$ where $a_\kappa = a$ and $a_\iota = 0$ for $\iota \neq \kappa$, we find that the map $a \mapsto \bar{a}$ represents $\mathfrak{A}_\kappa$ as an ideal of $\mathfrak{A}$. Now μ_κ is just the measure on $\mathfrak{A}_\kappa$ induced by μ. From this it is clear that $(\mathfrak{A}, \mu)$ is semi-finite iff $(\mathfrak{A}_\iota, \mu_\iota)$ is semi-finite for every $\iota\in I$.

51G Exercises (**a**) Let $(\mathfrak{A}, \mu)$ be a measure ring. Then there is a natural metric ρ on $\mathfrak{A}^f$ defined by

$$\rho(a, b) = \mu(a+b) \quad \forall\ a, b\in\mathfrak{A}^f.$$

(**b**) For any measure ring $(\mathfrak{A}, \mu)$, μ is order-continuous on the left. [Hint: use the technique of 51Cb, upside down.]

(**c**) Let $(\mathfrak{A}, \mu)$ be a measure ring. Then it is semi-finite iff

$$a = \sup\{b : b\in\mathfrak{A}^f,\ b \subseteq a\} \text{ for every } a\in\mathfrak{A}. \text{ [Cf. 41Lc.]}$$

(**d**) Let $(\mathfrak{A}, \mu)$ be a measure ring. Then $\boldsymbol{S}(\mathfrak{A}^f)$ and $\boldsymbol{L}^\infty(\mathfrak{A}^f)$ can be thought of as solid linear subspaces of $\boldsymbol{S}(\mathfrak{A})$ and $\boldsymbol{L}^\infty(\mathfrak{A})$ respectively [45Db]. Now the following are equivalent: (i) $(\mathfrak{A}, \mu)$ is semi-finite; (ii) $\boldsymbol{S}(\mathfrak{A}^f)$ is order-dense in $\boldsymbol{S}(\mathfrak{A})$; (iii) $\boldsymbol{L}^\infty(\mathfrak{A}^f)$ is order-dense in $\boldsymbol{L}^\infty(\mathfrak{A})$.

*(**e**) For any measure ring $(\mathfrak{A}, \mu)$, $\boldsymbol{S}(\mathfrak{A}^f)$ and $\boldsymbol{L}^\infty(\mathfrak{A}^f)$ have the countable sup property.

Notes and comments Although all the most important measure rings, being derived from measure spaces, are Dedekind σ-complete [61Dd], the work of the next section will appear in clearer perspective if we omit this condition for the time being.

51Cb echoes the corresponding result for L-spaces [26B]. The details of the proof have to be different for technical reasons, but the essential phenomenon is the same. In fact, the ring of elements of finite measure has a natural metric [51Ga], which behaves very like the norm of an L-space.

In 5XA are some elementary examples of measure rings. These are essentially trivial. For a substantive example we do need a 'non-atomic' measure like Lebesgue's [see the references in 6XAb]. 5XAb is an extreme example of a measure ring which is not semi-finite; every non-zero element is purely infinite.

52 The space $L^1(\mathfrak{A})$

In this section we study the properties of the spaces S, L^∞ and $L^\#$ from Chapter 4 when they are based on measure rings. The pivotal results are 52B and 52D, in which the measure is used to embed $S(\mathfrak{A}^f)$ as a Riesz subspace of $L^\#(\mathfrak{A})$ which, when $\mathfrak{A}$ is semi-finite, is order-dense. This allows us to think of $L^\#(\mathfrak{A})$ as a completion of $S(\mathfrak{A}^f)$ with respect to the appropriate norm. We now have a construction $L^1(\mathfrak{A})$ which is very like the classical L^1 spaces.

52A The embedding of $S(\mathfrak{A}^f)$ in $L^\#(\mathfrak{A})$ Let $(\mathfrak{A}, \mu)$ be a measure ring. For the moment, fix $a \in \mathfrak{A}^f$. Define $\nu_a: \mathfrak{A} \to \mathbf{R}$ by

$$\nu_a(b) = \mu(a \cap b) \quad \forall\ b \in \mathfrak{A}.$$

Then, because the map $b \to a \cap b : \mathfrak{A} \mapsto \mathfrak{A}^f$ is an order-continuous ring homomorphism, and $\mu: \mathfrak{A}^f \to \mathbf{R}$ is completely additive [51E], ν_a is completely additive. Moreover,

$$\sup\{\nu_a(b) : b \in \mathfrak{A}\} = \mu a < \infty,$$

so there is a unique $f_a \in L^\#(\mathfrak{A})$ such that

$$f_a(\chi b) = \nu_a(b) = \mu(a \cap b) \quad \forall\ b \in \mathfrak{A},$$

and $\|f_a\| = \mu a$ [44Bb].

Thus we have a map $a \mapsto f_a: \mathfrak{A}^f \to L^\#(\mathfrak{A})$. But clearly this map is additive and increasing, since if $a \cap c = 0$ in $\mathfrak{A}^f$,

$$f_{a+c}(\chi b) = \mu((a+c)\,b) = \mu(ab+cb) = \mu(ab)+\mu(cb)$$
$$= f_a(\chi b) + f_c(\chi b)$$

for every $b \in \mathfrak{A}$, and $f_{a+c} = f_a + f_c$. So it induces a canonical increasing linear map $\phi: S(\mathfrak{A}^f) \to L^\#(\mathfrak{A})$, given by $\phi(\chi a) = f_a$ for every $a \in \mathfrak{A}^f$.

The basic properties of ϕ are contained in the two propositions 52B and 52D.

52B Proposition For any measure ring $(\mathfrak{A}, \mu)$, the canonical map $\phi: S(\mathfrak{A}^f) \to L^\#(\mathfrak{A})$ defined in 52A is a one-to-one Riesz homomorphism. For any $x \in S(\mathfrak{A}^f)$, $\|\phi x\| = \hat{\mu}(|x|)$, where $\hat{\mu}$ is the linear functional on $S(\mathfrak{A}^f)$ corresponding to the additive functional $\mu: \mathfrak{A}^f \to \mathbf{R}$.

Proof (a) If $a, b \in \mathfrak{A}^f$ and $a \cap b = 0$, $f_a \wedge f_b = 0$. **P** For every $c \in \mathfrak{A}$, $\chi c = \chi(c \cap b) + \chi(c \backslash b)$ in $L^\infty(\mathfrak{A})$. So

$$0 \leqslant (f_a \wedge f_b)(\chi c) \leqslant f_a \chi(c \cap b) + f_b \chi(c \backslash b)$$
$$= \mu(a \cap c \cap b) + \mu(b \cap c \backslash b) = 0,$$

remembering that $f_a \wedge f_b$ is defined in

$$L^{\#}(\mathfrak{A}) = L^{\infty}(\mathfrak{A})^{\times} \subseteq L^{\infty}(\mathfrak{A})^{\sim} = L^{\sim}(L^{\infty}(\mathfrak{A}); \mathbf{R}),$$

and applying the formula of 16Eb. As c is arbitrary, $f_a \wedge f_b = 0$. **Q**

Thus $\phi(\chi a) \wedge \phi(\chi b) = 0$ whenever $a \cap b = 0$ in $\mathfrak{A}$. But now suppose that $x \wedge y = 0$ in $S(\mathfrak{A}^f)$. Then we can find $\alpha, \beta > 0$ in $\mathbf{R}$ and $a, b \in \mathfrak{A}^f$ such that

$$\alpha\chi a \leqslant x \leqslant \|x\|_{\infty}\chi a, \quad \beta\chi b \leqslant y \leqslant \|y\|_{\infty}\chi b$$

[42Ec]. Then $(\alpha \wedge \beta)\chi(a \cap b) \leqslant x \wedge y = 0$, so $a \cap b = 0$. Now, because ϕ is increasing,

$$0 \leqslant \phi x \wedge \phi y \leqslant \|x\|_{\infty} f_a \wedge \|y\|_{\infty} f_b = 0$$

because $f_a \wedge f_b = 0$ [14Kb]. Since x and y are arbitrary, this proves that ϕ is a Riesz homomorphism [14Eb].

(b) Now $\|\phi(\chi a)\| = \|f_a\| = \mu a = \hat{\mu}(\chi a)$ for any $a \in \mathfrak{A}^f$, recalling from 52A that $\|f_a\| = \mu a$. If next $x \in S(\mathfrak{A}^f)^+$, then x can be expressed as $\sum_{i<n} \alpha_i \chi a_i$, where each $\alpha_i \geqslant 0$; since $L^{\#}(\mathfrak{A})$ is an L-space,

$$\|\phi x\| = \|\textstyle\sum_{i<n} \alpha_i \phi(\chi a_i)\| = \sum_{i<n} \alpha_i \|\phi(\chi a_i)\|$$
$$= \textstyle\sum_{i<n} \alpha_i \mu a_i = \hat{\mu}(x).$$

Finally, for arbitrary $x \in S(\mathfrak{A}^f)$,

$$\|\phi x\| = \|\,|\phi x|\,\| = \|\phi(|x|)\| = \hat{\mu}(|x|).$$

(c) It follows that ϕ is one-to-one. **P** For if $x \neq 0$ in $S(\mathfrak{A}^f)$ then $|x| > 0$, so there is an $\alpha > 0$ and an $a \neq 0$ such that $\alpha\chi a \leqslant |x|$. Now

$$\|\phi x\| = \hat{\mu}(|x|) \geqslant \alpha\mu a > 0,$$

so $\phi x \neq 0$. **Q**

52C Lemma Let $\mathfrak{A}$ be a Boolean ring, and $\nu: \mathfrak{A} \to \mathbf{R}$ a completely additive functional. Suppose that $a \in \mathfrak{A}$ and that $\nu a > 0$. Then there is a non-zero $b \subseteq a$ such that $\nu c \geqslant 0$ for every $c \subseteq b$.

Proof Define $\lambda: \mathfrak{A} \to \mathbf{R}$ by $\lambda c = \nu(a \cap c)$ for every $c \in \mathfrak{A}$. Then it is easy to see that λ is completely additive, and $\lambda a > 0$. Now consider $\hat{\lambda} \in S(\mathfrak{A})^{\times}$ [42L]. Since

$$\hat{\lambda}^+(\chi a) \geqslant \hat{\lambda}(\chi a) = \lambda a > 0,$$

$\hat{\lambda}^+ > 0$. So by 32A there is an $x > 0$ in $S(\mathfrak{A})$ such that

$$\hat{\lambda}^+(x) > 0, \quad \hat{\lambda}^-(x) = 0$$

[for certainly $\hat{\lambda}^+ \wedge \hat{\lambda}^- = 0$]. Now by 42Ec there is a $b_0 \in \mathfrak{A}$ and a $\beta > 0$ such that

$$\beta\chi b_0 \leqslant x \leqslant \|x\|_\infty \chi b_0.$$

So $\hat{\lambda}^+(\chi b_0) > 0$ and $\hat{\lambda}^-(\chi b_0) = 0$.

Now set $b = a \cap b_0$. If $c \subseteq b$, then $0 \leqslant \chi c \leqslant \chi b_0$, so $\hat{\lambda}^-(\chi c) = 0$ and

$$\nu c = \lambda c = \hat{\lambda}(\chi c) = \hat{\lambda}^+(\chi c) \geqslant 0.$$

On the other hand,

$$\nu b = \lambda b_0 = \hat{\lambda}(\chi b_0) = \hat{\lambda}^+(\chi b_0) - \hat{\lambda}^-(\chi b_0) > 0,$$

so $b \neq 0$ as required.

***Exercise** Prove this lemma without using the construction $S(\ )$. [Hint: adapt the argument of 32A so that it refers directly to completely additive functionals on Boolean rings.]

52D Proposition Let $(\mathfrak{A}, \mu)$ be a semi-finite measure ring. Then the canonical map $\phi: S(\mathfrak{A}^f) \to L^\#(\mathfrak{A})$ defined in 52A embeds $S(\mathfrak{A}^f)$ as an order-dense Riesz subspace of $L^\#(\mathfrak{A})$.

Proof We already know that ϕ is a one-to-one Riesz homomorphism [52B], so I have only to prove that $\phi[S]$ is order-dense.

Suppose that $f > 0$ in $L^\#(\mathfrak{A})$. Then, because $(\mathfrak{A}, \mu)$ is semi-finite, there is an $a \in \mathfrak{A}^f$ such that $f(\chi a) > 0$. **P** Certainly there is an $a_0 \in \mathfrak{A}$ such that $f(\chi a_0) > 0$. Let

$$B = \{a_0 \backslash a : a \in \mathfrak{A}^f\}.$$

Then $B \downarrow$ in $\mathfrak{A}$. **?** Suppose, if possible, that B has a non-zero lower bound b_0 in $\mathfrak{A}$. Then there is a non-zero $c \subseteq b_0$ such that $c \in \mathfrak{A}^f$. So $c \subseteq a_0 \backslash c$, which is impossible. **X** Accordingly, $B \downarrow 0$ in $\mathfrak{A}$. Since $f\chi: \mathfrak{A} \to \mathbf{R}$ is order-continuous, $\inf_{b \in B} f(\chi b) = 0$. In particular, there is an $a \in \mathfrak{A}^f$ such that $f\chi(a_0 \backslash a) < f\chi a_0$. Now

$$f\chi a \geqslant f\chi(a \cap a_0) = f\chi a_0 - f\chi(a_0 \backslash a) > 0$$

as required. **Q**

Now let $\alpha > 0$ be such that $f(\chi a) > \alpha\mu a$. Set

$$g = f - \alpha\phi(\chi a)$$

in $L^\#(\mathfrak{A}) = L^\infty(\mathfrak{A})^\times$, and let $\nu = g\chi: \mathfrak{A} \to \mathbf{R}$, so that ν is a completely additive functional and

$$\nu a = g(\chi a) = f(\chi a) - \alpha\mu a > 0.$$

By 52C, there is a non-zero $b \subseteq a$ such that $\nu c \geqslant 0$ for every $c \subseteq b$, i.e.

$$f(\chi c) = g(\chi c) + \alpha\mu(a \cap c) = \nu c + \alpha\mu c \geqslant \alpha\mu c$$

for every $c \subseteq b$.

Consequently, for every $c \in \mathfrak{A}$,

$$f(\chi c) \geqslant f\chi(c \cap b) \geqslant \alpha\mu(c \cap b) = \alpha\phi(\chi b)(\chi c).$$

Thus $f \geqslant \alpha\phi(\chi b)$ in $L^\#$ [44Bb]. Since $\alpha > 0$ and $b \neq 0$ and ϕ is one-to one,

$$f \geqslant \phi(\alpha\chi b) > 0.$$

As f is arbitrary, this shows (using 15E) that $\phi[S]$ is order-dense in $L^\#$.

52E Proposition Let $(\mathfrak{A}, \mu)$ be a semi-finite measure ring. On $S(\mathfrak{A}^f)$, a Riesz norm $\| \ \|_1$ can be defined by writing $\|x\|_1 = \hat{\mu}(|x|)$, where $\hat{\mu}$ is the linear functional on $S(\mathfrak{A}^f)$ associated with the additive functional μ: $\mathfrak{A}^f \to \mathbf{R}$. Let $L^1(\mathfrak{A}, \mu)$ be the normed Riesz space completion of $S(\mathfrak{A}^f)$. Then the map ϕ: $S(\mathfrak{A}^f) \to L^\#(\mathfrak{A})$ defined in 52A extends to a normed Riesz space isomorphism between $L^1(\mathfrak{A}, \mu)$ and $L^\#(\mathfrak{A})$. Consequently $S(\mathfrak{A}^f)$ is super-order-dense in $L^1(\mathfrak{A}, \mu)$.

Proof (a) We know from 52B that $S(\mathfrak{A}^f)$, with $\| \ \|_1$, can be identified with the Riesz subspace $\phi[S]$ of $L^\#(\mathfrak{A})$. So $\| \ \|_1$ is a Riesz norm and by 22F the normed space completion $L^1(\mathfrak{A}, \mu)$ of $S(\mathfrak{A}^f)$ has a canonical Riesz space structure.

(b) Now $\phi[S]$ is dense in $L^\#(\mathfrak{A})$. **P** We know that $L^\#(\mathfrak{A})$ is an L-space [44Ba], so its topology is Lebesgue [26Ba]. Now $\phi[S]$ is order-dense in $L^\#(\mathfrak{A})$ [52D], that is, for any $f \geqslant 0$ in $L^\#$,

$$A = \{g : g \in \phi[S],\ 0 \leqslant g \leqslant f\} \uparrow f,$$

and $f \in \mathscr{I}\phi[S] \subseteq \overline{\phi[S]}$ [24Ba]. As $L^\# = L^{\#+} - L^{\#+}$, $\phi[S]$ is dense. **Q**

(c) So ϕ: $S(\mathfrak{A}^f) \to L^\#(\mathfrak{A})$, which is norm-preserving for the norms on S and $L^\#$, must extend to a normed space isomorphism between L^1 and $L^\#$, because $L^\#$ is complete. Because the lattice operations on both sides are continuous, and ϕ is a Riesz homomorphism on S, ϕ: $L^1 \to L^\#$ is also a Riesz space isomorphism. Now S is order-dense in L^1 because $\phi[S]$ is order-dense in $L^\#$. Consequently it is super-order-dense, because L^1 is an L-space [26Bd].

***52F $L^\#(\mathfrak{A})$ and $L^\#(\mathfrak{A}^f)$** If $(\mathfrak{A}, \mu)$ is any measure ring, then $(\mathfrak{A}^f, \mu_f)$ is a semi-finite measure ring, where μ_f is the restriction of μ to $\mathfrak{A}^f$, and of course $(\mathfrak{A}^f)^f = \mathfrak{A}^f$. So we may apply the results above to show that

$L^1(\mathfrak{A}, \mu)$, the completion of $S(\mathfrak{A}^f) = S(\mathfrak{A}^{ff})$, is the same as $L^1(\mathfrak{A}^f, \mu_f)$, and is isomorphic to $L^\#(\mathfrak{A}^f)$.

Of course $L^1(\mathfrak{A}, \mu)$ can also be identified with the closure of $\phi[S(\mathfrak{A}^f)]$ in $L^\#(\mathfrak{A})$. Thus $L^\#(\mathfrak{A}^f)$ is represented as a subspace of $L^\#(\mathfrak{A})$. The exercise 52Hc goes into this further.

52G I conclude this section with a remark on the duality between $L^\#(\mathfrak{A})$ and $L^\infty(\mathfrak{A})$ when $\mathfrak{A}$ carries a semi-finite measure.

Proposition Let $(\mathfrak{A}, \mu)$ be a semi-finite measure ring. Then the norms on $L^\infty(\mathfrak{A})$ and $L^\#(\mathfrak{A})$ are dual in the sense that

$$\|f\| = \sup\{|fx| : x \in L^\infty, \|x\|_\infty \leqslant 1\} \quad \forall\ f \in L^\#,$$

$$\|x\|_\infty = \sup\{|fx| : f \in L^\#, \|f\| \leqslant 1\} \quad \forall\ x \in L^\infty.$$

Proof The first part is just the definition of the norm on $L^\# = L^{\infty\times} \subseteq L^{\infty\prime}$. The second part is more interesting. Of course

$$\|x\|_\infty \geqslant \sup\{|fx| : f \in L^\#, \|f\| \leqslant 1\} = \|x\|'$$

for every $x \in L^\infty$, and $\|\ \|'$, so defined, is a seminorm. Now if x is a non-zero member of $S(\mathfrak{A})$, x can be expressed as $\sum_{i<n} \alpha_i \chi a_i$, where $\langle a_i \rangle_{i<n}$ is a disjoint sequence in $\mathfrak{A}$ and no a_i is 0; and in this case $\|x\|_\infty = |\alpha_j|$ for some $j < n$ [42Eb]. Let $a \in \mathfrak{A}^f$ be such that $a \subseteq a_j$ and $a \neq 0$. Then $f = f_a = \phi(\chi a) \in L^\#$ [notation as 52A], and $\|f\| = \mu a > 0$. But

$$|fx| = |\textstyle\sum_{i<n} \alpha_i \mu(a \cap a_i)| = |\alpha_j|\, \mu a = |\alpha_j|\, \|f\|,$$

because $a \cap a_i = 0$ if $i \neq j$ and $a \cap a_j = a$. From this it is clear that $\|x\|' \geqslant |\alpha_j| = \|x\|_\infty$. Thus $\|\ \|'$ and $\|\ \|_\infty$ agree on $S(\mathfrak{A})$; since $\|\ \|'$ is continuous with respect to $\|\ \|_\infty$, they agree on $L^\infty(\mathfrak{A})$.

***Remark** We observe that the above result is a property of $\mathfrak{A}$ alone; it is immaterial what the measure μ is, as long as it is semi-finite. It can be shown that this proposition characterizes those measure rings $\mathfrak{A}$ on which a semi-finite measure can be defined. [See 4XFj.]

***52H Exercises** (a) Let $\mathfrak{A}$ be a Dedekind σ-complete Boolean ring and $\nu: \mathfrak{A} \to \mathbf{R}$ a countably additive functional. Then for every $a \in \mathfrak{A}$ there is a $b \subseteq a$ such that $\nu b = \nu^+ a$, and $\nu^- b = 0$. (If $\langle b_n \rangle_{n \in \mathbf{N}}$ is a sequence such that $b_n \subseteq a$ and $\nu b_n \geqslant \nu^+ a - 2^{-n}$ for each $n \in \mathbf{N}$, set $b = \sup_{n \in \mathbf{N}} \inf_{i \geqslant n} b_i$.)

(b) Let $\boldsymbol{E}$ be a Riesz space and $\boldsymbol{F}$ a Riesz subspace of $\boldsymbol{E}^\times$ such that the map $x \mapsto \hat{x}: \boldsymbol{E} \to \boldsymbol{F}^*$ is a Riesz homomorphism from $\boldsymbol{E}$ to $\boldsymbol{F}^\times$. Use the method of 32B to show that the image of $\boldsymbol{E}$ is order-dense in $\boldsymbol{F}^\times$. Now apply this result with $\boldsymbol{E} = \boldsymbol{S}(\mathfrak{A}^f)$, $\boldsymbol{F} = \boldsymbol{L}^\infty(\mathfrak{A})$ to prove 52D.

(c) Let $(\mathfrak{A}, \mu)$ be any measure ring. Let

$$\boldsymbol{F} = \{g : g \in \boldsymbol{L}^\#(\mathfrak{A}),\ g(\chi a) = 0 \quad \forall\ a \in \mathfrak{A}^f\}.$$

Then $\boldsymbol{F}$ is a band in $\boldsymbol{L}^\#(\mathfrak{A})$. Let

$$\boldsymbol{E} = \boldsymbol{F}^d = \{f : f \in \boldsymbol{L}^\#(\mathfrak{A}),\ |f| \wedge |g| = 0 \quad \forall\ g \in \boldsymbol{F}\}$$

[cf. 15F]. Now the canonical map $\phi: \boldsymbol{S}(\mathfrak{A}^f) \to \boldsymbol{L}^\#(\mathfrak{A})$ embeds $\boldsymbol{S}$ as an order-dense Riesz subspace of $\boldsymbol{E}$ [use the method of 52D]. Consequently ϕ extends to an isomorphism between $\boldsymbol{L}^1(\mathfrak{A}, \mu)$ and $\boldsymbol{E}$, which sets up a canonical isomorphism $\theta: \boldsymbol{L}^\#(\mathfrak{A}^f) \to \boldsymbol{E}$ [see 52F].

At the same time, the order-continuous embedding of $\mathfrak{A}^f$ in $\mathfrak{A}$ induces a map $\hat{\pi}': \boldsymbol{L}^\#(\mathfrak{A}) \to \boldsymbol{L}^\#(\mathfrak{A}^f)$ [45E]. Now $\hat{\pi}'\theta$ is the identity on $\boldsymbol{L}^\#(\mathfrak{A}^f)$, and the kernel of $\hat{\pi}'$ is $\boldsymbol{F}$. Thus $\hat{\pi}'$ can be thought of as the projection of $\boldsymbol{L}^\#(\mathfrak{A})$ onto $\boldsymbol{E} \cong \boldsymbol{L}^\#(\mathfrak{A}^f) \cong \boldsymbol{L}^1(\mathfrak{A}, \mu)$, with kernel $\boldsymbol{F} = \boldsymbol{E}^d$ [15F].

$\theta: \boldsymbol{L}^\infty(\mathfrak{A}^f)^\times \to \boldsymbol{L}^\infty(\mathfrak{A})^\times$ can also be given by

$$(\theta f)(x) = \sup\{fy : y \in \boldsymbol{L}^\infty(\mathfrak{A}^f),\ 0 \leqslant y \leqslant x\}$$

for every $x \in \boldsymbol{L}^\infty(\mathfrak{A})^+$ and $f \in \boldsymbol{L}^\#(\mathfrak{A}^f)^+$, or by

$$(\theta f)(\chi a) = \sup\{f(\chi b) : b \in \mathfrak{A}^f, b \subseteq a\}$$

for every $a \in \mathfrak{A}$ and $f \in \boldsymbol{L}^\#(\mathfrak{A}^f)^+$.

Notes and comments The reason for maintaining a careful distinction between $\boldsymbol{L}^1(\mathfrak{A})$ and $\boldsymbol{L}^\#(\mathfrak{A})$, although they are canonically isomorphic in all significant cases, will appear in § 54. They are affected quite differently by homomorphisms between measure rings.

The properties of $\boldsymbol{L}^1$ can also be derived intrinsically, without considering $\boldsymbol{L}^\#$. It is easy to see that $\hat{\mu}: \boldsymbol{S}(\mathfrak{A}^f) \to \mathbf{R}$ is strictly positive, because $\mu: \mathfrak{A}^f \to \mathbf{R}$ is; so that $\| \ \|_1 = \hat{\mu}(|\ |)$ is a norm on $\boldsymbol{S}(\mathfrak{A}^f)$. Since of course

$$\| |x| + |y| \|_1 = \|x\|_1 + \|y\|_1$$

for all x, y in $\boldsymbol{S}(\mathfrak{A}^f)$, the same is true in the completion of $\boldsymbol{S}$, which is therefore an L-space. A less obvious point is the fact that $\boldsymbol{S}$ is order-dense in $\boldsymbol{L}^1$. This is clear from 52E, because $\phi[\boldsymbol{S}]$ is order-dense in $\boldsymbol{L}^\#$. But it is also a consequence of the fact that, because μ is completely additive on $\mathfrak{A}^f$ [51E], $\hat{\mu} \in \boldsymbol{S}^\times$ [42L], so that the $\| \ \|_1$-topology on $\boldsymbol{S}$ is

Lebesgue. It follows that $S' \subseteq S^\times$. Now S' is identified with $L^{1\prime}$ as linear space, and since the positive cone of L^1 is just the closure of the positive cone of S, S' and $L^{1\prime}$ have the same ordering. Also, $L^{1\prime} = L^{1\sim}$, so L^1 can be identified with a Riesz subspace of $(L^{1\prime})^\times = S'^\times$. The result now follows from 32B.

However, the isomorphism between $L^1(\mathfrak{A})$ and $L^\#(\mathfrak{A})$ is in itself one of the most important results of measure theory. In its more conventional forms, which will be touched on in § 63, it is the Radon–Nikodým theorem. The normal approach refers to countably additive functionals on σ-algebras of sets and uses 52Ha (the 'Hahn decomposition theorem') instead of 52C. Now an extra hypothesis must be added to ensure that, given a non-zero increasing countably additive functional $\nu: \mathfrak{A} \to \mathbf{R}$, there is an $a \in \mathfrak{A}^f$ such that $\nu a > 0$ (as in the first part of the proof of 52D). But such a hypothesis will always enable us to find a completely additive functional; see, for example, 63J.

53 Maharam algebras

If the underlying Boolean ring of a semi-finite measure ring is a Dedekind complete Boolean algebra, then its L^∞ space turns out to be 'perfect' in the sense of § 33, that is, L^∞ is isomorphic to $L^{\infty\times\times}$ or $L^{\#\times}$. Thus the duality between L^∞ and $L^\#$ or L^1 is symmetric in some respects. In many contexts, Maharam algebras appear to be the 'normal' measure rings, and all others 'unnatural'.

53A Definition A **measure algebra** is a measure ring $(\mathfrak{A}, \mu)$ in which $\mathfrak{A}$ is an algebra, i.e. has a 1. A **Maharam algebra** is a semi-finite measure algebra $(\mathfrak{A}, \mu)$ such that the Boolean algebra $\mathfrak{A}$ is Dedekind complete.

53B Theorem Let $(\mathfrak{A}, \mu)$ be a semi-finite measure ring. Then the following are equivalent:

(i) $(\mathfrak{A}, \mu)$ is a Maharam algebra;

(ii) $L^\infty(\mathfrak{A})$ is a Dedekind complete M-space with unit;

(iii) the duality between $L^\infty(\mathfrak{A})$ and $L^1(\mathfrak{A})$, derived from the isomorphism between $L^1(\mathfrak{A})$ and $L^\infty(\mathfrak{A})^\times$ [52E], represents $L^\infty(\mathfrak{A})$ as $L^1(\mathfrak{A})'$.

Proof (a) (i) $\Leftrightarrow$ (ii) We know that $\mathfrak{A}$ is Dedekind complete iff L^∞ is Dedekind complete [43Da]. If $\mathfrak{A}$ has a 1, then $\chi 1$ is the unit of

L^∞. Conversely if L^∞ has a unit e, then $\{a : \chi a \leqslant e\}$ is bounded above in $\mathfrak{A}$ [43Bd], and its upper bound must be the 1 of $\mathfrak{A}$.

(b) (ii) ⇒ (iii) We know that the canonical map from L^∞ to $L^{\infty\times\times}$ takes L^∞ onto a solid linear subspace of $L^{\infty\times\times}$, because L^∞ is Dedekind complete [32C]. Now $L^{\infty\times\times} = L^\#(\mathfrak{A})^\times$ is an M-space with unit [26C], and its unit is $\int$, given by $\int f = \|f\|$ whenever $f \geqslant 0$ in $L^\#$. But for $f \geqslant 0$ in $L^\#$, $\|f\| = f(\chi 1)$ [44Bb]. So $\int$ is the image of $\chi 1$ in $L^{\infty\times\times}$, and the image of L^∞ is the whole of $L^{\infty\times\times}$. Also, we have seen in 52G that $L^\#$ separates the points of L^∞, so L^∞ is isomorphic to $L^{\infty\times\times}$ as a Riesz space. At the same time, 52G shows that the norm of L^∞ is precisely that induced by its embedding in $L^{\#\prime}$; we recall from 26C that $L^{\#\prime} = L^{\#\times}$.

Thus L^∞ is identified with $L^{\#\prime}$. Since $L^\#$ and L^1 are canonically isomorphic, we can identify L^∞ with $L^{1\prime}$.

(c) (iii) ⇒ (ii) Finally, the dual of L^1 is certainly a Dedekind complete M-space with unit [26C], so if it is isomorphic to L^∞ the condition (ii) must be satisfied.

53C Definition A measure algebra $(\mathfrak{A}, \mu)$ has **finite magnitude** if $\mu 1 < \infty$.

53D Lemma A Dedekind σ-complete measure algebra of finite magnitude is a Maharam algebra.

Proof Of course any measure algebra $(\mathfrak{A}, \mu)$ of finite magnitude is semi-finite, as $\mathfrak{A} = \mathfrak{A}^f$. Now suppose that $\varnothing \subset B \uparrow$ in $\mathfrak{A}$. Then

$$C = \{a \backslash b : b \in B, a \text{ is an upper bound for } B\} \downarrow 0$$

[41Ha]. So by 51Cb there is a sequence $\langle c_n \rangle_{n \in \mathbf{N}}$ in C such that $\inf_{n \in \mathbf{N}} c_n = 0$. If $c_n = a_n \backslash b_n$, where $b_n \in B$ and a_n is an upper bound for B for each $n \in \mathbf{N}$, then $\inf_{n \in \mathbf{N}} a_n = \sup_{n \in \mathbf{N}} b_n$ must be the least upper bound of B. [Cf. 18B.]

53E Proposition Any product of Maharam algebras is a Maharam algebra.

Proof Obviously, any product of Boolean algebras is a Boolean algebra. But also a product of semi-finite measure rings is semi-finite [51F], and a product of Dedekind complete Boolean rings is Dedekind complete [13Hb]. Thus each of the defining properties of a Maharam algebra is preserved in the product.

53F Exercises (**a**) A measure algebra $(\mathfrak{A}, \mu)$ has **countable magnitude** if there is a sequence $\langle a_n \rangle_{n\in\mathbf{N}}$ in $\mathfrak{A}^f$ such that $\sup_{n\in\mathbf{N}} a_n = 1$. Show that any measure algebra of countable magnitude is semi-finite, and that it is a Maharam algebra iff it is Dedekind σ-complete.

(**b**) Let $\mathfrak{A}$ be any Dedekind complete Boolean algebra. For each $a \in \mathfrak{A}$, let $\mathfrak{A}_a = \{b : b \in \mathfrak{A},\, b \subseteq a\} \lhd \mathfrak{A}$. If $A \subseteq \mathfrak{A}$ is a disjoint set such that $\sup A = 1$, show that the map $b \mapsto \langle b \cap a \rangle_{a\in A} : \mathfrak{A} \to \prod_{a\in A} \mathfrak{A}_a$ is a Boolean ring isomorphism between $\mathfrak{A}$ and $\prod_{a\in A} \mathfrak{A}_a$.

Now let $(\mathfrak{A}, \mu)$ be a Maharam algebra. For $a \in \mathfrak{A}$, let $\mu_a : \mathfrak{A}_a \to [0, \infty]$ be the restriction of μ to $\mathfrak{A}_a$. Then if A satisfies the same conditions as before, $(\mathfrak{A}, \mu)$ is isomorphic to $\prod_{a\in A}(\mathfrak{A}_a, \mu_a)$ as measure ring.

(**c**) Any Maharam algebra is isomorphic to a product of Maharam algebras of finite magnitude. [Hint: in (b) above, use Zorn's lemma to find a maximal disjoint set $A \subseteq \mathfrak{A}^f$.]

*(**d**) Let $(\mathfrak{A}, \mu)$ be any semi-finite measure ring. For each $a \in \mathfrak{A}^f$, define $\rho_a : \mathfrak{A} \times \mathfrak{A} \to \mathbf{R}$ by

$$\rho_a(b, c) = \mu(a(b+c)) \quad \forall\ b, c \in \mathfrak{A}.$$

Then each ρ_a is a pseudo-metric [cf. 51Ga]. The collection $\{\rho_a : a \in \mathfrak{A}^f\}$ defines a Hausdorff uniformity on $\mathfrak{A}$ with respect to which the ring operations on $\mathfrak{A}$ and the map $\mu : \mathfrak{A} \to [0, \infty]$ are all uniformly continuous ($[0, \infty]$ being given its usual compact uniformity). If $\mathfrak{A}$ is a Maharam algebra, it is complete with respect to this uniformity [cf. Nakano's theorem, 23K]; otherwise, $\mathfrak{A}$ may be completed, and, with the natural extensions of μ and the ring operations, the completion is a Maharam algebra.

Notes and comments L^∞ and L^1 now constitute a dual pair of perfect Riesz spaces; L^∞ is isomorphic to $L^{1\times}$ and L^1 to $L^{\infty\times}$. The main value of this result lies in the fact that $L^{1\times} = L^{1\prime}$, so that L^∞ can be regarded as a representation of the normed space dual of L^1, or of S given the norm $\| \ \|_1$. From 53C–53Fa we see that this result will apply to any product of Dedekind σ-complete measure algebras of finite or countable magnitude.

The theorem 53B can be looked at from another direction. Suppose that $\mathfrak{A}$ is a Dedekind complete Boolean algebra; under what conditions is $L^\infty(\mathfrak{A})$ a perfect Riesz space? We have seen in 4XF that it need not be. Now 53B shows that it is sufficient that there should exist a strictly positive semi-finite measure on $\mathfrak{A}$. In fact it is possible to show that this condition is also necessary (cf. the remark at the end of 52G).

There is an effective structure theory known for Maharam algebras. The first step is to express an arbitrary Maharam algebra as a product of Maharam algebras of finite magnitude [53Fb/c]. A much deeper argument, due to MAHARAM, can now be used to express a Maharam algebra of finite magnitude as a product of homogeneous algebras, and to classify completely homogeneous Maharam algebras of finite magnitude. (Maharam's theorem may also be found in SEMADENI [§§ 26.4–5]). This classification must not be confused with the much easier representation theorem of Kakutani [61I].

54 Measure-preserving ring homomorphisms

When we come to examine homomorphisms between measure rings, we find that the most natural ones are the 'measure-preserving' ring homomorphisms, that is, ring homomorphisms π such that πa and a always have the same measure. On the ideal of elements of finite measure, such homomorphisms have to be order-continuous [54B], so we obtain linear maps between the $L^\#$ spaces as well as between the S and L^∞ spaces. Of course, a measure-preserving ring homomorphism has to be one-to-one, so in effect we are talking about subrings of measure rings. What is interesting is the fact that the map between the S spaces preserves the norm $\| \ \|_1$, so that we find ourselves with a natural map between the L^1 spaces. Now this canonical map between the L^1 spaces is in the opposite direction from the canonical map between the $L^\#_-$ spaces; so the isomorphism between the L^1 and $L^\#$ spaces has complex effects.

54A Definition Let $(\mathfrak{A}, \mu)$ and $(\mathfrak{B}, \nu)$ be measure rings. A ring homomorphism $\pi: \mathfrak{A} \to \mathfrak{B}$ is **measure-preserving** if $\nu(\pi a) = \mu a$ for every $a \in \mathfrak{A}$.

A measure-preserving ring homomorphism must be one-to-one, for

$$a \neq 0 \Rightarrow \mu a > 0 \Rightarrow \nu(\pi a) > 0 \Rightarrow \pi a \neq 0.$$

It need not be order-continuous, or even sequentially order-continuous; but we do have the following result.

54B Proposition Let $(\mathfrak{A}, \mu)$ and $(\mathfrak{B}, \nu)$ be measure rings, and $\pi: \mathfrak{A} \to \mathfrak{B}$ a measure-preserving ring homomorphism. Then $\pi[\mathfrak{A}^f] \subseteq \mathfrak{B}^f$

and $\pi: \mathfrak{A}^f \to \mathfrak{B}^f$ is an order-continuous measure-preserving ring homomorphism.

Proof Clearly $\pi[\mathfrak{A}^f] \subseteq \mathfrak{B}^f$. Now if $\varnothing \subset A \downarrow 0$ in $\mathfrak{A}^f$, $\inf_{a \in A} \mu a = 0$ [51Cb], so $\inf_{a \in A} \nu(\pi a) = 0$. As ν is strictly positive, $\inf \pi[A] = 0$. This proves that $\pi: \mathfrak{A}^f \to \mathfrak{B}^f$ is order-continuous, and of course it is a measure-preserving ring homomorphism.

Remark The effect of this is that we can apply 54E below, which refers to order-continuous measure-preserving ring homomorphisms between semi-finite measure rings, to any measure-preserving ring homomorphism, by considering its restriction to the ideal of elements of finite measure.

54C Proposition Let $(\mathfrak{A}, \mu)$ and $(\mathfrak{B}, \nu)$ be measure rings, and $\pi: \mathfrak{A} \to \mathfrak{B}$ a measure-preserving ring homomorphism. Then $\pi: \mathfrak{A}^f \to \mathfrak{B}^f$ induces a Riesz homomorphism $\hat{\pi}: \boldsymbol{S}(\mathfrak{A}^f) \to \boldsymbol{S}(\mathfrak{B}^f)$ [45B]. Now $\|\hat{\pi}x\|_1 = \|x\|_1$ for every $x \in \boldsymbol{S}(\mathfrak{A}^f)$.

Proof If $a \in \mathfrak{A}^f$,

$$\hat{\nu}\hat{\pi}(\chi a) = \hat{\nu}\chi(\pi a) = \nu(\pi a) = \mu a = \hat{\mu}(\chi a).$$

So $\hat{\nu}\hat{\pi} = \hat{\mu}: \boldsymbol{S}(\mathfrak{A}^f) \to \mathbf{R}$. Now

$$\|\hat{\pi}x\|_1 = \hat{\nu}(|\hat{\pi}x|) = \hat{\nu}\hat{\pi}(|x|) = \hat{\mu}(|x|) = \|x\|_1$$

for every $x \in \boldsymbol{S}(\mathfrak{A}^f)$, using the definition of $\| \ \|_1$ [52E] and the fact that $\hat{\pi}$ is a Riesz homomorphism.

54D Corollary $\hat{\pi}: \boldsymbol{S}(\mathfrak{A}^f) \to \boldsymbol{S}(\mathfrak{B}^f)$ extends to a norm-preserving Riesz homomorphism $\hat{\pi}: \boldsymbol{L}^1(\mathfrak{A}) \to \boldsymbol{L}^1(\mathfrak{B})$.

Proof By definition, $\boldsymbol{L}^1(\mathfrak{A})$ and $\boldsymbol{L}^1(\mathfrak{B})$ are the completions of $\boldsymbol{S}(\mathfrak{A}^f)$ and $\boldsymbol{S}(\mathfrak{B}^f)$ respectively for the norms $\| \ \|_1$. So $\hat{\pi}$ has an extension to a norm-preserving linear map from $\boldsymbol{L}^1(\mathfrak{A})$ to $\boldsymbol{L}^1(\mathfrak{B})$, which is a Riesz homomorphism because the lattice operations are continuous on both sides.

54E Proposition Let $(\mathfrak{A}, \mu)$ and $(\mathfrak{B}, \nu)$ be semi-finite measure rings. Let $\pi: \mathfrak{A} \to \mathfrak{B}$ be an order-continuous measure-preserving ring homomorphism. Let $\phi: \boldsymbol{L}^1(\mathfrak{A}) \to \boldsymbol{L}^\#(\mathfrak{A})$ and $\psi: \boldsymbol{L}^1(\mathfrak{B}) \to \boldsymbol{L}^\#(\mathfrak{B})$ be the canonical isomorphisms [52E]. Then by 54D and 45E we have a diagram:

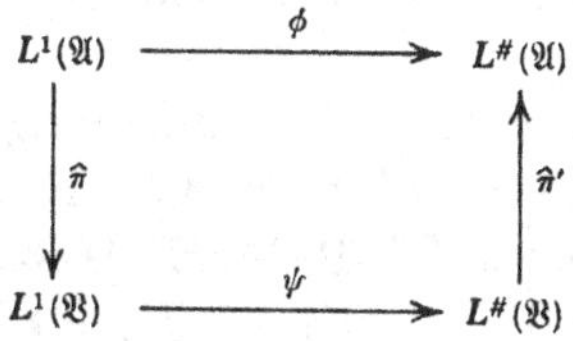

Now this diagram commutes, i.e.

$$\hat{\pi}'\psi\hat{\pi} = \phi.$$

Proof Suppose that $a \in \mathfrak{A}^f$. Then $(\hat{\pi}'\psi\hat{\pi})(\chi a)$ is given by:

$$\begin{aligned}(\hat{\pi}'\psi\hat{\pi}\chi a)(\chi b) &= (\psi\hat{\pi}\chi a)(\hat{\pi}\chi b)\\ &= (\psi\chi\pi a)(\chi\pi b) = \nu(\pi a \cap \pi b)\\ &= \nu\pi(a \cap b) = \mu(a \cap b)\\ &= (\phi\chi a)(\chi b)\end{aligned}$$

for every $b \in \mathfrak{A}$. So $\hat{\pi}'\psi\hat{\pi}\chi a = \phi\chi a$ in $L^\#(\mathfrak{A})$. But $\hat{\pi}'\psi\hat{\pi}$ and ϕ are both linear, so they agree on $S(\mathfrak{A}^f)$; and they are both continuous, so they agree on $L^1(\mathfrak{A})$.

54F Corollary Turning this diagram round, we can write

$$P_\pi = \phi^{-1}\hat{\pi}'\psi : L^1(\mathfrak{B}) \to L^1(\mathfrak{A})$$

and see that $P_\pi\hat{\pi}$ is the identity on $L^1(\mathfrak{A})$. Thus if the norm-preserving Riesz homomorphism $\hat{\pi}$ is taken as embedding $L^1(\mathfrak{A})$ as a closed Riesz subspace of $L^1(\mathfrak{B})$, then P_π is a projection from $L^1(\mathfrak{B})$ onto $L^1(\mathfrak{A})$. P_π can be defined by saying that, for any $y \in L^1(\mathfrak{B})$, $P_\pi y$ is that member of $L^1(\mathfrak{A})$ such that

$$\begin{aligned}\langle P_\pi y, \chi a\rangle &= (\phi P_\pi y)(\chi a) = (\hat{\pi}'\psi y)(\chi a)\\ &= (\psi y)(\hat{\pi}\chi a) = (\psi y)(\chi\pi a)\\ &= \langle y, \chi\pi a\rangle\end{aligned}$$

for every $a \in \mathfrak{A}$.

Notes and comments Since measure-preserving ring homomorphisms have to be one-to-one, we may regard all the work above as results about subrings of measure rings. If $\mathfrak{A}$ is a subring of $\mathfrak{B}$, then

$\mathfrak{A}^f = \mathfrak{A} \cap \mathfrak{B}^f$ is a subring of $\mathfrak{B}^f$, and $S(\mathfrak{A}^f)$ is a Riesz subspace of $S(\mathfrak{B}^f)$ [45Da]. Now 54C shows that the norm $\| \ \|_1$ on $S(\mathfrak{A}^f)$ agrees with the norm $\| \ \|_1$ on $S(\mathfrak{B}^f)$, so that $L^1(\mathfrak{A})$ will be just the closure of $S(\mathfrak{A}^f)$ in $L^1(\mathfrak{B})$.

Next, we see from 54B that the embedding of $\mathfrak{A}^f$ in $\mathfrak{B}^f$ is order-continuous; if $\varnothing \subset A \downarrow 0$ in $\mathfrak{A}^f$, then $A \downarrow 0$ in $\mathfrak{B}^f$ also. So we have a natural map from $L^{\#}(\mathfrak{B}^f)$ to $L^{\#}(\mathfrak{A}^f)$ [45E]. If we identify $L^{\#}(\mathfrak{B}^f)$ with $L^1(\mathfrak{B}^f) = L^1(\mathfrak{B})$ and $L^{\#}(\mathfrak{A}^f)$ with $L^1(\mathfrak{A}^f) = L^1(\mathfrak{A})$, we get a map $P: L^1(\mathfrak{B}) \to L^1(\mathfrak{A})$, which turns out to be a projection [54F]. These projections are particularly important in probability theory.

The point is that a map $\pi: \mathfrak{A} \to \mathfrak{B}$ leads naturally both to a map $\hat{\pi}: L^1(\mathfrak{A}) \to L^1(\mathfrak{B})$ and to a map $P_\pi: L^1(\mathfrak{B}) \to L^1(\mathfrak{A})$. Thus $L^1(\mathfrak{A})$, regarded as the completion of $S(\mathfrak{A}^f)$, is a covariant construction; but identified with $L^{\#}(\mathfrak{A}^f)$ it is contravariant.

5X Examples for Chapter 5

At this stage I restrict my examples to measure rings which are rings of sets. 5XB tries to show how, in a semi-finite measure ring $(\mathfrak{A}, \mu)$, $L^1(\mathfrak{A})$ depends on μ, even though it is always isomorphic to $L^{\#}(\mathfrak{A})$, which does not depend on μ. 5XC discusses a simple example of a measure-preserving ring homomorphism.

5XA Two simple types of measure ring (a) Let X be any set, and $\mathfrak{A}$ any subring of $\mathscr{P}X$. Define $\mu: \mathfrak{A} \to [0, \infty]$ by:

$$\mu E = n \quad \text{if } E \text{ has } n \text{ members, where } n \in \mathbf{N};$$

$$\mu E = \infty \quad \text{if } E \text{ is infinite.}$$

Then $(\mathfrak{A}, \mu)$ is a measure ring.

(b) If $\mathfrak{A}$ is any Boolean ring, define $\mu: \mathfrak{A} \to [0, \infty]$ by $\mu 0 = 0$, $\mu a = \infty$ if $a \neq 0$. Then $(\mathfrak{A}, \mu)$ is a measure ring.

5XB L^1 and $L^{\#}$ Let X be any set, and let $\mathfrak{A}$ be a subring of $\mathscr{P}X$ containing all finite subsets of X. Let $\zeta: X \to [0, \infty]$ be a function such that $\zeta(t) > 0$ for every $t \in X$. Define $\mu: \mathfrak{A} \to [0, \infty]$ by

$$\mu E = \textstyle\sum_{t \in E} \zeta(t) \quad \forall\ E \in \mathfrak{A}.$$

Then μ is a measure on $\mathfrak{A}$. (Note that, because $\{t\} \in \mathfrak{A}$ for every $t \in X$, the embedding $\mathfrak{A} \to \mathscr{P}X$ is order-continuous. If $A \subseteq \mathfrak{A}$ and $\sup A$ exists in $\mathfrak{A}$, then $\sup A = \bigcup A$.)

$\boldsymbol{S}(\mathfrak{A}^f)$ can be identified with a linear subspace of $\mathbf{R}^X$, as in 4XB. For $x \in \boldsymbol{S}(\mathfrak{A}^f)$,

$$\hat{\mu}(x) = \sum_{t \in X} x(t)\,\zeta(t),$$

the sum being absolutely convergent if we employ the convention that $0.\infty = 0$. So

$$\|x\|_1 = \sum_{t \in X} |x(t)|\,\zeta(t) \quad \forall\ x \in \boldsymbol{S}(\mathfrak{A}^f).$$

Now $\boldsymbol{L}^1(\mathfrak{A})$ can be identified with

$$\{x : x \in \mathbf{R}^X,\ \|x\|_1 = \sum_{t \in X} |x(t)|\,\zeta(t) < \infty\},$$

using the same arguments as those which show that $\boldsymbol{l}^1(X)$ is complete [2XB].

Let us regard $\boldsymbol{L}^\infty(\mathfrak{A})$ also as a Riesz subspace of $\mathbf{R}^X$. Then the duality between $\boldsymbol{L}^\infty(\mathfrak{A})$ and $\boldsymbol{S}(\mathfrak{A}^f)$ [52A] is given by

$$\langle x, y\rangle = \sum_{t \in X} x(t)\,y(t)\,\zeta(t) \quad \forall\ x \in \boldsymbol{S}(\mathfrak{A}^f),\ y \in \boldsymbol{L}^\infty(\mathfrak{A});$$

this formula is easily verified if $x = \chi E$ and $y = \chi F$, and is extended to general x and y by linearity and continuity. Also, because every finite subset of X belongs to $\mathfrak{A}$, $\boldsymbol{L}^\infty(\mathfrak{A})$ is order-dense in $\boldsymbol{l}^\infty(X)$. It follows from 17B and 2XA, or otherwise, that $\boldsymbol{L}^\#(\mathfrak{A}) = \boldsymbol{L}^\infty(\mathfrak{A})^\times$ can be identified with $\boldsymbol{l}^1(X)$. So the canonical map from $\boldsymbol{S}(\mathfrak{A}^f)$ to $\boldsymbol{L}^\#(\mathfrak{A})$ is given by $x \mapsto x \times \zeta$, where $(x \times \zeta)(t) = x(t)\,\zeta(t)$ for every $t \in X$.

Clearly μ is semi-finite iff $\zeta(t) < \infty$ for every $t \in X$. In this case, the isomorphism from $\boldsymbol{L}^1(\mathfrak{A})$ to $\boldsymbol{L}^\#(\mathfrak{A})$ [52E] is given by $x \mapsto x \times \zeta$ for every $x \in \boldsymbol{L}^1$, and this is now onto.

This example shows how $\boldsymbol{L}^1(\mathfrak{A})$ is subtly dependent on the measure μ, even though (as long as μ is semi-finite) it is isomorphic to $\boldsymbol{l}^1(X)$, which is entirely independent of μ.

Since $\mathfrak{A}$ contains all finite subsets of X, it will be a Dedekind complete Boolean algebra iff it is $\mathscr{P}X$. Now $\boldsymbol{L}^\infty(\mathfrak{A})$ can be identified with $\boldsymbol{l}^\infty(X)$ [4XCa], which is a perfect Riesz space [2XA]. Thus $(\mathfrak{A}, \mu)$ is a Maharam algebra iff $\mathfrak{A} = \mathscr{P}X$ and $\zeta(t) < \infty$ for every $t \in X$.

5XC Let X be the set $\mathbf{N} \times \mathbf{N}$; let $\mathfrak{B} = \mathscr{P}X$; let $\langle \zeta_{mn}\rangle_{m,n \in \mathbf{N}}$ be a double sequence such that

$$\zeta_{mn} > 0 \quad \forall\ m, n \in \mathbf{N},\ \sum_{m,n \in \mathbf{N}} \zeta_{mn} < \infty.$$

Let ν be the measure on $\mathfrak{B}$ given by

$$\nu E = \sum\{\zeta_{mn} : (m, n) \in E\} \quad \forall\ E \in \mathfrak{B}.$$

Now let $\mathfrak{A}$ be the algebra $\mathscr{P}\mathrm{N}$, and define

$$\mu A = \sum_{m\in A}\eta_m \quad \forall\ A\in\mathfrak{A},$$

where $\eta_m = \sum_{n\in\mathbf{N}}\zeta_{mn}$ for every $m\in\mathrm{N}$. Then $(\mathfrak{A},\mu)$ and $(\mathfrak{B},\nu)$ are Maharam algebras of the type described in 5XB; in fact they are of finite magnitude.

If we write

$$\pi A = A\times\mathrm{N} \quad \forall\ A\subseteq\mathrm{N},$$

then $\pi\colon \mathfrak{A}\to\mathfrak{B}$ is a measure-preserving ring homomorphism. Identifying $L^\infty(\mathfrak{A})$ and $L^\infty(\mathfrak{B})$ with $l^\infty(\mathrm{N})$ and $l^\infty(\mathrm{N}\times\mathrm{N})$ respectively, $\hat{\pi}\colon L^\infty(\mathfrak{A})\to L^\infty(\mathfrak{B})$ is given by

$$(\hat{\pi}x)(m,n) = x(m) \quad \forall\ m,n\in\mathrm{N},\ x\in l^\infty(\mathrm{N}).$$

So, now identifying $L^\#(\mathfrak{A})$ and $L^\#(\mathfrak{B})$ with $l^1(\mathrm{N})$ and $l^1(\mathrm{N}\times\mathrm{N})$ respectively, $\hat{\pi}'\colon L^\#(\mathfrak{B})\to L^\#(\mathfrak{A})$ is given by

$$(\hat{\pi}'z)(m) = \sum_{n\in\mathbf{N}}z(m,n) \quad \forall\ m\in\mathrm{N},\ z\in l^1(\mathrm{N}\times\mathrm{N}).$$

At the same time, 5XB shows how $L^1(\mathfrak{A})$ and $L^1(\mathfrak{B})$ can be identified with

$$\{x : x\in\mathbf{R}^{\mathbf{N}},\ \sum_{m\in\mathbf{N}}\eta_m|x(m)| < \infty\}$$

and

$$\{y : y\in\mathbf{R}^{\mathbf{N}\times\mathbf{N}},\ \sum_{m,n\in\mathbf{N}}\zeta_{mn}|y(m,n)| < \infty\}.$$

Clearly $\hat{\pi}\colon L^1(\mathfrak{A})\to L^1(\mathfrak{B})$ is still given by

$$(\hat{\pi}x)(m,n) = x(m) \quad \forall\ m,n\in\mathrm{N},\ x\in L^1(\mathfrak{A});$$

and $P_\pi\colon L^1(\mathfrak{B})\to L^1(\mathfrak{A})$ is given by

$$(P_\pi y)(m) = \eta_m^{-1}\sum_{n\in\mathbf{N}}\zeta_{mn}y(m,n) \quad \forall\ m\in\mathrm{N},\ y\in L^1(\mathfrak{B}).$$

Thus, for $y\in L^1(\mathfrak{B})$,

$$(\hat{\pi}P_\pi y)(m,n) = \eta_m^{-1}\sum_{i\in\mathbf{N}}\zeta_{mi}y(m,i) \quad \forall\ m,n\in\mathrm{N};$$

the projection $\hat{\pi}P_\pi\colon L^1(\mathfrak{B})\to L^1(\mathfrak{B})$ 'averages' y down each column.

6. Measure spaces

The object of this chapter is to show how the concepts of ordinary measure theory may be joined to the more abstract work done so far. The first step is to describe the measure algebra of a measure space [61D]; then § 62 and § 63 show how the L^∞ and L^1 spaces of this measure algebra can be identified with the usual function spaces. In view of the importance of Maharam measure algebras [§ 53], we are naturally concerned to identify the corresponding measure spaces; the search for useful criteria involves us in consideration of a number of special kinds of measure space, which are discussed in § 64. In §65 we examine an important class of function spaces, including the L^p spaces, which exemplify the results of Chapters 2 and 3 on locally convex topological Riesz spaces.

61 Definitions and basic properties

The most important idea here is the construction of the measure algebra of a measure space [61D]. The rest of the section comprises a miscellany of definitions and results which will be used later.

61A Definition A **measure space** is a triple (X, Σ, μ), where X is a set, Σ is a σ-subalgebra of $\mathscr{P}X$ [i.e. a subalgebra closed under countable unions and intersections], and $\mu: \Sigma \to [0, \infty]$ is a function such that:

(i) $\mu\varnothing = 0$;

(ii) $\mu(E \cup F) = \mu E + \mu F$ if $E, F \in \Sigma$ and $E \cap F = \varnothing$;

(iii) if $\langle E_n \rangle_{n \in \mathbf{N}}$ is an increasing sequence in Σ,

$$\mu(\bigcup_{n \in \mathbf{N}} E_n) = \lim_{n \to \infty} \mu E_n.$$

61B Notes (a) For a note on the use of '∞', see 51A.

(b) Observe that I allow $X = \varnothing$; in this case $\Sigma = \mathscr{P}X = \{\varnothing\}$ and μ must be the zero function.

(c) μ is called a **measure** on X, and the elements of Σ are called **measurable sets.**

61C Elementary results Let (X, Σ, μ) be a measure space.

(a) If $E, F \in \Sigma$ and $E \subseteq F$, then $\mu E \leqslant \mu F$.

(b) If $E, F \in \Sigma$, then $\mu(E \cup F) \leqslant \mu E + \mu F$.

(c) If $\langle E_n \rangle_{n \in \mathbf{N}}$ is any sequence in Σ, $\mu(\bigcup_{n \in \mathbf{N}} E_n) \leqslant \sum_{n \in \mathbf{N}} \mu E_n$.

Proof (a) $\mu F = \mu E + \mu(F \backslash E) \geqslant \mu E$.

(**b**) $\mu(E \cup F) = \mu E + \mu(F \backslash E) \leqslant \mu E + \mu F$.

(**c**) $\mu(\bigcup_{n \in \mathbf{N}} E_n) = \lim_{n \to \infty} \mu(\bigcup_{i \leqslant n} E_i)$

$$\leqslant \lim_{n \to \infty} \sum_{i \leqslant n} \mu E_i = \sum_{n \in \mathbf{N}} \mu E_n.$$

61D The measure algebra (a) Let (X, Σ, μ) be a measure space. Write Σ_0 for $\{E : E \in \Sigma, \mu E = 0\}$.

Consider Σ as a Boolean ring. Then Σ_0 is an ideal of Σ, by 61Ca and 61Cb [see 41J], and we can form the quotient $\mathfrak{A} = \Sigma / \Sigma_0$; $\mathfrak{A}$ is a Boolean algebra and the canonical map $E \mapsto E^\bullet : \Sigma \to \mathfrak{A}$ is a ring homomorphism. The 1 of $\mathfrak{A}$ is $X^\bullet$. We recall also that

$$E^\bullet \subseteq F^\bullet \Leftrightarrow E \backslash F \in \Sigma_0 \Leftrightarrow \mu(E \backslash F) = 0$$

[41J]. Now

$$E^\bullet = F^\bullet \Leftrightarrow \mu(E \backslash F) = \mu(F \backslash E) = 0$$

$$\Rightarrow \mu E = \mu(E \cap F) = \mu F.$$

So μ factors through $\mathfrak{A}$, that is, there is a function $\mu^\bullet : \mathfrak{A} \to [0, \infty]$ such that $\mu E = \mu^\bullet E^\bullet$ for every $E \in \Sigma$. Henceforth I shall allow myself to write μ instead of $\mu^\bullet$ for this function on $\mathfrak{A}$.

(**b**) Now the map $E \mapsto E^\bullet : \Sigma \to \mathfrak{A}$ is sequentially order-continuous. **P** Suppose that $\langle E_n \rangle_{n \in \mathbf{N}} \downarrow \varnothing$ in Σ. Since $\bigcap_{n \in \mathbf{N}} E_n$ belongs to Σ and is a lower bound for $\{E_n : n \in \mathbf{N}\}$ in Σ, $\bigcap_{n \in \mathbf{N}} E_n = \varnothing$. Let a be any lower bound for $\{E_n^\bullet : n \in \mathbf{N}\}$ in $\mathfrak{A}$. Let $E \in \Sigma$ be such that $E^\bullet = a$. Then $E = \bigcup_{n \in \mathbf{N}} E \backslash E_n$, so by 61Ca and 61Cc

$$\mu E \leqslant \sum_{n \in \mathbf{N}} \mu(E \backslash E_n) = 0$$

because $E^\bullet \subseteq E_n^\bullet$ for every $n \in \mathbf{N}$. Thus $\mu E = 0$ and $a = 0$. As a is arbitrary, $\langle E_n^\bullet \rangle_{n \in \mathbf{N}} \downarrow 0$ in $\mathfrak{A}$. Because the map $E \mapsto E^\bullet$ is a ring homomorphism, this is enough to show that it is sequentially order-continuous. **Q**

(**c**) It follows that $(\mathfrak{A}, \mu)$ is a measure algebra. **P** $\mathfrak{A}$ is a Boolean algebra and μ is a function from $\mathfrak{A}$ to $[0, \infty]$. (**i**) If $a, b \in \mathfrak{A}$ and $a \cap b = 0$, let $E, F \in \Sigma$ be such that $E^{\bullet} = a$ and $F^{\bullet} = b$. As $(E \cap F)^{\bullet} = E^{\bullet} \cap F^{\bullet} = 0$, $\mu(E \cap F) = 0$, so

$$\mu(a \cup b) = \mu((E \cup F)^{\bullet}) = \mu(E \cup F) = \mu E + \mu(F \backslash E)$$
$$= \mu E + \mu F = \mu a + \mu b.$$

(ii) If $a \in \mathfrak{A}$, let $E \in \Sigma$ be such that $E^{\bullet} = a$. Then

$$\mu a = 0 \Leftrightarrow \mu E = 0 \Leftrightarrow E \in \Sigma_0 \Leftrightarrow a = 0.$$

(iii) If $\langle a_n \rangle_{n \in \mathbf{N}} \uparrow a$ in $\mathfrak{A}$, choose for each $n \in \mathbf{N}$ an $E_n \in \Sigma$ such that $E_n^{\bullet} = a_n$. Set $G_n = \bigcup_{i \leqslant n} E_i$. Since the map $E \mapsto E^{\bullet}$ is a lattice homomorphism,

$$G_n^{\bullet} = \sup_{i \leqslant n} a_i = a_n \quad \forall\ n \in \mathbf{N}.$$

Let $G = \bigcup_{n \in \mathbf{N}} G_n$. By (b) above, $G^{\bullet} = \sup_{n \in \mathbf{N}} G_n^{\bullet} = a$. So

$$\mu a = \mu G = \lim_{n \to \infty} \mu G_n = \lim_{n \to \infty} \mu a_n.$$

Thus the three conditions of 51A are satisfied and $(\mathfrak{A}, \mu)$ is a measure algebra. **Q**

(d) It also follows from (b) that $\mathfrak{A}$ is Dedekind σ-complete. **P** Let $\langle a_n \rangle_{n \in \mathbf{N}}$ be any sequence in $\mathfrak{A}$. For each $n \in \mathbf{N}$, choose an $E_n \in \Sigma$ such that $E_n^{\bullet} = a_n$. Then $H = \bigcap_{n \in \mathbf{N}} E_n \in \Sigma$, and $H = \inf_{n \in \mathbf{N}} E_n$ in Σ. Because the map $E \mapsto E^{\bullet}$ is a sequentially order-continuous lattice homomorphism, $H^{\bullet} = \inf_{n \in \mathbf{N}} a_n$ in $\mathfrak{A}$. By 41I, this is enough to show that $\mathfrak{A}$ is Dedekind σ-complete. **Q**

(e) Now $(\mathfrak{A}, \mu)$ is the **measure algebra** of the measure space (X, Σ, μ).

***61E Inverse-measure-preserving functions** Given two measure spaces (X, Σ, μ) and (Y, T, ν), a natural class of functions to consider is given by the following definition. A function $f: X \to Y$ is **inverse-measure-preserving** if $f^{-1}[F] \in \Sigma$ and $\mu f^{-1}[F] = \nu F$ for every $F \in \mathrm{T}$.

Let $\mathfrak{A}$ and $\mathfrak{B}$ be the measure algebras of (X, Σ, μ) and (Y, T, ν) respectively. If $f: X \to Y$ is inverse-measure-preserving, then there is a sequentially order-continuous measure-preserving ring homomorphism $\pi: \mathfrak{B} \to \mathfrak{A}$ given by $\pi[F^{\bullet}] = (f^{-1}[F])^{\bullet}$ for every $F \in \mathrm{T}$.

P **(i)** We must verify first that π is well-defined. But this is a direct consequence of the fact that $f^{-1}: \mathrm{T} \to \Sigma$ is a ring homomorphism such that $f^{-1}[F] \in \Sigma_0$ for every $F \in \mathrm{T}_0$.

(ii) The same argument shows that π is a ring homomorphism. Now suppose that $\langle b_n \rangle_{n \in \mathbf{N}} \downarrow 0$ in $\mathfrak{B}$. Choose $F_n \in \mathrm{T}$ such that $F_n^{\bullet} = b_n$ for each

$n \in \mathbf{N}$. Now the map $F \mapsto F^{\bullet}: \mathrm{T} \to \mathfrak{B}$ is a sequentially order-continuous lattice homomorphism, so

$$(\bigcap_{n\in\mathbf{N}} F_n)^{\bullet} = \inf_{n\in\mathbf{N}} b_n = 0,$$

and $\nu(\bigcap_{n\in\mathbf{N}} F_n) = 0$. Similarly,

$$\inf_{n\in\mathbf{N}} \pi b_n = \inf_{n\in\mathbf{N}} (f^{-1}[F_n])^{\bullet} = (\bigcap_{n\in\mathbf{N}} f^{-1}[F_n])^{\bullet} = (f^{-1}[\bigcap_{n\in\mathbf{N}} F_n])^{\bullet}.$$

But $\mu f^{-1}[\bigcap_{n\in\mathbf{N}} F_n] = \nu(\bigcap_{n\in\mathbf{N}} F_n) = 0$, so $\inf_{n\in\mathbf{N}} \pi b_n = 0$. As $\langle b_n \rangle_{n\in\mathbf{N}}$ is arbitrary, π is sequentially order-continuous.

(iii) Finally, it is obvious that π is measure-preserving. **Q**

[See also 61Jc, e.]

61F Further definitions Let (X, Σ, μ) be a measure space.

(a) Following 51Ba, write

$$\Sigma^f = \{E : E \in \Sigma, \mu E < \infty\}.$$

If $\mathfrak{A}$ is the measure algebra of (X, Σ, μ), then $\mathfrak{A}^f = \Sigma^f/\Sigma_0$.

(b) (X, Σ, μ) is **semi-finite** if its measure algebra is semi-finite [51Bb], that is, whenever $E \in \Sigma$ and $\mu E = \infty$, there is an $F \subseteq E$ such that $F \in \Sigma$ and $0 < \mu F < \infty$.

*(**c**) A set $E \in \Sigma$ is **purely infinite** if $\mu E = \infty$, and, for every $F \in \Sigma$ such that $F \subseteq E$, μF is either 0 or ∞. Thus (X, Σ, μ) is semi-finite iff there are no purely infinite sets in Σ.

(d) A set $A \subseteq X$ is **negligible** if there is an $F \in \Sigma$ such that $A \subseteq F$ and $\mu F = 0$. Now (i) any subset of a negligible set is negligible, (ii) a countable union of negligible sets is negligible [using 61Cc].

(e) Suppose that Φ is any property applicable to points of X. I shall write '$\Phi(t)$ for almost every $t \in X$', or '$\Phi(t)$ p.p. (t)', or 'Φ almost everywhere on X', or 'Φ p.p.' to mean that $X \backslash \{t: \Phi(t)\}$ is negligible.

For instance, if $x: X \to \mathbf{R}$ is a function, then '$x = 0$ p.p.' means that $\{t : x(t) \neq 0\}$ is negligible. Or, if $\langle x_n \rangle_{n\in\mathbf{N}}$ is a sequence of real-valued functions on X, then '$\langle x_n(t) \rangle_{n\in\mathbf{N}} \to 0$ p.p.(t)' means that

$$\{t : \langle x_n(t) \rangle_{n\in\mathbf{N}} \not\to 0\}$$

is negligible.

(f) (X, Σ, μ) is called **complete** if, whenever $\mu F = 0$ and $E \subseteq F$, then $E \in \Sigma$ (so that $\mu E = 0$); i.e. if every negligible set is measurable.

(g) (X, Σ, μ) is **of finite magnitude,** or **totally finite**, if its measure algebra is [53C], i.e. if $\mu X < \infty$.

61G Direct sum of measure spaces Let $\langle(X_\iota, \Sigma_\iota, \mu_\iota)\rangle_{\iota\in I}$ be an indexed family of measure spaces. Let

$$X = \{(t,\iota) : \iota\in I, t\in X_\iota\},$$

the 'disjoint union' of $\langle X_\iota\rangle_{\iota\in I}$. For $E \subseteq X$ and $\iota\in I$ write

$$E_\iota = \{t : (t,\iota)\in E\} \subseteq X_\iota.$$

Let $\Sigma = \{E : E \subseteq X, E_\iota \in \Sigma_\iota \ \forall\ \iota\in I\}$, and define $\mu: \Sigma \to [0,\infty]$ by

$$\mu E = \Sigma_{\iota\in I}\mu_\iota E_\iota.$$

It is easy to see that (X,Σ,μ) is a measure space. I shall call it the **direct sum** of $\langle(X_\iota, \Sigma_\iota, \mu_\iota)\rangle_{\iota\in I}$.

In effect, the maps $t \mapsto (t,\iota) : X_\iota \to X$ embed each X_ι as a subset of X, and $\langle X_\iota\rangle_{\iota\in I}$ now appears as a partition of X. A set $E \subseteq X$ belongs to Σ iff $E_\iota = E \cap X_\iota \in \Sigma_\iota = \Sigma \cap \mathscr{P}X_\iota$ for every $\iota\in I$, and in this case

$$\mu E = \Sigma_{\iota\in I}\mu(E \cap X_\iota),$$

if we identify each μ_ι with the restriction of μ to Σ_ι. [See also 64Ga.]

61H Measure algebra of the direct sum Clearly, in this construction, Σ is isomorphic to the Boolean ring product $\prod_{\iota\in I}\Sigma_\iota$ [41K]. It follows that if $\mathfrak{A}_\iota$ is the measure algebra of $(X_\iota, \Sigma_\iota, \mu_\iota)$ for each $\iota\in I$, while $\mathfrak{A}$ is the measure algebra of (X,Σ,μ), then $\mathfrak{A} \cong \prod_{\iota\in I}\mathfrak{A}_\iota$. (For if $E\in\Sigma$, then $\mu E = 0$ iff $\mu_\iota E_\iota = 0$ for every $\iota\in I$.) And, given any $E\in\Sigma$,

$$\mu E^\bullet = \mu E = \Sigma_{\iota\in I}\mu_\iota E_\iota = \Sigma_{\iota\in I}\mu_\iota(E_\iota)^\bullet.$$

So $(\mathfrak{A},\mu)$ is the measure ring product of $\langle(\mathfrak{A}_\iota, \mu_\iota)\rangle_{\iota\in I}$, as defined in 51F.

Consequently, the direct sum of semi-finite measure spaces is semi-finite [61Fb, 51F].

***61I Representation of measure algebras** Some of the work of this chapter will appear in sharper relief if I give a general representation theorem. We have seen in 61D that the measure algebra of a measure space is always Dedekind σ-complete. But there is a converse result, as is shown by the following propositions.

Lemma Let $\mathfrak{A}$ be any Dedekind σ-complete Boolean algebra. Then there is a set X, a σ-subalgebra Σ of $\mathscr{P}X$, and a sequentially order-continuous ring homomorphism ψ from Σ onto $\mathfrak{A}$.

Proof See SIKORSKI, §29.1, or HALMOS L.B.A., p. 100, §23.

Proposition Let $(\mathfrak{A}, \mu)$ be any Dedekind σ-complete measure algebra. Then it is isomorphic to the measure algebra of some measure space.

Proof Choose X, Σ and ψ as in the lemma. Define $\nu: \Sigma \to [0, \infty]$ by $\nu E = \mu(\psi E)$ for every $E \in \Sigma$. Then it is easy to see that (because ψ is sequentially order-continuous) (X, Σ, ν) is a measure space. Since

$$\psi E = 0 \Leftrightarrow \mu(\psi E) = 0 \Leftrightarrow \nu E = 0,$$

$(\mathfrak{A}, \mu)$ is isomorphic to the measure algebra of (X, Σ, ν).

61J Exercises (**a**) Let (X, Σ, μ) be a semi-finite measure space. Then $\mu E = \sup\{\mu F : F \in \Sigma^f, F \subseteq E\}$ for every $E \in \Sigma$ [cf. 51Gc].

(**b**) Let (X, Σ, μ) be a measure space. Then $\mu: \Sigma^f \to \mathbf{R}$ is countably additive [cf. 51E].

*(**c**) Let (X, Σ, μ) be a measure space, Y any set, and $f: X \to Y$ any function. Let $\mathrm{T} = \{F : F \subseteq Y, f^{-1}[F] \in \Sigma\}$.

Then T is a σ-subalgebra of $\mathscr{P}Y$ and there is a unique measure on T for which f is inverse-measure-preserving.

(**d**) Let $\langle (X_\iota, \Sigma_\iota, \mu_\iota) \rangle_{\iota \in I}$ be an indexed family of measure spaces, with direct sum (X, Σ, μ). Then (X, Σ, μ) is complete iff $(X_\iota, \Sigma_\iota, \mu_\iota)$ is complete for every $\iota \in I$.

*(**e**) Let (X, Σ, μ) and (Y, T, ν) be measure spaces. Let $f: X \to Y$ be a function such that $f[E] \in \mathrm{T}$ and $\nu f[E] = \mu E$ for every $E \in \Sigma$. Then if $\mathfrak{A}$ and $\mathfrak{B}$ are the measure algebras of (X, Σ, μ) and (Y, T, ν) respectively, there is a measure-preserving ring homomorphism $\pi: \mathfrak{A}^f \to \mathfrak{B}^f$ given by $\pi(E^\bullet) = (f[E])^\bullet$ for every $E \in \Sigma^f$.

Notes and comments In 61D we have the first link between Chapter 5 and Chapter 6; others will appear in 62H and 63A; it is my purpose in this book to emphasize these connections. In every construction involving measure spaces, the effect on the measure algebras is one of the most important things to look for. Thus 61E, 61H and 61Je are typical investigations.

One of the oddities of measure theory is the lack of a completely convincing class of morphisms. The inverse-measure-preserving maps of 61E are probably the most important ones; but many curious results [e.g. 61Je] attend other functions. Inverse-measure-preserving functions constitute a category for which the measure algebra is a contravariant construction; but for the functions of 61Je it is covariant.

62 Measurable functions; the space L^0

Given a measure space (X, Σ, μ) ,we call a real-valued function on X *measurable* if the inverse image of an open set in $\mathbf{R}$ is always in Σ; we have already encountered bounded measurable functions in 4XD. The space $M(\Sigma)$ of all real-valued measurable functions on X is now a Riesz subspace of $\mathbf{R}^X$, and also a σ-sublattice (i.e. it is closed under countable suprema and infima) and a subring (for ordinary pointwise multiplication in $\mathbf{R}^X$). If we say that two measurable functions are equivalent if they are equal almost everywhere, we find that the set L^0 of equivalence classes inherits most of the structure of $M(\Sigma)$ to become a Dedekind σ-complete Riesz space with an order-continuous multiplication [62G, 62I].

The main importance of the construction L^0 lies in its solid linear subspaces. If $\mathfrak{A}$ is the measure algebra of (X, Σ, μ), then $L^\infty(\mathfrak{A})$ has a canonical representation as a solid linear subspace of L^0 [62H]. In the next section I shall show how $L^1(\mathfrak{A})$ can also be embedded naturally in L^0, and § 65 deals with a large class of subspaces of which L^1 and L^∞ are archetypes. L^0 is in itself a remarkable space. Although its most dramatic properties depend on certain conditions being imposed on (X, Σ, μ), and will therefore be held back for the moment, 62K and 62Mh give a foretaste of what can be proved.

62A Definition Let (X, Σ, μ) be a measure space and $(Y, \mathfrak{S})$ a topological space. A function $f: X \to Y$ is **measurable** if $f^{-1}[G] \in \Sigma$ for every $G \in \mathfrak{S}$.

***Remark** We must beware of conflicting definitions when Y is endowed with a σ-subalgebra $\mathrm{T} \subseteq \mathscr{P}Y$ as well as a topology $\mathfrak{S}$. In this case (especially in view of 61E), we are tempted to call a function $f: X \to Y$ 'measurable' if $f^{-1}[F] \in \Sigma$ for every $F \in \mathrm{T}$. This may not coincide with the definition above. Consider, for instance, the case $X = Y = \mathbf{R}$, with $\mathfrak{S}$ the usual topology and $\Sigma = \mathrm{T}$ the class of Lebesgue measurable sets. Then there exist functions $f: X \to Y$ which are $(\Sigma, \mathfrak{S})$-measurable, indeed continuous, but not (Σ, T)-measurable.

62B Lemma Let (X, Σ, μ) be a measure space, and $(Y, \mathfrak{S})$ and $(Z, \mathfrak{T})$ topological spaces. If $f: X \to Y$ is measurable and $g: Y \to Z$ is continuous, then $gf: X \to Z$ is measurable.

Proof For if $G \in \mathfrak{T}$, $g^{-1}[G] \in \mathfrak{S}$, so

$$(gf)^{-1}[G] = f^{-1}[g^{-1}[G]] \in \Sigma.$$

62C Lemma Let (X, Σ, μ) be a measure space, and $x: X \to \mathbf{R}$ and $y: X \to \mathbf{R}$ measurable functions, where $\mathbf{R}$ is given its usual topology. Then the function

$$t \mapsto (x(t), y(t)): X \to \mathbf{R}^2$$

is measurable.

Proof Let $G \subseteq \mathbf{R}^2$ be open. Then G can be expressed as $\bigcup_{n \in \mathbf{N}} I_n \times J_n$, where I_n and J_n are open subsets of $\mathbf{R}$ for each $n \in \mathbf{N}$. (In fact, unless $G = \varnothing = \varnothing \times \varnothing$, each I_n and J_n can be taken to be an open interval with rational end-points.) Now

$$\{t: (x(t), y(t)) \in G\} = \bigcup_{n \in \mathbf{N}} (x^{-1}[I_n] \cap y^{-1}[J_n]) \in \Sigma.$$

62D Lemma Let (X, Σ, μ) be a measure space. Then a function $x: X \to \mathbf{R}$ is measurable iff $\{t: x(t) > \alpha\} \in \Sigma$ for every $\alpha \in \mathbf{R}$.

Proof Obviously the condition is necessary. Conversely, suppose it is satisfied, and let $G \subseteq \mathbf{R}$ be open. Then G can be expressed as

$$\bigcup_{n \in \mathbf{N}} \{\alpha : \alpha_n < \alpha \leqslant \beta_n\}$$

for suitably chosen sequences $\langle \alpha_n \rangle_{n \in \mathbf{N}}$ and $\langle \beta_n \rangle_{n \in \mathbf{N}}$, and now

$$x^{-1}[G] = \bigcup_{n \in \mathbf{N}} (\{t: x(t) > \alpha_n\} \backslash \{t: x(t) > \beta_n\}) \in \Sigma.$$

62E Proposition Let (X, Σ, μ) be a measure space. Let M be the set of all measurable functions from X to $\mathbf{R}$. Then M is a Riesz subspace and a σ-sublattice of $\mathbf{R}^X$. Moreover, $x \times y \in M$ for all $x, y \in M$, where

$$(x \times y)(t) = x(t)\, y(t) \quad \forall\ t \in X.$$

Proof If $x, y \in M$, then $x + y \in M$, by 62B, for $+: \mathbf{R}^2 \to \mathbf{R}$ is continuous and $t \mapsto (x(t), y(t)): X \to \mathbf{R}^2$ is measurable by 62C. Similarly, αx and $|x|$ belong to M for every $\alpha \in \mathbf{R}$, because $\beta \mapsto \alpha\beta: \mathbf{R} \to \mathbf{R}$ and $|\ |: \mathbf{R} \to \mathbf{R}$ are continuous. Thus M is a Riesz subspace of $\mathbf{R}^X$.

If $A \subseteq M$ is countable and $x = \sup A$ exists in $\mathbf{R}^X$, then

$$\{t: x(t) > \alpha\} = \bigcup_{y \in A} \{t: y(t) > \alpha\}$$

for every $\alpha \in \mathbf{R}$. Using 62D in both directions, and the fact that Σ is closed under countable unions, we see that $x \in M$. Equally, if $\inf A$ exists in $\mathbf{R}^X$, then

$$\inf A = -\sup(-A) \in M.$$

Thus M is a σ-sublattice of $\mathbf{R}^X$.

Finally, M is closed under $\times$ because multiplication, like addition, is a continuous function from $\mathbf{R}^2$ to $\mathbf{R}$.

62F Definitions (a) Let (X, Σ, μ) be a measure space. I shall write $M = M(\Sigma)$ for the space of real-valued measurable functions on X, described in 62E above.

(b) Consider

$$M_0 = \{x : x \in M, x = 0 \text{ p.p.}\};$$

that is, $$x \in M_0 \Leftrightarrow \mu\{t : x(t) \neq 0\} = 0.$$

Then M_0 is a solid linear subspace of M, and also a σ-sublattice. **P** Suppose, for instance, that $x, y \in M_0$. Then

$$\{t : (x+y)(t) \neq 0\} \subseteq \{t : x(t) \neq 0\} \cup \{t : y(t) \neq 0\};$$

so

$$\mu\{t : (x+y)(t) \neq 0\} \leqslant \mu\{t : x(t) \neq 0\} + \mu\{t : y(t) \neq 0\} = 0.$$

Thus M_0 is closed under addition. And it is easy to see that it is solid and closed under scalar multiplication. If $A \subseteq M_0$ is countable and $x = \sup A$ exists in M, then $x(t) = \sup_{y \in A} y(t)$ for every $t \in X$ [see 62E]; so

$$\{t : x(t) \neq 0\} \subseteq \bigcup_{y \in A}\{t : y(t) \neq 0\}$$

is negligible. Thus M_0 is a σ-sublattice of M. **Q**

(c) Now let us write $L^0 = L^0(\Sigma, \mu)$ for the Riesz space quotient M/M_0, described in 14G.

(d) We observe that, for $x, y \in M$,

$$x^{\cdot} \leqslant y^{\cdot} \Leftrightarrow (x-y)^+ \in M_0 \Leftrightarrow \mu\{t \colon x(t) > y(t)\} = 0,$$

where $x^{\cdot}$ is the image of x in L^0. Thus

$$x^{\cdot} = y^{\cdot} \Leftrightarrow x = y \quad \text{p.p.}$$

62G Proposition Let (X, Σ, μ) be any measure space. Then $L^0 = L^0(\Sigma, \mu)$, defined in 62F above, is a Dedekind σ-complete Riesz space, and the canonical map from $M = M(\Sigma)$ to L^0 is sequentially order-continuous.

Proof In the construction of quotient spaces in 14G it was remarked that the canonical map is always a Riesz homomorphism. Now suppose that $\langle x_n \rangle_{n \in \mathbf{N}} \downarrow 0$ in M, and that u is a lower bound of $\{x_n^{\cdot} : n \in \mathbf{N}\}$ in L^0. Let $x \in M$ be such that $x^{\cdot} = u$. Then $x^{\cdot} \leqslant x_n^{\cdot}$, so $(x - x_n)^+ \in M_0$ for every

$n \in \mathbf{N}$ [62Fd]; but $\langle (x - x_n)^+ \rangle_{n \in \mathbf{N}} \uparrow x^+$, so $x^+ \in \boldsymbol{M}_0$, as $\boldsymbol{M}_0$ is a σ-sublattice of $\boldsymbol{M}$ [62Fb]. It follows that $u = x^{\cdot} \leqslant 0$ in $\boldsymbol{L}^0$. As u is arbitrary, $\inf_{n \in \mathbf{N}} x_n^{\cdot} = 0$; as $\langle x_n \rangle_{n \in \mathbf{N}}$ is arbitrary, the canonical map is sequentially order-continuous. [See 14Lb.]

Now let $\langle u_n \rangle_{n \in \mathbf{N}}$ be any sequence in $\boldsymbol{L}^0$ bounded above by $u \in \boldsymbol{L}^0$. For each $n \in \mathbf{N}$, choose an $x_n \in \boldsymbol{M}$ such that $x_n^{\cdot} = u_n$, and also an $x \in \boldsymbol{M}$ such that $x^{\cdot} = u$. Set

$$y = \sup_{n \in \mathbf{N}} x \wedge x_n,$$

which exists in $\boldsymbol{M}$ because $\boldsymbol{M}$ is a σ-sublattice of $\mathbf{R}^X$, and is therefore Dedekind σ-complete. Because the canonical map from $\boldsymbol{M}$ to $\boldsymbol{L}^0$ is a sequentially order-continuous Riesz homomorphism,

$$y^{\cdot} = \sup_{n \in \mathbf{N}} x^{\cdot} \wedge x_n^{\cdot} = \sup_{n \in \mathbf{N}} u_n.$$

As $\langle u_n \rangle_{n \in \mathbf{N}}$ is arbitrary, $\boldsymbol{L}^0$ is Dedekind σ-complete.

62H We come now to the identification of the $\boldsymbol{L}^\infty$ space of the measure algebra of (X, Σ, μ).

Theorem Let (X, Σ, μ) be any measure space, and $\mathfrak{A}$ its measure algebra [61D]. Then $\boldsymbol{L}^\infty(\mathfrak{A})$ can be identified, as Riesz space, with the super-order-dense solid linear subspace of $\boldsymbol{L}^0(\Sigma, \mu)$ generated by $e = (\chi X)^{\cdot}$. For each $E \in \Sigma$, this embedding identifies $\chi(E^{\cdot})$ in $\boldsymbol{L}^\infty(\mathfrak{A})$ with $(\chi E)^{\cdot}$ in $\boldsymbol{L}^0$. In particular, e becomes the unit of the M-space $\boldsymbol{L}^\infty(\mathfrak{A})$.

Proof In 4XB we have seen that $\boldsymbol{L}^\infty(\Sigma)$ can be identified with a space of real-valued functions on X, which in 4XD (using the fact that Σ is a σ-subalgebra of $\mathscr{P}X$) were identified as those bounded functions which are measurable by the criterion of 62D. Thus $\boldsymbol{L}^\infty(\Sigma)$ can be thought of as the solid linear subspace of $\boldsymbol{M}(\Sigma)$ generated by χX.

Now $\mathfrak{A}$ is defined as the Boolean ring quotient Σ/Σ_0, where Σ_0 is the ideal of sets of measure 0 [61D]. So, by 45Dc, $\boldsymbol{L}^\infty(\mathfrak{A}) \simeq \boldsymbol{L}^\infty(\Sigma)/\boldsymbol{L}^\infty(\Sigma_0)$. But it was pointed out in 4XE that $\boldsymbol{L}^\infty(\Sigma_0)$ is precisely

$$\{x : x \in \boldsymbol{L}^\infty(\Sigma), \quad \{t : x(t) \neq 0\} \in \Sigma_0\},$$

because Σ_0 is a σ-ideal of Σ. Thus $\boldsymbol{L}^\infty(\Sigma_0) = \boldsymbol{L}^\infty(\Sigma) \cap \boldsymbol{M}_0$, where $\boldsymbol{M}_0$ is the space of functions which are zero almost everywhere, as in 62F. It follows at once that $\boldsymbol{L}^\infty(\Sigma)/\boldsymbol{L}^\infty(\Sigma_0)$ can be identified, as Riesz space, with the image of $\boldsymbol{L}^\infty(\Sigma)$ in $\boldsymbol{L}^0 = \boldsymbol{M}/\boldsymbol{M}_0$. And because $\boldsymbol{L}^\infty(\Sigma)$ is the solid linear subspace of $\boldsymbol{M}$ generated by χX, its image in $\boldsymbol{L}^0$ will be the solid linear subspace generated by $e = (\chi X)^{\cdot}$.

On examining this embedding, we see that, for any $E \in \Sigma$, $\chi(E^{\cdot})$ in $L^\infty(\mathfrak{A})$ is identified with the image of χE in $L^\infty(\Sigma)/L^\infty(\Sigma_0)$; that is, with $(\chi E)^{\cdot}$. The unit of $L^\infty(\mathfrak{A})$ is of course $\chi(X^{\cdot})$, since $X^{\cdot}$ is the identity of the algebra $\mathfrak{A}$, and this is identified with $(\chi X)^{\cdot} = e$.

Finally, suppose that $u \geqslant 0$ in L^0. Then there is an $x \in M^+$ such that $x^{\cdot} = u$; now $\langle x \wedge n\chi X\rangle_{n\in\mathbf{N}} \uparrow x$ in M, so $\langle u \wedge ne\rangle_{n\in\mathbf{N}} \uparrow u$ in L^0. Thus $L^\infty(\mathfrak{A})$ is super-order-dense in L^0.

62I The multiplication on L^0. In 62E it was pointed out that $M(\Sigma)$ always has a natural multiplicative structure. Since M_0, as defined in 62F, is clearly an ideal of M, we can define a multiplication on $L^0 = M/M_0$ by writing

$$x^{\cdot} \times y^{\cdot} = (x \times y)^{\cdot} \quad \forall\ x, y \in M.$$

χX is the multiplicative identity of M, so $e = (\chi X)^{\cdot}$ is the multiplicative identity of L^0.

If $x \in M^+$, the map $y \mapsto x \times y$: $M \to M$ is a Riesz homomorphism. From this it follows easily that the map $v \mapsto u \times v$: $L^0 \to L^0$ is a Riesz homomorphism for every $u \in L^{0+}$. In fact, this map is order-continuous. **P** Suppose that $\varnothing \subset A \downarrow 0$ in L^0. Let $w \geqslant 0$ be a lower bound for $\{u \times v : v \in A\}$. Take some $v_0 \in A$ and let $x, y_0, z \in M^+$ be such that $x^{\cdot} = u$, $y_0^{\cdot} = v_0$ and $z^{\cdot} = w$. Then if $z_1 = z \wedge (x \times y_0)$ we shall have $z_1^{\cdot} = w \wedge (u \times v_0) = w$. And

$$\{t : x(t) \neq 0\} \supseteq \{t : z_1(t) \neq 0\}.$$

Define x_1: $X \to \mathbf{R}$ by

$$x_1(t) = x(t)^{-1} \quad \text{if} \quad x(t) \neq 0,$$
$$= 0 \qquad \text{if} \quad x(t) = 0.$$

Using 62D, or otherwise, it is easy to see that $x_1 \in M$. Now

$$0 \leqslant x_1 \times x \leqslant \chi X,$$

so $u_1 \times u \leqslant e$, where $u_1 = x_1^{\cdot}$, and

$$0 \leqslant u_1 \times w \leqslant u_1 \times u \times v \leqslant v \quad \forall\ v \in A.$$

So $0 = u_1 \times w = (x_1 \times z_1)^{\cdot}$.

But now $\qquad \{t : z_1(t) \neq 0\} = \{t : (x_1 \times z_1)(t) \neq 0\},$

so $z_1 \in M_0$ and $w = z_1^{\cdot} = 0$. As w is arbitrary, $\inf_{v\in A} u \times v = 0$; as A is arbitrary, the map $v \mapsto u \times v$ is order-continuous. **Q**

(Note also the results in 62Mc, which will be used later.)

62J The next theorem describes one of the most remarkable properties of the space L^0. In its strongest form it requires the measure algebra to be Dedekind complete, and will appear in 64D. To begin with I give a sequential form which is always true.

First we need a lemma.

Lemma Let (X, Σ, μ) be a measure space, and let $\langle u_n \rangle_{n \in \mathbf{N}}$ be a sequence in $L^0(\Sigma, \mu)^+$. Then *either* $\{u_n : n \in \mathbf{N}\}$ is bounded above *or* there is some $v > 0$ in L^0 such that

$$kv = \sup_{n \in \mathbf{N}} u_n \wedge kv \quad \forall\ k \in \mathbf{N}.$$

Proof For each $n \in \mathbf{N}$, choose $x_n \in M(\Sigma)$ such that $x_n^{\cdot} = u_n$. Set

$$E = \{t : \sup_{n \in \mathbf{N}} x_n(t) < \infty\}.$$

(i) If $\mu(X \backslash E) = 0$,

$$x = \sup_{n \in \mathbf{N}} (x_n \times \chi E)$$

exists in M. Now

$$x^{\cdot} = \sup_{n \in \mathbf{N}} (x_n \times \chi E)^{\cdot} = \sup_{n \in \mathbf{N}} x_n^{\cdot} = \sup_{n \in \mathbf{N}} u_n$$

in L^0, so $\{u_n : n \in \mathbf{N}\}$ has an upper bound.

(ii) If $\mu(X \backslash E) > 0$, set $y = \chi(X \backslash E)$ and $v = y^{\cdot}$. Since

$$ky = \sup_{n \in \mathbf{N}} x_n \wedge ky \quad \forall\ k \in \mathbf{N},$$

$$kv = \sup_{n \in \mathbf{N}} u_n \wedge kv \quad \forall\ k \in \mathbf{N}.$$

As $\mu(X \backslash E) > 0$, $v > 0$ in L^0, as required.

62K Theorem Let (X, Σ, μ) be a measure space and E an Archimedean Riesz space. Suppose that F is a super-order-dense Riesz subspace of E and that $T: F \to L^0(\Sigma, \mu)$ is an order-continuous Riesz homomorphism. Then T has a unique extension to an order-continuous Riesz homomorphism from E to $L^0(\Sigma, \mu)$.

Proof I mean to apply 17B. To do so, I must show that

$$\{Ty : y \in F,\ 0 \leqslant y \leqslant x\}$$

has a least upper bound in L^0 for every $x \in E^+$.

(a) I show first that if $\langle y_n \rangle_{n \in \mathbf{N}}$ is a sequence in F^+ which is bounded above by $x \in E$, then $\{Ty_n : n \in \mathbf{N}\}$ is bounded above in L^0.

P? For suppose otherwise. By the last lemma, there is a $v > 0$ in L^0 such that

$$kv = \sup_{n \in \mathbf{N}} Ty_n \wedge kv \quad \forall\ k \in \mathbf{N}.$$

In particular, setting $k = 1$, there is an $m \in \mathbf{N}$ such that $Ty_m \wedge v > 0$.

Now, because E is Archimedean,

$$y_m = \sup_{\alpha>0}(y_m - \alpha x)^+$$

in E [use 15D]. So, because F is order-dense in E, $y_m = \sup A$, where

$$A = \{y : y \in F, \quad \exists\ \alpha > 0, y \leqslant (y_m - \alpha x)^+\}.$$

(For any upper bound for A must be an upper bound for

$$\{(y_m - \alpha x)^+ : \alpha > 0\}$$

and therefore greater than or equal to y.) Now $T: F \to L^0$ is order-continuous and $A \uparrow y_m$, so $Ty_m = \sup_{y \in A} Ty$ and there is a $y \in A$ such that

$$w = Ty \wedge v > 0.$$

Let $\alpha > 0$ be such that $y \leqslant (y_m - \alpha x)^+$.

Let us now examine, for k, $n \in \mathbf{N}$,

$$\begin{aligned} y_n \wedge ky \leqslant x \wedge ky \leqslant x \wedge k(y_m - \alpha x)^+ &= x \wedge \alpha k(\alpha^{-1}y_m - x)^+ \\ &\leqslant \alpha^{-1}y_m + (x - \alpha^{-1}y_m)^+ \wedge \alpha k(\alpha^{-1}y_m - x)^+ \\ &= \alpha^{-1}y_m, \end{aligned}$$

using 14Kb and 14Bl. So

$$\begin{aligned} kw = kw \wedge kv &= kw \wedge \sup_{n \in \mathbf{N}} Ty_n \wedge kv \\ &= \sup_{n \in \mathbf{N}} Ty_n \wedge kv \wedge kw \leqslant \sup_{n \in \mathbf{N}} Ty_n \wedge kTy \\ &= \sup_{n \in \mathbf{N}} T(y_n \wedge ky) \leqslant \alpha^{-1}Ty_m. \end{aligned}$$

But L^0 is also Archimedean, so this is impossible, since $w > 0$. **XQ**

(b) Now let $x \in E^+$. Then there is a sequence $\langle y_n \rangle_{n \in \mathbf{N}}$ in F^+ such that $x = \sup_{n \in \mathbf{N}} y_n$. By (a) above, $\{Ty_n : n \in \mathbf{N}\}$ is bounded above; as L^0 is Dedekind σ-complete, $u = \sup_{n \in \mathbf{N}} Ty_n$ exists in L^0. Now if y is any member of F such that $0 \leqslant y \leqslant x$, $y = \sup_{n \in \mathbf{N}} y \wedge y_n$, so

$$Ty = \sup_{n \in \mathbf{N}} Ty \wedge Ty_n \leqslant u.$$

Thus
$$u = \sup\{Ty : y \in F, 0 \leqslant y \leqslant x\}.$$

(c) Now by 17B there is a unique extension of T to an order-continuous increasing linear map $U: E \to L^0$, and by 17Ca U is a Riesz homomorphism.

***62L** A corollary of this theorem throws an interesting light on the real nature of the construction L^0.

Corollary Let (X, Σ, μ) and (Y, T, ν) be measure spaces, with measure algebras $\mathfrak{A}$ and $\mathfrak{B}$ respectively. Let $\pi: \mathfrak{A} \to \mathfrak{B}$ be a Boolean ring isomorphism. Then $\hat{\pi}: \boldsymbol{L}^\infty(\mathfrak{A}) \to \boldsymbol{L}^\infty(\mathfrak{B})$ is a Riesz space isomorphism which extends uniquely to a Riesz space isomorphism from $\boldsymbol{L}^0(\Sigma, \mu)$ to $\boldsymbol{L}^0(\mathrm{T}, \nu)$.

Proof We know from 45C that $\hat{\pi}: \boldsymbol{L}^\infty(\mathfrak{A}) \to \boldsymbol{L}^\infty(\mathfrak{B})$ is onto and norm-preserving; so of course it is a Riesz space isomorphism. Since $\boldsymbol{L}^\infty(\mathfrak{B})$ is solid in $\boldsymbol{L}^0(\mathrm{T}, \nu)$, $\hat{\pi}: \boldsymbol{L}^\infty(\mathfrak{A}) \to \boldsymbol{L}^0(\mathrm{T}, \nu)$ is order-continuous [17E]. Now $\boldsymbol{L}^\infty(\mathfrak{A})$ is super-order-dense in $\boldsymbol{L}^0(\Sigma, \mu)$ [62H], so we can apply the theorem to obtain an order-continuous Riesz homomorphism

$$U: \boldsymbol{L}^0(\Sigma, \mu) \to \boldsymbol{L}^0(\mathrm{T}, \nu)$$

extending $\hat{\pi}$.

To show that U is the required Riesz space isomorphism, we can use the following device. $\pi^{-1}: \mathfrak{B} \to \mathfrak{A}$ is also an isomorphism, so

$$(\pi^{-1})^\wedge: \boldsymbol{L}^\infty(\mathfrak{B}) \to \boldsymbol{L}^\infty(\mathfrak{A})$$

has an extension to an order-continuous Riesz homomorphism $V: \boldsymbol{L}^0(\mathrm{T}, \nu) \to \boldsymbol{L}^0(\Sigma, \mu)$. Now $VU: \boldsymbol{L}^0(\Sigma, \mu) \to \boldsymbol{L}^0(\Sigma, \mu)$ is an order-continuous increasing linear map extending $(\pi^{-1})^\wedge\hat{\pi}: \boldsymbol{L}^\infty(\mathfrak{A}) \to \boldsymbol{L}^\infty(\mathfrak{B})$. By 45F, $(\pi^{-1})^\wedge\hat{\pi} = (\pi^{-1}\pi)^\wedge$, which is of course the identity on $\boldsymbol{L}^\infty(\mathfrak{A})$. Since $\boldsymbol{L}^\infty(\mathfrak{A})$ is order-dense in $\boldsymbol{L}^0(\Sigma, \mu)$, VU is the identity on $\boldsymbol{L}^0(\Sigma, \mu)$. Similarly, UV is the identity on $\boldsymbol{L}^0(\mathrm{T}, \nu)$. Thus U is indeed an isomorphism.

62M Exercises (a) If (X, Σ, μ) is a measure space and $\langle x_n \rangle_{n \in \mathbf{N}}$ is a sequence in $\boldsymbol{M}(\Sigma)$ such that $\langle x_n(t) \rangle_{n \in \mathbf{N}} \to x(t)$ for every $t \in X$, then $x: X \to \mathbf{R}$ is measurable.

(b) (i) Let $\mathfrak{A}$ be any Boolean ring, and Z its Stone space. Then $\boldsymbol{S}(\mathfrak{A})$, regarded as a linear subspace of $\mathbf{R}^Z$, is closed under the multiplication in $\mathbf{R}^Z$. Consequently $\boldsymbol{L}^\infty(\mathfrak{A})$ is also closed under multiplication. (ii) Let (X, Σ, μ) be a measure space and $\mathfrak{A}$ its measure algebra. Show that $\boldsymbol{L}^\infty(\mathfrak{A})$ is closed under the multiplication in $\boldsymbol{L}^0(\Sigma, \mu)$, and that this is the unique norm-continuous bilinear multiplication on $\boldsymbol{L}^\infty(\mathfrak{A})$ such that $\chi a \times \chi b = \chi(a \cap b)$ for all $a, b \in \mathfrak{A}$. Consequently it agrees with the multiplication on $\boldsymbol{L}^\infty(\mathfrak{A})$ defined in part (i) above.

(c) Let (X, Σ, μ) be a measure space. (i) Show that $|u \times v| = |u| \times |v|$ for all $u, v \in \boldsymbol{L}^0(\Sigma, \mu)$. (ii) Show that a set $A \subseteq \boldsymbol{L}^0$ is solid iff $u \times v \in A$ whenever $u \in A$ and $|v| \leqslant e$. (iii) Show that $u \times v = 0$ iff $|u| \wedge |v| = 0$. *(iv) Show that an element u of $\boldsymbol{L}^0$ has a multiplicative inverse iff it is a weak order unit.

***(d)** Show that in 62L the Riesz homomorphism

$$U: L^0(\Sigma, \mu) \to L^0(\mathrm{T}, \nu)$$

is multiplicative.

***(e)** Let (X, Σ, μ) be a measure space and E an Archimedean Riesz space. Let F be a super-order-dense Riesz subspace of E and

$$T: F \to L^0(\Sigma, \mu)$$

a sequentially order-continuous Riesz homomorphism. Then T has a unique extension to a sequentially order-continuous Riesz homomorphism from E to L^0. [Use 17Gc.]

***(f)** Let (X, Σ, μ) and (Y, T, ν) be measure spaces, with measure algebras $\mathfrak{A}$ and $\mathfrak{B}$ respectively. Let $\pi: \mathfrak{A} \to \mathfrak{B}$ be a sequentially order-continuous ring homomorphism. Then $\hat{\pi}: S(\mathfrak{A}) \to S(\mathfrak{B})$ is sequentially order-continuous [hint: use the technique of 42J/42O], so extends uniquely to a multiplicative sequentially order-continuous Riesz homomorphism from $L^0(\Sigma, \mu)$ to $L^0(\mathrm{T}, \nu)$. [Use (e) above.]

(g) Let (X, Σ, μ) be a measure space. Suppose that $\langle u_n \rangle_{n \in \mathbf{N}} \downarrow 0$ in $L^0(\Sigma, \mu)$. Show that for any $\alpha > 0$ in $\mathbf{R}$, $\{n(u_n - \alpha u_0)^+ : n \in \mathbf{N}\}$ is bounded above in L^0.

(h) Let (X, Σ, μ) be a measure space. Let E be any Archimedean Riesz space and $T: L^0(\Sigma, \mu) \to E$ an increasing linear map. Then T is sequentially order-continuous. [Use (g) above.]

Notes and comments The most important result above is the representation of the L^∞ space of the measure algebra of a measure space as a solid linear subspace of the L^0 space. Of course it is this solid linear subspace that is normally thought of as the L^∞ space of a measure space; so from the ordinary point of view it is this result which justifies the emphasis I have laid on the construction $L^\infty(\mathfrak{A})$, and also its name.

If we look closer at the construction of L^0, se see that it does not really depend on the measure. The original space $M(\Sigma)$, of course, is defined by the pair (X, Σ). Now the solid linear subspace M_0 in 62F depends only on the σ-ideal Σ_0 of Σ, not on anything else about μ. Thus it is not surprising that $L^0 = M/M_0$ is determined by the algebra $\mathfrak{A} = \Sigma/\Sigma_0$, without further reference to the measure [62L].

The same result is the reason for the notation I use. Since L^0 is determined by the algebra Σ and the function $\mu: \Sigma \to [0, \infty]$, regardless of how Σ is embedded in $\mathscr{P}X$, I write $L^0 = L^0(\Sigma, \mu)$. Similarly, M is determined, up to a multiplicative Riesz space isomorphism, by the algebra Σ. For it is easy to prove 62J and 62K with M replacing L^0, and therefore to prove an analogue of 62L showing that if Σ and T are σ-algebras of sets which are isomorphic as Boolean algebras, then $M(\Sigma) \cong M(\mathrm{T})$.

Indeed, every space $M(\Sigma)$ can be regarded as an L^0. For suppose that X is a set and Σ a σ-subalgebra of $\mathscr{P}X$. Define $\mu: \Sigma \to [0, \infty]$ by

$$\begin{aligned} \mu E &= n \quad \text{if } E \text{ has } n \text{ members, where } n \in \mathbf{N}; \\ &= \infty \quad \text{otherwise.} \end{aligned}$$

Then (X, Σ, μ) is a measure space. Since $\Sigma_0 = \{\varnothing\}$, $\Sigma/\Sigma_0 \cong \Sigma$ and $L^0(\Sigma, \mu) \cong M(\Sigma)$. Thus it is unsurprising that most of the results of this section [62G, 62I, 62J] consist in showing that L^0 spaces have properties which are a matter of course for $M(\Sigma)$ spaces.

It is clear that the theorem 62K relies only on the lemma 62J and the Dedekind σ-completeness of L^0. A more powerful result along the same lines is given in 64C/D.

*Note that we have seen in 5XAb that any Dedekind σ-complete Boolean algebra $\mathfrak{A}$ can be given a measure, so by 61I it is isomorphic to the measure algebra of a measure space. This gives us a construction of $L^0(\mathfrak{A})$, which by 62L is well-defined up to isomorphism. The functorial nature of the construction L^0 is given by 62Mf.

63 Integration

The theory of integration on an arbitrary measure space (X, Σ, μ) can be developed in many ways. All the customary approaches, however, proceed by defining a space $\mathfrak{L}^1(\mu)$ of 'integrable' real-valued functions, together with an 'integral', a sequentially order-continuous increasing linear functional $\int$ which extends the functional $\hat{\mu}: S(\Sigma^f) \to \mathbf{R}$. On $\mathfrak{L}^1$ a semi-norm $\| \;\|_1$ is defined by writing $\|x\|_1 = \int |x|$ for every $x \in \mathfrak{L}^1$, and $\mathfrak{L}^1$ is found to be complete, in the sense that every Cauchy sequence has at least one limit. It can then be pointed out that

$$H = \{x : x \in \mathfrak{L}^1, \|x\|_1 = 0\}$$

is a linear subspace of $\mathfrak{L}^1$, so that the quotient $\mathfrak{L}^1/H$ is a Banach space, in fact an L-space.

In most systems, all integrable functions are measurable; so, in the language of §62 above, $\mathfrak{L}^1 \subseteq M(\Sigma)$ and $H = \mathfrak{L}^1 \cap M_0$. So the quotient $\mathfrak{L}^1/H$ can be identified with the image of $\mathfrak{L}^1$ in $L^0(\Sigma, \mu)$. It is usually easy to show that $S(\Sigma^f)$ is $\| \;\|_1$-dense in $\mathfrak{L}^1$, so that its image (identified in 62H with $S(\mathfrak{A}^f)$) becomes dense in $\mathfrak{L}^1/H$; it follows at once that $\mathfrak{L}^1/H$ is isomorphic to $L^1(\mathfrak{A})$ as defined in §52, where $\mathfrak{A}$ is, as usual, the measure algebra of (X, Σ, μ).

In this book, however, I propose to use this work to demonstrate the power of the general theorems I have introduced so far. We know that $\boldsymbol{S}(\mathfrak{A}^f)$ is a super-order-dense Riesz subspace of $\boldsymbol{L}^1(\mathfrak{A})$ [52E]; it is easy to see that the natural embedding of $\boldsymbol{S}(\mathfrak{A}^f)$ in $\boldsymbol{L}^0$ [62H] is an order-continuous Riesz homomorphism. So by 62K it extends to an embedding of $\boldsymbol{L}^1$ in $\boldsymbol{L}^0$. I define $\mathfrak{L}^1$ in reverse as

$$\{x : x \in \boldsymbol{M}(\Sigma),\ x^{\cdot} \in \boldsymbol{L}^1\};$$

the characterization of members of $\mathfrak{L}^1$ requires care, but proceeds satisfactorily [63D]. We can accordingly regard the integral on $\mathfrak{L}^1$ as the linear functional defined by the integral on the L-space $\boldsymbol{L}^1$, i.e.

$$\int x = \int x^{\cdot} = \|x^{\cdot}\|_1 \quad \text{whenever} \quad x \geqslant 0 \text{ in } \mathfrak{L}^1.$$

The famous convergence theorems are now consequences of the fact that $\boldsymbol{L}^1$ is an L-space [26Hj–l, 63Ma–d].

I prove without difficulty that $\boldsymbol{L}^1$ is solid in $\boldsymbol{L}^0$ [63E]; it follows that $u \times v \in \boldsymbol{L}^1$ whenever $u \in \boldsymbol{L}^1$ and $v \in \boldsymbol{L}^\infty$. Now, writing $\langle u, v\rangle = \int u \times v$, we have a duality between $\boldsymbol{L}^1$ and $\boldsymbol{L}^\infty$; this is of course the usual duality taken between these spaces; we have to check that it agrees with the identification of $\boldsymbol{L}^1$ with $\boldsymbol{L}^{\infty\times}$ in 52E. On the basis of this, I can use previous results to establish a form of the Radon–Nikodým theorem [63J].

63A Theorem Let (X, Σ, μ) be any measure space, with measure algebra $(\mathfrak{A}, \mu)$. The embedding of $\boldsymbol{S}(\mathfrak{A}^f)$ as a Riesz subspace of $\boldsymbol{L}^0(\Sigma, \mu)$, described in 62H, extends uniquely to an order-continuous embedding of $\boldsymbol{L}^1(\mathfrak{A})$ as a Riesz subspace of $\boldsymbol{L}^0(\Sigma, \mu)$.

Proofs **(a)** It will help if we regard the embedding of $\boldsymbol{S}(\mathfrak{A}^f)$ in $\boldsymbol{L}^0$ as induced by a one-to-one Riesz homomorphism $T: \boldsymbol{S}(\mathfrak{A}^f) \to \boldsymbol{L}^0$. Since $\boldsymbol{L}^\infty(\mathfrak{A})$ is solid in $\boldsymbol{L}^0$ [62H], $\boldsymbol{L}^\infty(\mathfrak{A}^f)$ is solid in $\boldsymbol{L}^\infty(\mathfrak{A})$ [45Db], and $\boldsymbol{S}(\mathfrak{A}^f)$ is order-dense in $\boldsymbol{L}^\infty(\mathfrak{A}^f)$ [43Bc], the embeddings

$$\boldsymbol{S}(\mathfrak{A}^f) \subseteq \boldsymbol{L}^\infty(\mathfrak{A}^f) \subseteq \boldsymbol{L}^\infty(\mathfrak{A}) \subseteq \boldsymbol{L}^0$$

are all order-continuous [17E], so $T: \boldsymbol{S}(\mathfrak{A}^f) \to \boldsymbol{L}^0$ is order-continuous.

(b) Now 62K shows at once that T has a unique order-continuous extension $U: \boldsymbol{L}^1(\mathfrak{A}) \to \boldsymbol{L}^0$, because by 52E $\boldsymbol{S}(\mathfrak{A}^f)$ is super-order-dense in $\boldsymbol{L}^1$. We see also from 17Cb that $U: \boldsymbol{L}^1 \to \boldsymbol{L}^0$ is a one-to-one Riesz homomorphism. So we can think of U as an actual embedding of $\boldsymbol{L}^1$ as a Riesz subspace of $\boldsymbol{L}^0$.

63B Integration of real-valued functions: the space $\mathfrak{L}^1$ We have now found a canonical representation of L^1 as a Riesz subspace of L^0. Let us write $\mathfrak{L}^1$, or $\mathfrak{L}^1(\mu)$, for

$$\{x : x \in M(\Sigma),\ x^{\cdot} \in L^1\}.$$

Then $\mathfrak{L}^1$ is a Riesz subspace of $M(\Sigma)$. Since L^1 is an L-space, it has a natural integral $\int$ given by

$$\int u = \|u^+\|_1 - \|u^-\|_1 \quad \forall\ u \in L^1$$

[see 26C]. This induces an increasing linear functional on $\mathfrak{L}^1$, writing

$$\int x = \int x^{\cdot}.$$

Members of $\mathfrak{L}^1$ are the **integrable** real-valued functions, and $\int x$ or $\int x d\mu$ or $\int x(t)\,\mu(dt)$ is the **integral** of x with respect to μ.

63C In order to describe $\mathfrak{L}^1$ more adequately, we must first look again at the embedding of $S(\mathfrak{A}^f)$ in L^0.

Lemma Let (X, Σ, μ) be a measure space and $\mathfrak{A}$ its measure algebra.

(a) $S(\Sigma^f) \subseteq \mathfrak{L}^1(\mu)$ and $\int x d\mu = \hat{\mu}(x)$ for every $x \in S(\Sigma^f)$, where $\mathfrak{L}^1$ and $\int$ are defined in 63B, and $\hat{\mu}$: $S(\Sigma^f) \to \mathbf{R}$ is the linear functional associated with the additive functional μ on Σ^f.

(b) If $u \in S(\mathfrak{A}^f)$, then there is an $x \in S(\Sigma^f)$ such that $x^{\cdot} = u$.

Proof (a) For suppose that $E \in \Sigma^f$. Then $E^{\cdot} \in \mathfrak{A}^f$ and $\chi(E^{\cdot})$, in $S(\mathfrak{A}^f)$, is identified with $(\chi E)^{\cdot}$ in L^0 [62H]. So $\chi E \in \mathfrak{L}^1$ and

$$\int \chi E = \|(\chi E)^{\cdot}\|_1 = \|\chi(E^{\cdot})\|_1 = \mu E^{\cdot} = \mu E = \hat{\mu}(\chi E).$$

Now the result follows because $S(\Sigma^f)$ is the linear space generated by $\{\chi E : E \in \Sigma^f\}$ and $\int$ and $\hat{\mu}$ are both linear functionals.

(**b**) follows directly from 45Cb, as $\mathfrak{A}^f = \{E^{\cdot} : E \in \Sigma^f\}$.

63D Proposition Let (X, Σ, μ) be a measure space, and let $x \in M(\Sigma)$. Then $x \in \mathfrak{L}^1(\mu)$ iff

(i) $\alpha = \sup\{\hat{\mu}y : y \in S(\Sigma^f),\ 0 \leqslant y \leqslant |x|\} < \infty$;

(ii) $\{t : t \in X,\ x(t) \neq 0\}$ has no purely infinite measurable subset [for definition, see 61Fc]. In this case, $\alpha = \int |x|\, d\mu = \|x^{\cdot}\|_1$.

Proof (a) Suppose that $x \in \mathfrak{L}^1$, i.e. that $x^{\cdot} \in L^1$, embedded in L^0 by 63A. Then $\int |x| = \int |x^{\cdot}| = \|x^{\cdot}\|_1$. If $y \in S(\Sigma^f)$ and $0 \leqslant y \leqslant |x|$, then $0 \leqslant y^{\cdot} \leqslant |x^{\cdot}|$ in L^0 and

$$\hat{\mu}y = \int y = \int y^{\cdot} = \|y^{\cdot}\|_1 \leqslant \|x^{\cdot}\|_1,$$

using 63Ca. So $\alpha \leqslant \|x^{\cdot}\|_1 < \infty$. On the other hand, let $F \in \Sigma$ be any purely infinite set. Then $\mu(F \cap E) = 0$ for every $E \in \Sigma^f$, because $\mu(F \cap E)$ must be either 0 or ∞. So $(\chi F)^{\cdot} \wedge (\chi E)^{\cdot} = \chi(E \cap F)^{\cdot} = 0$ for every $E \in \Sigma^f$. It follows that $(\chi F)^{\cdot} \wedge |u| = 0$ for every $u \in S(\mathfrak{A}^f)$ [using 63Cb]. Now $S(\mathfrak{A}^f)$ is order-dense in L^1, so

$$A = \{u : u \in S(\mathfrak{A}^f), \quad 0 \leqslant u \leqslant |x^{\cdot}|\} \uparrow |x^{\cdot}|$$

in L^1; as the embedding of L^1 in L^0 is order-continuous, $A \uparrow |x^{\cdot}|$ in L^0; so by the distributive law [14D],

$$(\chi F \wedge |x|)^{\cdot} = \chi F^{\cdot} \wedge |x^{\cdot}| = \sup_{u \in A} \chi F^{\cdot} \wedge u = 0.$$

Thus $$\mu\{t : t \in F, \ |x(t)| > 0\} = 0,$$

and F cannot be a subset of $\{t : |x(t)| > 0\}$. Thus x satisfies both conditions.

(b) Conversely, suppose that $x \in M(\Sigma)^+$ satisfies the conditions. Let A be the set

$$\{y^{\cdot} : y \in S(\Sigma^f), \ 0 \leqslant y \leqslant x\}.$$

Then $A \uparrow$ and $\|u\|_1 \leqslant \alpha$ for every $u \in A$ [using 63Ca again], so $v = \sup A$ exists in L^1 [because the L-space L^1 is Levi and Dedekind complete, by 26B]. Again, the embedding of L^1 in L^0 being order-continuous, $v = \sup A$ in L^0. Since $u \leqslant x^{\cdot}$ for every $u \in A$, $v \leqslant x^{\cdot}$ in L^0. Let $z \in M(\Sigma)^+$ be such that $z^{\cdot} = v$.

Now consider $$F = \{t : z(t) < x(t)\} \in \Sigma.$$

? Suppose, if possible, that $\mu F > 0$. For $\gamma \in \mathbf{Q}$, set

$$F_\gamma = \{t : z(t) < \gamma \leqslant x(t)\};$$

since $F = \bigcup_{\gamma \in \mathbf{Q}} F_\gamma$, there is a $\gamma \in \mathbf{Q}$ such that $\mu F_\gamma > 0$. Now $\gamma > 0$ as $z \geqslant 0$, so $F_\gamma \subseteq \{t : x(t) \neq 0\}$. By the condition (ii), F_γ is not purely infinite, and there is a measurable set $H \subseteq F_\gamma$ such that $0 < \mu H < \infty$. Let $y = \gamma \chi H$. Then $y \in S(\Sigma^f)$ and $y \leqslant x$, so $y^{\cdot} \in A$. Thus

$$y^{\cdot} \leqslant \sup A = z^{\cdot}.$$

But $$\mu\{t : z(t) < y(t)\} = \mu H > 0,$$

which is impossible. **X**

So we see that $\mu F = 0$ and $x^{\cdot} \leqslant z^{\cdot}$; thus $x^{\cdot} = z^{\cdot} \in L^1$ and $x \in \mathfrak{L}^1$. Also, because the norm on the L-space L^1 is Fatou,

$$\|x^{\cdot}\|_1 = \|z^{\cdot}\|_1 = \sup\{\|u\|_1 : u \in A\}$$
$$= \sup\{\hat{\mu} y : y \in S(\Sigma^f), \ 0 \leqslant y \leqslant x\} = \alpha.$$

Thus the result is true for $x \geqslant 0$. In general, it is clear that $|x|$ and

x^+ satisfy the conditions if x does, and now $x = 2x^+ - |x| \in \mathfrak{L}^1$, while $\alpha = \| |x|^{\cdot} \|_1 = \| x^{\cdot} \|_1$.

63E Corollaries (a) $\mathfrak{L}^1$ is solid in $\boldsymbol{M}(\Sigma)$ and $\boldsymbol{L}^1$ is solid in $\boldsymbol{L}^0$.

(b) For any $x \geqslant 0$ in $\mathfrak{L}^1$,

$$\int x = \sup \{ \int y : y \in \boldsymbol{S}(\Sigma^f), 0 \leqslant y \leqslant x \}.$$

Proof (a) It is obvious from the criterion in 63D that $\mathfrak{L}^1$ is solid in $\boldsymbol{M}$, and it follows that $\boldsymbol{L}^1$ is solid in $\boldsymbol{L}^0$. (Or use 17Gd.) (**b**) This is just a restatement of the last sentence in 63D, identifying $\int y$ with $\hat{\mu} y$ by 63Ca.

63F Lemma If (X, Σ, μ) is a semi-finite measure space, then $\boldsymbol{L}^1$ is order-dense in $\boldsymbol{L}^0(\Sigma, \mu)$.

Proof For, if $\mathfrak{A}$ is the measure algebra of (X, Σ, μ) then $\boldsymbol{L}^\infty(\mathfrak{A})$ is order-dense in $\boldsymbol{L}^0$ [62H], and $\boldsymbol{S}(\mathfrak{A})$ is order-dense in $\boldsymbol{L}^\infty(\mathfrak{A})$ [43Bc]. But, if (X, Σ, μ) is semi-finite, then $\boldsymbol{S}(\mathfrak{A}^f)$ is order-dense in $\boldsymbol{S}(\mathfrak{A})$ [51Gd]. So [using 15E, or otherwise], $\boldsymbol{S}(\mathfrak{A}^f)$ is order-dense in $\boldsymbol{L}^0$ and $\boldsymbol{L}^1$ is order-dense in $\boldsymbol{L}^0$.

63G The duality between $\boldsymbol{L}^1$ and $\boldsymbol{L}^\infty$ Let (X, Σ, μ) be a semi-finite measure space, with measure algebra $\mathfrak{A}$. In 62H and 63A we have identified both $\boldsymbol{L}^\infty(\mathfrak{A})$ and $\boldsymbol{L}^1(\mathfrak{A})$ with subspaces of $\boldsymbol{L}^0(\Sigma, \mu)$. If $u \in \boldsymbol{L}^1$ and $v \in \boldsymbol{L}^\infty$, then $|v| \leqslant \|v\|_\infty e$, where $e = \chi X^{\cdot}$ [see 62H]. So

$$|u \times v| = |u| \times |v| \leqslant |u| \times \|v\|_\infty e = \|v\|_\infty |u|.$$

As $\boldsymbol{L}^1$ is solid [63E], $u \times v \in \boldsymbol{L}^1$ and

$$|\int u \times v| \leqslant \|u \times v\|_1 \leqslant \|v\|_\infty \|u\|_1.$$

Now we already have a duality between $\boldsymbol{L}^1$ and $\boldsymbol{L}^\infty$ given by the identification of $\boldsymbol{L}^1$ with $\boldsymbol{L}^{\infty\times}$ in 52E. This duality is continuous for the norms of $\boldsymbol{L}^1$ and $\boldsymbol{L}^\infty$ and is defined by saying that

$$\langle \chi a, \chi b \rangle = \mu(a \cap b) \quad \forall \; a \in \mathfrak{A}^f, b \in \mathfrak{A}.$$

But, given $a \in \mathfrak{A}^f$ and $b \in \mathfrak{A}$,

$$\chi a \times \chi b = \chi(a \cap b) \quad \text{in} \quad \boldsymbol{L}^0,$$

because $\chi E \times \chi F = \chi(E \cap F)$ in $\boldsymbol{M}(\Sigma)$ for all $E, F \in \Sigma$. So

$$\langle \chi a, \chi b \rangle = \int \chi a \times \chi b.$$

But now it follows that $\langle u, v\rangle = \int u \times v$ for every $u \in \boldsymbol{S}(\mathfrak{A}^f)$ and $v \in \boldsymbol{S}(\mathfrak{A})$, and therefore for every $u \in \boldsymbol{L}^1$ and $v \in \boldsymbol{L}^\infty$, by continuity.

Thus the duality between $\boldsymbol{L}^1$ and $\boldsymbol{L}^\infty$, under which $\boldsymbol{L}^1$ is a representation of $\boldsymbol{L}^{\infty\times}$, is given by

$$\langle u, v\rangle = \int u \times v \quad \forall\ u \in \boldsymbol{L}^1, v \in \boldsymbol{L}^\infty,$$

and this is the way in which I shall generally regard it from now on.

***63H** For any measure space (X, Σ, μ), we can identify $\boldsymbol{L}^1(\mathfrak{A})$ with $\boldsymbol{L}^\infty(\mathfrak{A}^f)^\times$ [52F]. Exactly the same arguments show that the same formula $\langle u, v\rangle = \int u \times v$ gives this duality also.

63I Notation Let (X, Σ, μ) be a measure space. Suppose that $x \in \mathfrak{L}^1(\mu)$ and that $E \in \Sigma$. Then $|x \times \chi E| \leqslant |x|$ so $x \times \chi E \in \mathfrak{L}^1$. We write

$$\int_E x = \int x \times \chi E = \langle x^\cdot, \chi E^\cdot\rangle,$$

regarding $\chi E^\cdot$ as a member of $\boldsymbol{L}^\infty(\mathfrak{A})$.

63J The Radon–Nikodým theorem I can now bring this famous theorem into the light of day.

First form Let $(\mathfrak{A}, \mu)$ be a measure ring. Let $\nu: \mathfrak{A}^f \to \mathbf{R}$ be bounded and completely additive. Then there is a unique $u \in \boldsymbol{L}^1(\mathfrak{A})$ such that $\nu a = \langle u, \chi a\rangle$ for every $a \in \mathfrak{A}^f$.

Second form Let (X, Σ, μ) be a measure space. Let $\nu: \Sigma^f \to \mathbf{R}$ be bounded, countably additive, and such that $\nu E = 0$ whenever $\mu E = 0$. Then there is an integrable function $x: X \to \mathbf{R}$ such that $\nu E = \int_E x$ for every $E \in \Sigma^f$.

Proof The 'first form' is just 44Bb (which finds u in $\boldsymbol{L}^\#(\mathfrak{A}^f)$) and 52E/F (which is the isomorphism between $\boldsymbol{L}^1(\mathfrak{A})$ and $\boldsymbol{L}^\#(\mathfrak{A}^f)$). The 'second form' follows. You see at once that the condition

$$\text{‘}\mu E = 0 \Rightarrow \nu E = 0\text{’}$$

is simply to ensure that ν descends to a function $\nu^\cdot: \mathfrak{A}^f \to \mathbf{R}$ given by $\nu^\cdot(E^\cdot) = \nu E$ for every $E \in \Sigma^f$. (As usual, $\mathfrak{A}$ is the measure algebra of (X, Σ, μ).) Then $\nu^\cdot$ is bounded and additive; in fact it is completely additive. **P** By 51D, it is enough to show that $\nu^\cdot$ is countably additive.

Suppose that $\langle a_n \rangle_{n\in\mathbf{N}} \downarrow 0$ in $\mathfrak{A}^f$. For each $n \in \mathbf{N}$, choose $E_n \in \Sigma^f$ such that $E_n^\cdot = a_n$. Set $E = \bigcap_{n\in\mathbf{N}} E_n$; then $E^\cdot = 0$, so

$$0 = \nu(\bigcap_{n\in\mathbf{N}} E_n) = \lim_{n\to\infty} \nu(\bigcap_{i\leqslant n} E_i) = \lim_{n\to\infty} \nu^\cdot(\inf_{i\leqslant n} a_i)$$
$$= \lim_{n\to\infty} \nu^\cdot a_n,$$

which is what we need. **Q**

So there is a $u \in \mathbf{L}^1(\mathfrak{A})$ such that $\langle u, \chi E^\cdot \rangle = \nu^\cdot E^\cdot = \nu E$ for every $E \in \Sigma^f$. Now let $x \in \mathfrak{L}^1(\mu)$ be such that $x^\cdot = u$. Then

$$\int_E x = \int x \times \chi E = \int u \times \chi E^\cdot = \langle u, \chi E^\cdot \rangle = \nu E$$

for every $E \in \Sigma^f$, using the notation of 63I and the identification of the duality in 63G/H.

***63K The topology on L^0** For any measure space (X, Σ, μ), there is a 'natural' topology on $\mathbf{L}^0(\Sigma, \mu)$, which may be defined as follows. For each $u \geqslant 0$ in $\mathbf{L}^1$, define

$$\rho_u(v) = \int u \wedge |v| \quad \forall \ v \in \mathbf{L}^0.$$

Because $\mathbf{L}^1$ is solid in $\mathbf{L}^0$, $\rho_u : \mathbf{L}^0 \to \mathbf{R}^+$ is well-defined. Now it is a Fatou pseudo-norm. **P** (i) Of course, if $|v| \leqslant |w|$, then $\rho_u(v) \leqslant \rho_u(w)$. (ii) If v and w belong to $\mathbf{L}^0$,

$$u \wedge |v+w| \leqslant u \wedge (|v|+|w|) \leqslant u \wedge |v| + u \wedge |w|$$

[14Bn, 14Ja]. So $\rho_u(v+w) \leqslant \rho_u(v) + \rho_u(w)$. (iii) If $\varnothing \subset A \downarrow 0$ in $\mathbf{L}^0$, then $\{u \wedge v : v \in A\} \downarrow 0$ in $\mathbf{L}^1$, so $\inf_{v\in A} \rho_u(v) = 0$ (because $\int : \mathbf{L}^1 \to \mathbf{R}$ is order-continuous). (iv) Consequently, if $\varnothing \subset A \uparrow w$ in $(\mathbf{L}^0)^+$,

$$\inf_{v\in A} \rho_u(w-v) = 0,$$

so $\rho_u(w) = \sup_{v\in A} \rho_u(v)$. (v) Also, if $v \in \mathbf{L}^0$,

$$\{\alpha |v| : \alpha > 0\} \downarrow 0 \quad \text{in} \quad \mathbf{L}^0,$$

so $\inf_{\alpha>0} \rho_u(\alpha v) = \inf_{\alpha>0} \rho_u(\alpha |v|) = 0$. **Q**

Since, given $u, v \in (\mathbf{L}^1)^+$,

$$\rho_{u+v} \geqslant \max(\rho_u, \rho_v),$$

the system $\{\rho_u : u \in (\mathbf{L}^1)^+\}$ defines a Fatou topology $\mathfrak{T}$ on $\mathbf{L}^0$ which by (iii) above is also Lebesgue.

When $\mu X < \infty$, $\mathfrak{T}$ (or the topology it induces on $\mathbf{M}(\Sigma)$) is commonly called the topology of **convergence in measure**; see 63Mk. I regret that space does not allow me to discuss at length its remarkable properties, but some of them are in 63L, 63M j–k, 64E, 64Jc(v) and 64Je. For two examples see 6XAa and 6XBc.

***63L Lemma** If (X, Σ, μ) is semi-finite, the topology $\mathfrak{T}$ described above is Hausdorff.

Proof Let v be a non-zero member of L^0 and let $y \in M(\Sigma)$ be such that $y^{\cdot} = v$. Then $\mu\{t : y(t) \neq 0\} \neq 0$, so there is an $n \in \mathrm{N}$ such that $\mu\{t : |y(t)| \geqslant 2^{-n}\} \neq 0$. Let H be a measurable subset of $\{t : |y(t)| \geqslant 2^{-n}\}$ such that $0 < \mu H < \infty$; such a set exists as (X, Σ, μ) is semi-finite. Set $u = (\chi H)^{\cdot} \in L^1$. Then

$$\rho_u(v) = \int u \wedge |v| = \int \chi H \wedge |y| \geqslant 2^{-n}\mu H > 0.$$

As v is arbitrary, this shows that $\mathfrak{T}$ is Hausdorff. [See also 63Mj.]

63M Exercises (a) (B. Levi's theorem.) Let (X, Σ, μ) be a measure space and $\langle x_n \rangle_{n \in \mathbf{N}}$ a sequence of integrable real-valued functions on X such that $\alpha = \sup_{n \in \mathbf{N}} \int x_n < \infty$ and, for each n, $x_n \leqslant x_{n+1}$ p.p. Then (i) there is an $x \in M(\Sigma)$ such that $x(t) = \sup_{n \in \mathbf{N}} x_n(t)$ p.p. (t), (ii) any such x is integrable, with $\int x = \alpha$. [See 26Hj.]

(b) Let (X, Σ, μ) be a measure space. Let $\langle x_n \rangle_{n \in \mathbf{N}}$ be a sequence of integrable real-valued functions on X such that $\sum_{n \in \mathbf{N}} \int |x_n| < \infty$. Then there is an integrable function x such that $x(t) = \sum_{n \in \mathbf{N}} x_n(t)$ p.p. (t).

(c) (Fatou's lemma.) Let (X, Σ, μ) be a measure space, and $\langle x_n \rangle_{n \in \mathbf{N}}$ a sequence of non-negative integrable real-valued functions on X such that $\sup_{n \in \mathbf{N}} \int x_n < \infty$. Then $\liminf_{n \to \infty} x_n(t) < \infty$ p.p. (t), and if $y: X \to \mathbf{R}$ is defined by

$$y(t) = \liminf_{n \to \infty} x_n(t) \quad \text{when this is finite,} \quad 0 \text{ otherwise,}$$

then y is integrable and $\int y \leqslant \liminf_{n \to \infty} \int x_n$. [See 26Hk.]

(d) (Lebesgue's Dominated Convergence Theorem.) Let (X, Σ, μ) be a measure space and x a non-negative integrable real-valued function on X. Let $\langle x_n \rangle_{n \in \mathbf{N}}$ be a sequence of measurable real-valued functions on X such that (i) $|x_n| \leqslant x$ for every $n \in \mathrm{N}$ (ii) $\langle x_n(t) \rangle_{n \in \mathbf{N}} \to y(t)$ for every $t \in X$. Then $\int y = \lim_{n \to \infty} \int x_n$. [See 26Hl.]

(e) Let (X, Σ, μ) be a measure space and x an integrable real-valued function on X. Let $F \in \Sigma$ be such that $\mu F > 0$ and $x(t) > 0$ for every $t \in F$. Then $\int_F x > 0$.

(f) Let (X, Σ, μ) be a measure space, and $x: X \to \mathbf{R}$ an integrable function. (i) The map $F \mapsto \int_F x: \Sigma \to \mathbf{R}$ is countably additive. (ii) For every $\epsilon > 0$, there is a $y \in S(\Sigma^f)$ such that $\int |x - y| \leqslant \epsilon$. (iii) For every $\epsilon > 0$ there is a set $E \in \Sigma^f$ and a $\delta > 0$ such that $|\int_F x| \leqslant \epsilon$ whenever $F \in \Sigma$ and $\mu(F \cap E) \leqslant \delta$.

(g) Let (X, Σ, μ) be a measure space of finite magnitude. Then $L^\infty(\Sigma) \subseteq \mathfrak{L}^1(\mu)$, and $\int : L^\infty(\Sigma) \to \mathbf{R}$ is continuous for $\| \ \|_\infty$.

(h) Let (X, Σ, μ) be a measure space, and $x: X \to \mathbf{R}$ a non-negative integrable function. (i) For every $\alpha > 0$, $\mu\{t : x(t) > \alpha\} < \infty$. (ii) There is a sequence $\langle x_n \rangle_{n \in \mathbf{N}}$ in $S(\Sigma^f)$ such that $\langle x_n \rangle_{n \in \mathbf{N}} \uparrow x$ in $M(\Sigma)$; i.e. $S(\Sigma^f)$ is super-order-dense in $\mathfrak{L}^1(\mu)$.

*(i) Let (X, Σ, μ) and (Y, T, ν) be measure spaces. Let $f: X \to Y$ be an inverse-measure-preserving function [61E]. (i) If $x: Y \to \mathbf{R}$ is measurable, so is $xf: X \to \mathbf{R}$. In this case, $\int x\, d\nu$ exists iff $\int xf\, d\mu$ exists, and they are then equal. (ii) The map $x \mapsto xf: M(\mathrm{T}) \to M(\Sigma)$ induces a sequentially order-continuous Riesz homomorphism from $L^0(\mathrm{T}, \nu)$ to $L^0(\Sigma, \mu)$, which includes norm-preserving maps from L^1 to L^1 and from L^∞ to L^∞. These maps are precisely the maps $\hat{\pi}$ described in 54D and 45B, where π is the homomorphism between the measure algebras described in 61E. See also 62Mf.

*(j) Let (X, Σ, μ) be a measure space. For $E \in \Sigma^f$ and $\epsilon > 0$, let

$$U(E, \epsilon) = \{x^\cdot : x \in M(\Sigma),\ \mu\{t : t \in E,\ |x(t)| \geq \epsilon\} \leq \epsilon\} \subseteq L^0(\Sigma, \mu).$$

Then $\{U(E, \epsilon) : E \in \Sigma^f, \epsilon > 0\}$ is a base of neighbourhoods of 0 for the topology $\mathfrak{T}$ on L^0 described in 63K. Hence, or otherwise, show that $\mathfrak{T}$ is Hausdorff iff (X, Σ, μ) is semi-finite.

*(k) Let (X, Σ, μ) be a measure space of finite magnitude, and suppose that $\langle x_n \rangle_{n \in \mathbf{N}}$ is a sequence in $M(\Sigma)$. (i) Show that $\langle x_n^\cdot \rangle_{n \in \mathbf{N}} \to 0$ in $L^0(\Sigma, \mu)$ for the topology $\mathfrak{T}$ of 63K iff

$$\langle \mu\{t : |x_n(t)| \geq \epsilon\} \rangle_{n \in \mathbf{N}} \to 0 \quad \forall\ \epsilon > 0.$$

(ii) Let $f: \mathbf{R} \to \mathbf{R}$ be continuous and bounded. Show that there is a function $\phi: L^0 \to \mathbf{R}$ given by

$$\phi(x^\cdot) = \int fx\, d\mu \quad \forall\ x \in M(\Sigma),$$

and that ϕ is continuous for $\mathfrak{T}$.

Notes and comments In Chapters 4 and 5, the relationships $S(\mathfrak{A}^f) \subseteq L^1(\mathfrak{A})$ and $S(\mathfrak{A}^f) \subseteq S(\mathfrak{A}) \subseteq L^\infty(\mathfrak{A})$ were established for any measure ring $\mathfrak{A}$. Now the effect of 62H and 63A is to show that when $\mathfrak{A}$ is the measure algebra of a measure space (X, Σ, μ), then all these spaces can be embedded in $L^0(\Sigma, \mu)$. This is, of course, the ordinary way of looking at L^1 and L^∞. So the work of 63G/H is essential, since we must be sure not only that the abstract spaces of Chapters 4 and 5 are isomorphic to the usual function spaces, but also that they are in the familiar duality.

Most of the results in this section are, like 63G, a matter of checking that we really are in the situation we expected. But we now find our labour beginning to bear fruit, in a quick proof of the Radon–Nikodým theorem [63J]. The 'first form' is just a restatement of earlier results; the 'second form', the conventional Radon–Nikodým theorem, is a straightforward corollary.

It is instructive to take a conventional proof of the Radon–Nikodým theorem [e.g. WILLIAMSON p. 100, theorem 6.3d, or ROYDEN chap. 11, § 5] for contrast with the proof here. The 'Jordan decomposition theorem' is 42N; the 'Hahn decomposition theorem' is 52Ha. The proof proceeds very much on the lines of 52D, reading Σ^f for $\mathfrak{A}$ throughout. It finishes with an argument based essentially on the fact that the space of bounded countably additive functionals on Σ^f is an L-space, just as in 52E. You will see that the fact that Σ is a σ-algebra of sets is vital, for the sake of the Hahn decomposition; whereas no such condition is required in § 52. In fact it is used, not in this part of the argument, but in the embedding of L^1 in L^0 [62K/63A].

*It is clear from 63Mj that the topology of L^0 can be described without reference to the theory of integration; but this seems an unnecessary refinement. It has usually been studied for measure spaces of finite magnitude [see 63Mk], and various notions of convergence have been introduced in other cases. But I think there can be no doubt that the topology of 63K is the right one for ordinary purposes.

64 Maharam measure spaces

Now that we have identified the function spaces L^1 and L^∞, the work of § 53 can be seen in a new light. I shall define a Maharam measure space to be one with Maharam measure algebra, so that by 53B the dual of L^1 is identified with L^∞. We find that, because L^∞ is Dedekind complete, so is L^0, and that this leads to striking consequences [64D, 64E].

The brief discussion in 53C–53E showed that a product of measure algebras of finite magnitude is Maharam; so a direct sum of measure spaces of finite magnitude is Maharam. Such a measure space I will call decomposable. All important Maharam measure spaces are of this kind. A useful sufficient condition for a measure space to be decomposable is in 64I.

64A Definition A measure space is **Maharam** (sometimes called **localizable**) if its measure algebra is Maharam, i.e. semi-finite and Dedekind complete [53A].

64B Theorem If (X, Σ, μ) is a semi-finite measure space, with measure algebra $\mathfrak{A}$, then these are equivalent:

(i) (X, Σ, μ) is Maharam;

(ii) $L^\infty(\mathfrak{A})$ is Dedekind complete;

(iii) $L^0(\Sigma, \mu)$ is Dedekind complete;

(iv) the natural duality between $L^1(\mathfrak{A})$ and $L^\infty(\mathfrak{A})$ represents L^∞ as $(L^1)'$.

Proof In 53B it was shown that (i), (ii), and (iv) are equivalent. (Note that of course $L^\infty(\mathfrak{A})$ has a unit, $e = (\chi X)^\cdot$.) Obviously (iii) $\Rightarrow$ (ii), as L^∞ is a solid linear subspace of L^0 [62H]. Conversely, suppose that L^∞ is Dedekind complete and that $A \subseteq (L^0)^+$ is non-empty and bounded above. Then for each $n \in \mathbf{N}$.

$$v_n = \sup \{u \wedge ne : u \in A\}$$

exists in L^∞, because L^∞ is Dedekind complete; as L^∞ is solid in L^0, $v_n = \sup\{u \wedge ne : u \in A\}$ in L^0. Now $\sup A = \sup_{n \in \mathbf{N}} v_n$ exists in L^0 because L^0 is Dedekind σ-complete [62G]. Thus (ii) $\Rightarrow$ (iii).

64C We can now prove a stronger version of 62K, which will be useful later, as well as being intrinsically remarkable. As before, I begin with a lemma.

Lemma Let (X, Σ, μ) be a Maharam measure space, and let $A \subseteq L^0(\Sigma, \mu)^+$. Then *either* A is bounded above *or* there is a $v > 0$ such that

$$kv = \sup_{u \in A} u \wedge kv \quad \forall\ k \in \mathbf{N}.$$

Proof If A is not bounded above, set

$$u_n = \sup_{u \in A} u \wedge ne \quad \forall\ n \in \mathbf{N},$$

where as usual $e = (\chi X)^\cdot$; these all exist because L^0 is Dedekind complete [64B]. Then $\{u_n : n \in \mathbf{N}\}$ is not bounded above, so by 62J there is a $v > 0$ such that

$$kv = \sup_{n \in \mathbf{N}} u_n \wedge kv \leqslant \sup_{u \in A} u \wedge kv \leqslant kv$$

for every $k \in \mathbf{N}$.

64D Theorem Let (X, Σ, μ) be a Maharam measure space and E an Archimedean Riesz space. Suppose that F is an order-dense Riesz subspace of E and that $T: F \to L^0(\Sigma, \mu)$ is an order-continuous Riesz homomorphism. Then T has a unique extension to an order-continuous Riesz homomorphism from $\boldsymbol{E}$ to $\boldsymbol{L}^0$.

Proof Just as in 62K, I use 17B. Suppose that $x \geqslant 0$ in $\boldsymbol{E}$, and set

$$C = \{z : z \in \boldsymbol{F}, 0 \leqslant z \leqslant x\}.$$

? Suppose, if possible, that $T[C]$ is not bounded above in $\boldsymbol{L}^0$. Then by 64C there is a $v > 0$ in $\boldsymbol{L}^0$ such that

$$kv = \sup \{kv \wedge Tz : z \in C\} \quad \forall \ k \in \mathbf{N}.$$

In particular, there is a $y_0 \in C$ such that $v \wedge Ty_0 > 0$. Now, because $\boldsymbol{E}$ is Archimedean, $y_0 = \sup A$ where

$$A = \{y : y \in \boldsymbol{F}^+, \ \exists \ \alpha > 0, y \leqslant (y_0 - \alpha x)^+\}.$$

As T is an order-continuous Riesz homomorphism, $Ty_0 = \sup T[A]$, and there is a $y \in A$ with $w = v \wedge Ty > 0$. Let $\alpha > 0$ be such that $y \leqslant (y_0 - \alpha x)^+$.

Now, just as in 62K, consider, for $z \in C$ and $k \in \mathbf{N}$,

$$\begin{aligned} z \wedge ky \leqslant x \wedge ky &\leqslant x \wedge k(y_0 - \alpha x)^+ \\ &= x \wedge \alpha k(\alpha^{-1} y_0 - x)^+ \\ &\leqslant \alpha^{-1} y_0 + (x - \alpha^{-1} y_0)^+ \wedge \alpha k(\alpha^{-1} y_0 - x)^+ \\ &= \alpha^{-1} y_0. \end{aligned}$$

So

$$\begin{aligned} kw = kw \wedge kv &= kw \wedge \sup_{z \in C} kv \wedge Tz \\ &= \sup_{z \in C} kw \wedge kv \wedge Tz \leqslant \sup_{z \in C} kTy \wedge Tz \\ &= \sup_{z \in C} T(ky \wedge z) \leqslant \alpha^{-1} Ty_0; \end{aligned}$$

contradicting the fact that $w > 0$. **X**

Since $\boldsymbol{L}^0$ is Dedekind complete [64B], $\sup T[C]$ exists. By 17B, T has a unique extension to an order-continuous increasing linear map from $\boldsymbol{E}$ to $\boldsymbol{L}^0$, which by 17Ca is a Riesz homomorphism.

***64E Theorem** Let (X, Σ, μ) be a semi-finite measure space. Then it is Maharam iff the usual topology $\mathfrak{T}$ on $\boldsymbol{L}^0(\Sigma, \mu)$ [63K] is complete, and in this case $\mathfrak{T}$ is Levi.

Proof By 63L, $\mathfrak{T}$ is Hausdorff. We know it is Lebesgue and locally solid [63K], so if it is complete, $\boldsymbol{L}^0$ must be Dedekind complete [24E], and (X, Σ, μ) is Maharam [64B].

Conversely, if (X, Σ, μ) is Maharam, $\mathfrak{T}$ is Levi. **P** Suppose that $A \subseteq (L^0)^+$ is directed upwards and bounded for $\mathfrak{T}$. **?** If A is not bounded above, let $v > 0$ in L^0 be such that

$$kv = \sup_{u \in A} u \wedge kv \quad \forall\ k \in \mathbb{N}$$

[64C] Then $v = \sup_{u \in A} k^{-1}u \wedge v$ for every $k \geqslant 1$. Let ρ be a continuous Fatou pseudo-norm on L^0 such that $\gamma = \rho v > 0$ [23B or 63K]. Let $k \geqslant 1$ be such that $k^{-1}A \subseteq \{u : \rho u \leqslant \frac{1}{2}\gamma\}$, i.e. $\rho(k^{-1}u) \leqslant \frac{1}{2}\gamma$ for every $u \in A$. Then

$$\gamma = \rho u = \rho(\sup_{u \in A} k^{-1}u \wedge v) = \sup_{u \in A} \rho(k^{-1}u \wedge v)$$
$$\leqslant \sup_{u \in A} \rho(k^{-1}u) \leqslant \tfrac{1}{2}\gamma,$$

using the fact that $\{k^{-1}u \wedge v : u \in A\} \uparrow v$. **X** Thus A is bounded above. As A is arbitrary, $\mathfrak{T}$ is Levi. **Q**

But we know that $\mathfrak{T}$ is Fatou [63K] and that L^0 is Dedekind complete [64B], so by Nakano's theorem [23K], $\mathfrak{T}$ is complete.

64F Proposition (a) A measure space of finite magnitude is Maharam.

(b) A direct sum of Maharam measure spaces is Maharam.

Proof (a) follows directly from 53D, since the measure algebra of any measure space is Dedekind σ-complete [61Dd]. Now (b) follows from 53E since the measure algebra of a direct sum is the product of the measure algebras [61H].

64G Definition (a) The last proposition shows the importance of the following concept. A measure space (X, Σ, μ) is **decomposable** (sometimes called **strictly localizable**) if it is (isomorphic to) a direct sum of measure spaces of finite magnitude.

From the discussion in 61G, it is clear that (X, Σ, μ) is decomposable iff there is a partition $\langle X_\iota \rangle_{\iota \in I}$ of X such that: (i) $X_\iota \in \Sigma^f$ for every $\iota \in I$; (ii) given a set $E \subseteq X$, $E \in \Sigma$ iff $E \cap X_\iota \in \Sigma$ for every $i \in I$; (iii) for every $E \in \Sigma$, $\mu E = \sum_{\iota \in I} \mu(E \cap X_\iota)$.

(b) A special case of the above [see 64Ha] is the following. A measure space (X, Σ, μ) is **σ-finite** or **of countable magnitude** if there is a sequence $\langle E_n \rangle_{n \in \mathbb{N}}$ in Σ^f such that $X = \bigcup_{n \in \mathbb{N}} E_n$.

(c) A weaker property is important to us because of its use in 64I below. A measure space (X, Σ, μ) is **locally determined** if it is semi-finite and, for any set $E \subseteq X$,

$$E \cap F \in \Sigma \quad \forall\ F \in \Sigma^f \Rightarrow E \in \Sigma.$$

64H Proposition (a) A measure space of countable magnitude is decomposable.

(b) A decomposable measure space is Maharam and locally determined.

Proof (a) Let (X, Σ, μ) be a measure space of countable magnitude. Let $\langle E_n \rangle_{n \in \mathbf{N}}$ be a sequence in Σ^f such that $X = \bigcup_{n \in \mathbf{N}} E_n$. Set

$$X_n = E_n \backslash \bigcup_{i<n} E_i$$

for each $n \in \mathbf{N}$. Then $\langle X_n \rangle_{n \in \mathbf{N}}$ is a partition of X, and $X_n \in \Sigma^f$ for every $n \in \mathbf{N}$. If $E \subseteq X$, then

$$E \in \Sigma \Rightarrow E \cap X_n \in \Sigma \quad \forall \ n \in \mathbf{N}$$

$$\Rightarrow \bigcup_{n \in \mathbf{N}} (E \cap X_n) \in \Sigma \Rightarrow E \in \Sigma.$$

Finally, if $E \in \Sigma$,

$$\mu E = \lim_{n \to \infty} \mu(\bigcup_{i \leqslant n} E \cap X_i) = \lim_{n \to \infty} \sum_{i \leqslant n} \mu(E \cap X_i)$$

$$= \sum_{n \in \mathbf{N}} \mu(E \cap X_n).$$

So $\langle X_n \rangle_{n \in \mathbf{N}}$ has the required properties, and (X, Σ, μ) is decomposable.

(b) By 64F, a decomposable measure space is Maharam; it is therefore semi-finite. So it is obviously locally determined.

***64I Proposition** Let (X, Σ, μ) be a locally determined measure space which is complete in the sense of 61Ff. Suppose that there is a disjoint family $\mathscr{A} \subseteq \Sigma$ such that

$$\sup \{H^{\bullet} : H \in \mathscr{A}\} = 1 = X^{\bullet}$$

in the measure algebra $\mathfrak{A}$ of (X, Σ, μ). Then (X, Σ, μ) is decomposable.

Proof (a) Observe first that if $F \in \Sigma^f$, there is a countable set $\mathscr{A}_0 \subseteq \mathscr{A}$ such that $\mu(F \backslash \bigcup \mathscr{A}_0) = 0$. **P** Since $\mathscr{A}$ is disjoint, the set

$$\{H : H \in \mathscr{A}, \mu(H \cap F) \geqslant 2^{-n}\}$$

must be finite for each $n \in \mathbf{N}$. So

$$\mathscr{A}_0 = \{H : H \in \mathscr{A}, \mu(H \cap F) > 0\}$$

is countable. Consequently $\bigcup \mathscr{A}_0$ and $E = F \backslash \bigcup \mathscr{A}_0$ both belong to Σ. Now $\mu(H \cap E) = 0$ for every $H \in \mathscr{A}$, so $E^{\bullet} \cap H^{\bullet} = 0$ in $\mathfrak{A}$ for every $H \in \mathscr{A}$. As $\sup \{H^{\bullet} : H \in \mathscr{A}\} = 1$, $E^{\bullet} = 0$, i.e. $\mu E = 0$, as required. **Q**

(b) Now suppose that $E \subseteq X$ and that $E \cap H \in \Sigma$ for every $H \in \mathscr{A}$.

Then $E \in \Sigma$. **P** Let $F \in \Sigma^f$. By (a) above, there is a countable set $\mathscr{A}_0 \subseteq \mathscr{A}$ such that $\mu(F \backslash \bigcup \mathscr{A}_0) = 0$. Now $E \cap H \cap F \in \Sigma$ for every $H \in \mathscr{A}_0$, so $E \cap F \cap \bigcup \mathscr{A}_0 \in \Sigma$. On the other hand,

$$E \cap F \backslash \bigcup \mathscr{A}_0 \subseteq F \backslash \bigcup \mathscr{A}_0,$$

which is of zero measure; so $E \cap F \backslash \bigcup \mathscr{A}_0 \in \Sigma$ as (X, Σ, μ) is complete. Thus $E \cap F \in \Sigma$. But F was an arbitrary member of Σ^f; as (X, Σ, μ) is locally determined, $E \in \Sigma$. **Q**

(**c**) Moreover, if $E \in \Sigma$, then $\mu E = \sum_{H \in \mathscr{A}} \mu(E \cap H)$. **P** (i) Suppose first that $E \in \Sigma^f$. Then there is a countable $\mathscr{A}_0 \subseteq \mathscr{A}$ such that $\mu(E \backslash \bigcup \mathscr{A}_0) = 0$. Since $\mathscr{A}_0$ is countable,

$$\mu E = \mu(E \cap \bigcup \mathscr{A}_0) \leqslant \sum_{H \in \mathscr{A}_0} \mu(E \cap H) \leqslant \sum_{H \in \mathscr{A}} \mu(E \cap H).$$

(ii) In general, we know that $\mu E = \sup\{\mu F : F \in \Sigma^f, F \subseteq E\}$ because (X, Σ, μ) is semi-finite [61Ja]. So

$$\begin{aligned} \mu E &= \sup\{\mu F : F \in \Sigma^f, F \subseteq E\} \\ &\leqslant \sup\{\textstyle\sum_{H \in \mathscr{A}} \mu(F \cap H) : F \in \Sigma^f, F \subseteq E\} \\ &\leqslant \textstyle\sum_{H \in \mathscr{A}} \mu(E \cap H) \\ &= \sup\{\textstyle\sum_{H \in \mathscr{I}} \mu(E \cap H) : \mathscr{I} \subseteq \mathscr{A}, \mathscr{I} \text{finite}\} \\ &\leqslant \mu E, \end{aligned}$$

because $\mathscr{A}$ is disjoint. **Q**

(**d**) So if we set $H_0 = X \backslash \bigcup \mathscr{A}$, $\mathscr{B} = \mathscr{A} \cup \{H_0\}$ is a suitable partition of X. For by (b) above, $H_0 \in \Sigma$, and by (c) $\mu H_0 = 0$. Now if $E \subseteq X$ and $E \cap H \in \Sigma$ for every $H \in \mathscr{B}$, then $E \in \Sigma$ by (b) and

$$\mu E = \sum_{H \in \mathscr{A}} \mu(E \cap H) = \sum_{H \in \mathscr{B}} \mu(E \cap H)$$

by (c). Thus (X, Σ, μ) is decomposable.

64J Exercises (a) Let (X, Σ, μ) be any measure space. Let I be the σ-ideal of negligible sets in $\mathscr{P}X$ [61Fd]. Show that

$$\begin{aligned} \Sigma' &= \{E \triangle A : E \in \Sigma, A \in I\} \\ &= \{G : \exists\, E, F \in \Sigma, E \subseteq G \subseteq F, \mu(F \backslash E) = 0\} \end{aligned}$$

is a σ-subalgebra of $\mathscr{P}X$, and that there is a unique extension of μ to a measure μ' on Σ'. Show that: (i) (X, Σ', μ') is a complete measure space;

(ii) (X, Σ, μ) and (X, Σ', μ') have isomorphic measure algebras;

(iii) $M(\Sigma') = \{x : x \in \mathbf{R}^X, \ \exists\, y \in M(\Sigma), y = x \text{ p.p.}\}$;

(iv) $L^0(\Sigma,\mu) \cong L^0(\Sigma',\mu')$;

(v) $\mathfrak{L}^1(\mu') = \{x : x \in \mathbf{R}^X, \ \exists \ y \in \mathfrak{L}^1(\mu), y = x \text{ p.p.}\}$;

(vi) (X,Σ',μ') is semi-finite or Maharam iff (X,Σ,μ) is;

(vii) if (X,Σ,μ) is decomposable, so is (X,Σ',μ').

(b) Let (X,Σ,μ) be any measure space. Let

$$\Sigma' = \{E : E \cap F \in \Sigma \quad \forall \ F \in \Sigma^f\}.$$

Define $\mu' : \Sigma' \to [0,\infty]$ by

$$\mu' E = \sup\{\mu(E \cap F) : F \in \Sigma^f\} \quad \forall \ E \in \Sigma'.$$

Show that: (i) (X,Σ',μ') is a locally determined measure space; (ii) if $\mathfrak{A}$ and $\mathfrak{B}$ are the measure algebras of (X,Σ,μ) and (X,Σ',μ') respectively, then $\mathfrak{A}^f \cong \mathfrak{B}^f$; (iii) if (X,Σ,μ) is semi-finite, then μ' is an extension of μ; (iv) if (X,Σ,μ) is Maharam, then $\mathfrak{A} \cong \mathfrak{B}$, so that $L^0(\Sigma,\mu) \cong L^0(\Sigma',\mu')$; (v) if (X,Σ,μ) is complete, so is (X,Σ',μ'). [Hint for (iv): if $E \in \Sigma'$, let $H \in \Sigma$ be such that

$$H^{\cdot} = \sup\{(E \cap F)^{\cdot} : F \in \Sigma^f\} \text{ in } \mathfrak{A}.]$$

*(**c**) Let (X,Σ,μ) be a semi-finite measure space, with measure algebra $\mathfrak{A}$. Show that these are equivalent: (i) (X,Σ,μ) is of countable magnitude; (ii) $\mathfrak{A}$ is of countable magnitude [53Fa]; (iii) $L^\infty(\mathfrak{A})$ has the countable sup property; (iv) $L^0(\Sigma,\mu)$ has the countable sup property; (v) the topology $\mathfrak{T}$ on $L^0(\Sigma,\mu)$ [63K] is metrizable.

(d) Let (X,Σ,μ) be a measure space of countable magnitude, and $\nu : \Sigma \to \mathbf{R}$ a countably additive functional such that $\nu E = 0$ whenever $\mu E = 0$. Then there is an integrable function $x : X \to \mathbf{R}$ such that $\int_E x = \nu E$ for every $E \in \Sigma$.

*(**e**) Let (X,Σ,μ) be a measure space of countable magnitude and $\langle x_n\rangle_{n\in\mathbf{N}}$ a sequence in $M(\Sigma)$. Show that $\langle x_n^{\cdot}\rangle_{n\in\mathbf{N}} \to 0$ for the usual topology on $L^0(\Sigma,\mu)$ [63K] iff every subsequence $\langle y_n\rangle_{n\in\mathbf{N}}$ of $\langle x_n\rangle_{n\in\mathbf{N}}$ has in turn a subsequence $\langle z_n\rangle_{n\in\mathbf{N}}$ such that $\langle z_n(t)\rangle_{n\in\mathbf{N}} \to 0$ p.p. (t).

(f) A direct sum of locally determined measure spaces is locally determined. A direct sum of decomposable measure spaces is decomposable.

*(**g**) Any Maharam algebra is the measure algebra of a decomposable measure space. [Use 53Fc, 61I.]

*(**h**) Let (X,Σ,μ) be a Maharam measure space such that whenever $E \subseteq X$ is such that $E \cap F \in \Sigma$ and $\mu(E \cap F) = 0$ for every $F \in \Sigma^f$, then $E \in \Sigma$. Show that (X,Σ,μ) is locally determined.

Notes and comments This section demonstrates the power of the abstract theory we have been studying. It shows that, for a semi-finite measure space, Dedekind completeness of the measure algebra is equivalent to a series of very important properties of the function spaces [64B].

There is an application of 64D in the next section [65A]. It is clear that 64C/D use only the Dedekind completeness of the measure algebra, not the fact that it is semi-finite. This is significant because we know that any Dedekind complete Boolean algebra can be expressed as the measure algebra of some measure space [see the note at the end of § 62]. In fact 64D is characteristic of Dedekind complete Riesz spaces satisfying 64C; these include the C_∞ spaces of VULIKH [chapter V, § 2] or LUXEMBURG & ZAANEN R.S. [§ 47].

We can see from 64Ja and 64Jb that any semi-finite measure space can be converted into a complete measure space and thence into a complete locally determined space, by adding new measurable sets. In fact some important constructions for measure spaces can be adjusted so as to yield complete locally determined spaces directly [see, for example, 71A].

There do exist Maharam measure spaces which are not decomposable but which are complete and locally determined (and are therefore unaffected by the transformations in 64Ja/b), but these are all thoughly unnatural. This is why 64I is so important; it is clearly a necessary and sufficient condition for a complete locally determined measure space to be decomposable, and it is in practice nearly a necessary condition for a properly constructed measure space to be Maharam.

Measures of countable magnitude, which include, for instance, all the ordinary Lebesgue–Stieltjes measures on finite-dimensional Euclidean spaces, are the simplest non-trivial decomposable measures. They have a number of special properties such as those in 64Jc–e.

65 Banach function spaces

In §§ 62 and 63 we have identified two of the most important subspaces of L^0. A great many other spaces have attracted interest, principally L^2 and the other L^p spaces. In this section, I show how an important class of normed spaces can be discussed in terms of the concepts I have introduced; specific examples are in 6XD–6XH. I begin with a powerful general result which enables us to identify the dual space $E^\times$ for a large proportion of naturally arising Riesz spaces.

65A Theorem Let (X, Σ, μ) be a Maharam measure space. Let E be an order-dense solid linear subspace of $L^0(\Sigma, \mu)$. Let

$$F = \{w : w \in L^0, u \times w \in L^1 \quad \forall \; u \in E\}.$$

Then F is a solid linear subspace of L^0. Define a duality between E and F by writing.

$$\langle u, w \rangle = \int u \times w \quad \forall \; u \in E, \quad w \in F.$$

Then this duality induces a Riesz space isomorphism between F and $E^\times$.

Proof (a) Because L^1 is a linear subspace of L^0 [63A] and $\times$ is bilinear, F is a linear subspace of L^0. Because L^1 is solid [63Ea] and E is a Riesz subspace, F is solid. **P** Suppose that $w \in F$ and that $|v| \leqslant |w|$. Then for any $u \in E$

$$|u \times v| = |u| \times |v| \leqslant |u| \times |w| = |u \times w| \in L^1$$

[using 62Mc(i)]. So $u \times v \in L^1$. As u is arbitrary, $v \in F$. **Q**

(b) Clearly the duality given is bilinear, and therefore induces a linear map $T: F \to E^*$. Now $T[F] \subseteq E^\times$. **P** Suppose first that $w \in F^+$ and that $\varnothing \subset A \downarrow 0$ in E. Then $\{u \times w : u \in A\} \downarrow 0$ in L^1 [62I]. So

$$\inf_{u \in A} (Tw)(u) = \inf_{u \in A} \int u \times w = 0,$$

because $\int$ is order-continuous on L^1. Thus $Tw \in E^\times$. Now, for any $w \in F$, $Tw = Tw^+ - Tw^- \in E^\times$. **Q**

(c) In fact $T: F \to E^\times$ is a Riesz homomorphism. **P** It is clear that T is increasing, as

$$(Tw)(u) = \int u \times w \geqslant 0 \quad \forall \; u, w \geqslant 0.$$

So suppose that $v \wedge w = 0$ in F, and that $u \in E^+$. Let $x, y, z \in M(\Sigma)^+$ be such that $x^\cdot = u$, $y^\cdot = v$ and $z^\cdot = w$. Examine

$$x_1 = \sup_{n \in \mathbf{N}} x \wedge n(y-z)^+$$

in $M(\Sigma)$. Clearly $x_1(t) = x(t)$ if $y(t) > z(t)$, and $x_1(t) = 0$ otherwise. So $x_1 \times (y-z)^+ = x \times (y-z)^+$, while $x_1 \times (z-y)^+ = 0$. Now set $u_1 = x_1^\cdot$ in L^0. Then $0 \leqslant u_1 \leqslant u$ and

$$\begin{aligned} u_1 \times v = u_1 \times (v-w)^+ &= [x_1 \times (y-z)^+]^\cdot = [x \times (y-z)^+]^\cdot \\ &= u \times (v-w)^+ = u \times v, \end{aligned}$$

while similarly

$$u_1 \times w = u_1 \times (w-v)^+ = [x_1 \times (z-y)^+]^\cdot = 0.$$

So
$$(Tv \wedge Tw)(u) \leqslant (Tw)(u_1) + (Tv)(u - u_1)$$
$$= \int u_1 \times w + \int (u - u_1) \times v = 0.$$

As u is arbitrary, $Tv \wedge Tw = 0$. But this shows that T is a Riesz homomorphism [14Eb]. **Q**

(**d**) It follows (because $\boldsymbol{E}$ is order-dense in $\boldsymbol{L}^0$) that T is one-to-one. **P** If $w \in \boldsymbol{F}$ and $w \neq 0$, then $|w| > 0$. So there is a $u \in \boldsymbol{E}$ such that $0 < u \leqslant |w|$. Now $u \times |w| > 0$, so

$$|Tw|(u) = T(|w|)(u) = \int u \times |w| > 0,$$

and $Tw \neq 0$. **Q**

(**e**) Next, $\boldsymbol{G} = T[\boldsymbol{F}]$ is order-dense in $\boldsymbol{E}^\times$. **P** Suppose that $f > 0$ in $\boldsymbol{E}^\times$. Let $u_0 > 0$ in $\boldsymbol{E}$ be such that $fu_0 > 0$. Now since $\boldsymbol{L}^\infty$ is order-dense in $\boldsymbol{L}^0$ [62H], we know that

$$A = \{u : u \in \boldsymbol{L}^\infty, 0 \leqslant u \leqslant u_0\} \uparrow u_0.$$

So there is a $u \in A$ such that $fu > 0$. Let $u_1 = \|u\|_\infty^{-1} u$, so that $fu_1 > 0$ and $\|u_1\|_\infty = 1$.

Consider the functional $g: \boldsymbol{L}^\infty \to \mathbf{R}$ given by

$$g(v) = f(u_1 \times v) \quad \forall\ v \in \boldsymbol{L}^\infty.$$

Since the maps $v \mapsto u_1 \times v: \boldsymbol{L}^\infty \to \boldsymbol{E}$ and $f: \boldsymbol{E} \to \mathbf{R}$ are both increasing and order-continuous, $g \in (\boldsymbol{L}^\infty)^\times$. So by 63G there is a $w \in \boldsymbol{L}^1$ with

$$\int w \times v = g(v) = f(u_1 \times v) \quad \forall\ v \in \boldsymbol{L}^\infty;$$

and, setting $v = e$, $\int w = f(u_1) > 0$, so $w > 0$.

Now suppose that $u \in \boldsymbol{E}^+$. Then once again

$$B = \{v : v \in \boldsymbol{L}^\infty, 0 \leqslant v \leqslant u\} \uparrow u,$$

so
$$\{w \times v : v \in B\} \uparrow w \times u.$$

But if $v \in B$,
$$\int w \times v = f(u_1 \times v) \leqslant f(v) \leqslant f(u),$$

since $u_1 \leqslant e = (\chi X)^\bullet$. As $\boldsymbol{L}^1$ is an L-space, $\{w \times v : v \in B\}$ is bounded above in $\boldsymbol{L}^1$, i.e. $w \times u \in \boldsymbol{L}^1$, and

$$\int w \times u = \sup_{v \in B} \int w \times v \leqslant f(u).$$

So for any $u \in \boldsymbol{E}$, $|w \times u| = w \times |u| \in \boldsymbol{L}^1$, and $w \in \boldsymbol{F}$. As $w > 0$, $Tw > 0$ in $\boldsymbol{E}^\times$, by (d) above. Finally, we have seen that, for any $u \in \boldsymbol{E}^+$,

$$(Tw)(u) = \int w \times u \leqslant f(u),$$

so $0 < Tw \leqslant f$. Thus [using 15E] $T[\boldsymbol{F}]$ is order-dense in $\boldsymbol{E}^\times$.

(f) Thus T is a Riesz space isomorphism between $\boldsymbol{F}$ and the order-dense Riesz subspace $\boldsymbol{G}$ of $\boldsymbol{E}^{\times}$. Define $S: \boldsymbol{G} \to \boldsymbol{L}^0$ by $S = T^{-1}$. Now S is a one-to-one Riesz homomorphism, and $S[\boldsymbol{G}] = \boldsymbol{F}$ is solid in $\boldsymbol{L}^0$ [(a) above], so S is order-continuous [17E]. Therefore, S extends to a one-to-one Riesz homomorphism $U: \boldsymbol{E}^{\times} \to \boldsymbol{L}^0$ [64D, using 17Cb].

Now $U[\boldsymbol{E}^{\times}] \subseteq \boldsymbol{F}$. **P** Suppose that $f \geqslant 0$ in $\boldsymbol{E}^{\times}$, and consider

$$\begin{aligned} A &= \{v : v \in \boldsymbol{F}^+, Tv \leqslant f\} \\ &= \{Sg : g \in \boldsymbol{G}, 0 \leqslant g \leqslant f\} \end{aligned}$$

Then, because $\boldsymbol{G}$ is order-dense in $\boldsymbol{E}^{\times}$ and S is order-continuous, $A \uparrow Uf$ in $\boldsymbol{L}^0$. But suppose that $u \in \boldsymbol{E}$. Then $\{|u| \times v : v \in A\} \uparrow |u| \times Uf$. But also

$$\int |u| \times v = (Tv)(|u|) \leqslant f(|u|) \quad \forall \ v \in A.$$

So once again $|u \times Uf| = |u| \times Uf \in \boldsymbol{L}^1$. As u is arbitrary, $Uf \in \boldsymbol{F}$. Now for any $f \in \boldsymbol{E}^{\times}$, $Uf = Uf^+ - Uf^- \in \boldsymbol{F}$. **Q**

(g) But now $T: \boldsymbol{F} \to \boldsymbol{E}^{\times}$ and $U: \boldsymbol{E}^{\times} \to \boldsymbol{F}$ are both one-to-one Riesz homomorphisms, and UT is the identity on $\boldsymbol{F}$. It follows at once that T is an isomorphism between $\boldsymbol{F}$ and $\boldsymbol{E}^{\times}$.

***65B Corollary** The same result applies for any order-dense linear subspace $\boldsymbol{E}$ of $\boldsymbol{L}^0$, whether solid or not.

Sketch of proof Just as in part (a) of the proof above, $\boldsymbol{F}$ is certainly solid. Now let $\boldsymbol{G}$ be the solid hull of $\boldsymbol{E}$ in $\boldsymbol{L}^0$. Then it is easy to see that

$$\boldsymbol{F} = \{w : w \in \boldsymbol{L}^0, u \times w \in \boldsymbol{L}^1 \quad \forall \ u \in \boldsymbol{G}\},$$

so that the duality between $\boldsymbol{F}$ and $\boldsymbol{G}$ sets up an isomorphism between $\boldsymbol{F}$ and $\boldsymbol{G}^{\times}$. Now using 17B it is easy to see that every increasing order-continuous functional $f: \boldsymbol{E} \to \mathbf{R}$ extends to an increasing order-continuous functional on $\boldsymbol{G}$, and therefore corresponds to an element of $\boldsymbol{F}$. The remainder of the verification that $\boldsymbol{F}$ is isomorphic to $\boldsymbol{E}^{\times}$ is relatively straightforward [cf. 17Gb].

65C Definition Let (X, Σ, μ) be a semi-finite measure space. An **extended Fatou norm** on $\boldsymbol{L}^0(\Sigma, \mu)$ is a function $\rho: \boldsymbol{L}^0 \to [0, \infty]$ such that:

(i) $\rho(u+v) \leqslant \rho(u) + \rho(v) \quad \forall \ u, v \in \boldsymbol{L}^0$;

(ii) $\rho(\alpha u) = |\alpha| \rho(u) \quad \forall \ u \in \boldsymbol{L}^0, \alpha \in \mathbf{R}$;

(iii) if $|u| \leqslant |v|$ in L^0, then $\rho(u) \leqslant \rho(v)$;

(iv) if $\varnothing \subset A \uparrow v$ in L^0, then $\rho(v) \leqslant \sup_{u \in A} \rho(u)$;

(v) $L^\rho = \{u : u \in L^0, \rho(u) < \infty\}$ is order-dense in L^0;

(vi) if $\rho(u) = 0$, then $u = 0$.

Notes For the use of '∞', see 51A; in particular, recall that $0 . \infty = 0$. The idea is that ρ should be a Fatou norm on L^ρ [see 65E below]. The conditions (v) and (vi) are designed to eliminate uninstructive complications. For examples of extended Fatou norms, see 6XE–6XG.

65D Theorem Let (X, Σ, μ) be a semi-finite measure space and ρ an extended Fatou norm on $L^0(\Sigma, \mu)$. Then ρ has an **associate** θ, given by
$$\theta(w) = \sup\{\|u \times w\|_1 : u \in L^0, \rho(u) \leqslant 1\}.$$
θ is also an extended Fatou norm, and the associate of θ is now ρ, i.e.
$$\rho(u) = \sup\{\|w \times u\|_1 : \theta(w) \leqslant 1\} \quad \forall\ u \in L^0.$$

Note Here I am employing the convention of 6XE, that if $v \in L^0 \backslash L^1$ then $\|v\|_1 = \infty$.

Proof It is easy to verify, directly from the fact that $\| \ \|_1$ is an extended Fatou norm, that θ satisfies the conditions (i)–(iv) of 65C. If $w > 0$, then, because L^ρ is order-dense, there is a $u \in L^\rho$ such that $0 < u \leqslant w$; if $\alpha = \rho(u)$, then
$$\theta(w) \geqslant \|w \times \alpha^{-1} u\|_1 > 0.$$
Thus θ satisfies condition (vi) of 65C.

Before showing that θ satisfies 65C(v), I shall show that ρ is the associate of θ. Of course, if $\rho(u_0) \leqslant 1$, then
$$\phi(u_0) = \sup\{\|u_0 \times w\|_1 : \theta(w) \leqslant 1\} \leqslant 1.$$
On the other hand, suppose that $u_0 \in L^0$ and that $\rho(u_0) > 1$. Consider
$$U = \{u : u \in L^1, \rho(u) \leqslant 1\}.$$
Then by conditions (i)–(iii) of 65C, U is solid and convex. Moreover, if $\varnothing \subset A \subseteq U$ and $A \uparrow v$ in L^1, then by 65C (iv)
$$\rho(v) \leqslant \sup_{u \in A} \rho(u) \leqslant 1.$$
It follows from 23L (or otherwise) that U is closed for the norm topology on L^1. Also, since L^1 is order-dense in L^0 [63F],
$$B = \{u : u \in L^1, 0 \leqslant u \leqslant |u_0|\} \uparrow |u_0|,$$

and there is a $u_1 \in B$ such that $\rho(u_1) > 1$, i.e. $u_1 \notin U$. Now by the Hahn–Banach theorem there is a continuous linear functional $f: L^1 \to \mathbf{R}$ such that $f(u_1) > 1$ but $|fu| \leqslant 1$ for every $u \in U$.

Recall that $(L^1)' = (L^1)^\times$ [26C]. Consider $|f|$ in $(L^1)^\times$. As U is solid,

$$|f|(|u|) = \sup\{|fv| : |v| \leqslant |u|\} \leqslant 1 \quad \forall\ u \in U,$$

while of course $|f|(u_1) \geqslant |f(u_1)| > 1$. Now the natural duality between L^1 and L^∞ represents L^1 as $(L^\infty)^\times$ [63G], so it must map L^∞ onto an order-dense Riesz subspace of $(L^1)^\times = (L^1)'$ [32B]. Thus if

$$C = \{w : w \in (L^\infty)^+,\ \hat{w} \leqslant |f|\},$$

we know that $\{\hat{w} : w \in C\} \uparrow |f|$ in $(L^1)^\times$, so by 16Db there is a $w \in C$ such that

$$\int u_1 \times w = \hat{w}(u_1) > 1.$$

We now find that $\theta(w) \leqslant 1$. **P** Suppose that $u \in L^0$ and that $\rho(u) \leqslant 1$. Once again, because L^1 is order-dense in L^0,

$$\{v \times w : v \in L^1,\ 0 \leqslant v \leqslant |u|\} \uparrow |u| \times w = |u \times w|,$$

so

$$\begin{aligned} \|u \times w\|_1 &\leqslant \sup\{\textstyle\int v \times w : v \in L^1,\ 0 \leqslant v \leqslant |u|\} \\ &= \sup\{\hat{w}(v) : v \in L^1,\ 0 \leqslant v \leqslant |u|\} \\ &\leqslant \sup\{|f|(v) : v \in L^1,\ 0 \leqslant v \leqslant |u|\} \\ &\leqslant \sup\{|f|(|v|) : v \in U\} \leqslant 1. \end{aligned}$$

As u is arbitrary, this shows that $\theta(w) \leqslant 1$. **Q**

Consequently,

$$\phi(u_0) \geqslant \phi(u_1) \geqslant \|u_1 \times w\|_1 \geqslant \textstyle\int u_1 \times w > 1.$$

Thus we see that, for $u_0 \in L^0$,

$$\rho(u_0) \leqslant 1 \Leftrightarrow \phi(u_0) \leqslant 1.$$

Since both ρ and ϕ satisfy the positive-homogeneity condition 65C(ii), $\rho = \phi$; i.e. ρ is the associate of θ.

Finally, if $v > 0$ in L^0, $\phi(v) = \rho(v) > 0$; so there is a $w \in L^0$ such that $\theta(w) \leqslant 1$ and $\|v \times w\|_1 > 0$. Then $0 < v \wedge |w| \leqslant v$ and $\theta(v \wedge |w|) < \infty$, so θ satisfies the condition (v) of 65C, and is an extended Fatou norm.

65E Proposition Let (X, Σ, μ) be a Maharam measure space and ρ an extended Fatou norm on $L^0(\Sigma, \mu)$. Then ρ, restricted to L^ρ, is a Fatou norm, and the topology it induced on L^ρ is Levi and complete.

Proof The conditions (i)–(iv) and (vi) of 65C make it clear that ρ is a Fatou norm on L^ρ. If A is a non-empty set in $(L^\rho)^+$ which is bounded and directed upwards, then it must be bounded above in L^0. **P ?** For otherwise, by 64C, there is a $v > 0$ in L^0 such that

$$\{u \wedge kv : u \in A\} \uparrow kv \quad \forall \ k \in \mathbf{N}.$$

Now it follows that, for any $k \in \mathbf{N}$,

$$k\rho(v) = \rho(kv) \leqslant \sup_{u \in A} \rho(u \wedge kv) \leqslant \sup_{u \in A} \rho(u) < \infty,$$

so $\rho(v) = 0$; which contradicts 65C(vi). **XQ** As L^0 is Dedekind complete [64B] it follows that $w_0 = \sup A$ exists in L^0. Now $A \uparrow w_0$, so $\rho(w_0) \leqslant \sup_{u \in A} \rho(u) < \infty$, and $w_0 \in L^\rho$. Thus A is bounded above in L^ρ. As A is arbitrary, the norm topology on L^ρ is Levi.

It follows by Nakano's theorem [23K], or otherwise [see 65Ia], that L^ρ is complete.

65F Proposition Let (X, Σ, μ) be a Maharam measure space and ρ an extended Fatou norm on $L^0(\Sigma, \mu)$. Let θ be the associate of ρ [65D]. Then

$$L^\theta = \{w : w \in L^0, \quad w \times u \in L^1 \quad \forall u \in L^\rho\},$$

so L^θ may be identified with $(L^\rho)^\times$.

Proof (**a**) If $u \in L^\rho$ and $w \in L^\theta$, let $\alpha = \rho(u)$. If $\alpha = 0$, then $u = 0$ and $u \times w = 0 \in L^1$. If $\alpha > 0$, then

$$\infty > \theta(w) \geqslant \|\alpha^{-1} u \times w\|_1 = \alpha^{-1} \|u \times w\|_1.$$

Thus $u \times w \in L^1$ and $\|u \times w\|_1 \leqslant \rho(u)\,\theta(w)$.

(**b**) Conversely, suppose that $w \in L^0$ and that $w \times u \in L^1$ for every $u \in L^\rho$. Then $|w| \times |u| = |w \times u| \in L^1$ for every $u \in L^\rho$. Set

$$B = \{v : v \in L^\theta, \ 0 \leqslant v \leqslant |w|\}.$$

For every $v \in B$, $\hat{v} \in (L^\rho)'$, where

$$\hat{v}(u) = \int u \times v \quad \forall \ u \in L^\rho,$$

and

$$\|\hat{v}\| = \sup\{|\hat{v}(u)| : \rho(u) \leqslant 1\} = \sup\{\|u \times v\|_1 : \rho(u) \leqslant 1\} = \theta(v),$$

using the fact that $\{u : \rho(u) \leqslant 1\}$ is solid. Morever, for any $u \in L^\rho$,

$$\sup_{v \in B} |\hat{v}(u)| \leqslant \sup_{v \in B} \int |u| \times v \leqslant \int |u| \times |w| < \infty.$$

So by the uniform boundedness theorem (since L^ρ is a Banach space, 65E),

$$\infty > \sup_{v \in B} \|\hat{v}\| = \sup_{v \in B} \theta(v) = \theta(|w|) = \theta(w),$$

since $B \uparrow |w|$ because L^θ is order-dense in L^0. Thus $w \in L^\theta$. As w is arbitrary, this shows that

$$L^\theta = \{w : w \times u \in L^1 \quad \forall\ u \in L^\rho\},$$

as required.

(c) Now by 65A above the natural duality between L^ρ and L^θ, writing $\langle u, v\rangle = \int u \times v$, represents L^θ as $(L^\rho)^\times$.

65G Corollary If (X, Σ, μ) is a Maharam measure space and ρ an extended Fatou norm on $L^0(\Sigma, \mu)$, then L^ρ is perfect in the sense of 33A.

Proof For if θ is the associate of ρ, then ρ is the associate of θ [65D], so $(L^\rho)^{\times\times} = (L^\theta)^\times = L^\rho$.

65H Proposition Let (X, Σ, μ) be a Maharam measure space and ρ an extended Fatou norm on $L^0(\Sigma, \mu)$. Then $(L^\rho)' = (L^\rho)^\sim$, and the following are equivalent:

(i) the norm topology on L^ρ is Lebesgue;

(ii) whenever $\varnothing \subset A \downarrow 0$ in L^ρ, $\inf_{u \in A} \rho(u) = 0$;

(iii) $(L^\rho)' = (L^\rho)^\times$;

(iv) whenever $\langle u_n\rangle_{n\in\mathbf{N}}$ is a disjoint sequence in $(L^0)^+$ and

$$\sup_{n\in\mathbf{N}} \rho(\textstyle\sum_{i<n} u_i) < \infty,$$

then $\langle\rho(u_n)\rangle_{n\in\mathbf{N}} \to 0$.

Proof (a) $(L^\rho)' = (L^\rho)^\sim$ by 25G, because by 65E L^ρ is a Banach lattice.

(b) Of course (i) and (ii) are equivalent by definition, and (i) and (iii) are equivalent by 25M.

Because the norm topology on L^ρ is Levi [65E], we see that, for a disjoint sequence $\langle u_n\rangle_{n\in\mathbf{N}}$ in $(L^\rho)^+$, $\sup_{n\in\mathbf{N}} \rho(\sum_{i<n} u_i) < \infty$ iff

$$\{u_n : n \in \mathbf{N}\}$$

is bounded above in L^ρ. So (iv) is equivalent to 'whenever $\langle u_n\rangle_{n\in\mathbf{N}}$ is a disjoint sequence in $(L^\rho)^+$ which is bounded above, then

$$\langle\rho(u_n)\rangle_{n\in\mathbf{N}} \to 0\text{'},$$

which by 24J is equivalent to (i).

65I Exercises **(a)** Let (X, Σ, μ) be any semi-finite measure space, and ρ an extended Fatou norm on $L^0(\Sigma, \mu)$. Show that (i) ρ is a Fatou norm on L^ρ; (ii) if $\langle u_n\rangle_{n\in\mathbf{N}}$ is a bounded increasing sequence in L^ρ, then $\sup_{n\in\mathbf{N}} u_n$ exists in L^0 and belongs to L^ρ [use 62J]; (iii) L^ρ is complete [use 25Na]; (iv) if θ is the associate of ρ,

$$L^\theta = \{w : w \times u \in L^1 \quad \forall \ u \in L^\rho\},$$

just as in 65F; (v) the whole of 65H is still true [use (ii) and (iii) above instead of 65E].

***(b)** Let (X, Σ, μ) be a semi-finite measure space, and ρ an extended Fatou norm on $L^0(\Sigma, \mu)$. Give L^0 its usual topology [63K]. Then the embedding $L^\rho \subseteq L^0$ is continuous.

Notes and comments The theorem 65A/B gives a general method of describing order-continuous linear functionals on most of the usual function and sequence spaces [see 6XD]; it includes, for instance, the identifications $(l^\infty)^\times = l^1$, $(l^1)^\times = l^\infty$, and $(c_0)^\times = l^1$ in 2XA–2XC, as well as $(L^\infty)^\times = L^1$ [63G] and $(L^1)^\times = L^\infty$ [64B]. Moreover, it can be shown that if E is any Riesz space such that $E^\times$ separates the points of E, then E can be represented as an order-dense Riesz subspace of $L^0(\Sigma, \mu)$ for some Maharam measure space (X, Σ, μ) [FREMLIN A.K.S. II]; so that 65B gives a picture of the duality between E and $E^\times$ which is in some sense complete.

The examples 6XE–6XH give some idea of the kind of spaces that can be generated by extended Fatou norms. They include most of the solid normed function spaces that have attracted interest. In fact, using the representation theorem mentioned above and the method of 6XH, it is not hard to show that if E is any Dedekind complete Riesz space with a Fatou norm inducing a Levi topology, and if $E^\times$ separates the points of E, then E is an L^ρ space.

When the underlying measure space is Maharam, then 65E–65H give a description of these function spaces which in its own terms is fairly complete. They are all perfect and their topologies are Fatou and Levi. The question of when they are Lebesgue is tackled in 65H.

For the sake of simplicity, the underlying measure space is assumed Maharam all through this work. But in 65Ia I list those results which can be extended to arbitrary semi-finite measure spaces. They give a good idea of the way in which, for metric spaces, it is often enough to be able to handle sequences.

In 6XI I give a example of a class of spaces which is very similar to the function spaces of this section except that the defining functions are not norms and the topologies are consequently not locally convex.

6X Examples for Chapter 6

I begin with simple examples of measure spaces; the most important one is of course Lebesgue measure [6XAb, 6XB]. Because Lebesgue measure is diffuse [6XBa], the associated L^0 space is not locally convex [6XBb]. The next paragraph [6XC] is an examination of a natural inverse-measure-preserving function, showing how the ideas of § 54 are linked to Fubini's theorem. The rest of the section is a discussion of some simple function spaces, based on the ideas of § 65, showing how the results of Chapters 2 and 3 can be applied.

6XA Measure spaces (a) Let X be any set, and set $\Sigma = \mathscr{P}X$. Define $\mu: \Sigma \to [0, \infty]$ by

$$\mu E = n \quad \text{if } E \text{ has } n \text{ members, where } n \in \mathbf{N};$$
$$= \infty \quad \text{otherwise.}$$

Then (X, Σ, μ) is a complete decomposable measure space. As

$$\mu E = 0 \ \text{ iff } \ E = \varnothing,$$

its measure algebra $\mathfrak{A}$ is isomorphic to Σ, and $L^0(\Sigma, \mu) \cong M(\Sigma) = \mathbf{R}^X$. $L^\infty(\mathfrak{A}) \cong L^\infty(\Sigma) = l^\infty(X)$ [4XCa], and $L^1(\mathfrak{A}) \cong \mathfrak{L}^1(\mu) = l^1(X)$ [cf. 5XB]. The usual topology on L^0 [63K] is precisely the product topology on $\mathbf{R}^X$ [cf. 1XD].

(b) **Lebesgue measure** For the construction of Lebesgue measure on $\mathbf{R}$, I refer you to WIDOM, chapter 1, or BARTLE, p. 96, chapter 9; an alternative method is suggested in 7XA. Its basic property is that $\mu[\alpha, \beta] = \beta - \alpha$ whenever $\alpha \leqslant \beta$ in $\mathbf{R}$. It is complete and of countable magnitude.

(c) For further examples, see MUNROE [chapter III] and Chapter 7 below; also HALMOS M.T., chapter XI, §§ 57–60, for a description of Haar measure.

6XB Lebesgue measure on [0, 1] Now let us take X to be the unit interval [0, 1], and Σ the class of Lebesgue measurable subsets of X; let μ be the restriction of Lebesgue measure to Σ. Then (X, Σ, μ) is a complete measure space and $\mu X = 1$.

(a) Suppose that $E \in \Sigma$ and that $\mu E > 0$. Let $n \geqslant 1$ be such that $\alpha = n^{-1} < \mu E$. Since $[0, 1] = \bigcup_{i<n}[\alpha i, \alpha(i+1)]$, there is an $i < n$ such that

$$0 < \mu(E \cap [\alpha i, \alpha(i+1)]) \leqslant \alpha < \mu E.$$

Thus (X, Σ, μ) is **diffuse**, i.e. for every non-negligible $E \in \Sigma$, there is an $F \in \Sigma$ such that $F \subseteq E$ and neither F nor $E \backslash F$ is negligible. Another way of putting this is to say that the measure algebra of (X, Σ, μ) has no minimal non-zero elements or 'atoms'.

(b) Consider now $L^0(\Sigma, \mu)$. If $f: L^0 \to \mathbf{R}$ is increasing and linear, $f = 0$. **P ?** Otherwise, let $g: M(\Sigma) \to \mathbf{R}$ be given by $gx = f(x^{\cdot})$ for every $x \in M(\Sigma)$. Let $x_0 > 0$ in $M(\Sigma)$ be such that $gx_0 > 0$. For each $n \in \mathrm{N}$, observe that

$$x_0 \leqslant \sum_{i<2^n} x_0 \times \chi([2^{-n}i, 2^{-n}(i+1)]),$$

so there must be an $i < 2^n$ such that

$$\alpha_n = g(y_n) > 0,$$

where $$y_n = x_0 \times \chi([2^{-n}i, 2^{-n}(i+1)]).$$

Now let $$z_n = 2^n \alpha_n^{-1} y_n,$$

so that $g(z_n) = 2^n$ and $\mu\{\gamma: z_n(\gamma) \neq 0\} \leqslant 2^{-n}$. If we set

$$E = \{\gamma: \sup_{n\in\mathbf{N}} z_n(\gamma) = \infty\},$$

we see that $$E \subseteq \bigcap_{n\in\mathbf{N}} \bigcup_{i\geqslant n} \{\gamma : z_i(\gamma) \neq 0\},$$

so that $$\mu E \leqslant \inf_{n\in\mathbf{N}} \sum_{i\geqslant n} 2^{-i} = 0.$$

Let $$z = \sup_{n\in\mathbf{N}} z_n \times \chi(X\backslash E).$$

Then $z^{\cdot} \geqslant z_n^{\cdot}$ in L^0, so

$$g(z) = f(z^{\cdot}) \geqslant f(z_n^{\cdot}) = g(z_n) = 2^n$$

for every $n \in \mathrm{N}$, which is impossible. **XQ**

*(c) Consequently, if $\mathfrak{T}$ is the usual topology on L^0 [63K], then the dual of L^0 for $\mathfrak{T}$ is $\{0\}$ [22D]. In particular, $\mathfrak{T}$ is not locally convex, though it is complete and metrizable [64E, 64Jc]. For further striking properties of $\mathfrak{T}$, see PRYCE U.S.

***6XC An inverse-measure-preserving function** Let I be the unit interval $[0, 1]$, and let $X = I^2 \subseteq \mathbf{R}^2$; let μ be the restriction of Lebesgue planar measure to subsets of X, and Σ the domain of μ. Let (Y, T, ν) be Lebesgue linear measure on I, as in 6XB above. Define $f: X \to Y$ by writing $f(\alpha, \beta) = \alpha$ for all $\alpha, \beta \in I$. Then f is inverse-measure-preserving.

Let $\mathfrak{A}$ be the measure algebra of (X, Σ, μ) and $\mathfrak{B}$ that of (Y, T, ν). Then the measure-preserving ring homomorphism $\pi: \mathfrak{B} \to \mathfrak{A}$ of 61E is given by

$$\pi(E^{\cdot}) = (f^{-1}[E])^{\cdot} = (E \times I)^{\cdot} \quad \forall\ E \in \mathrm{T}.$$

So $\hat{\pi}: L^\infty(\mathfrak{B}) \to L^\infty(\mathfrak{A})$ is given by

$$\hat{\pi}(x^\cdot) = (x \otimes 1)^\cdot \quad \forall \ x \in L^\infty(\mathrm{T}),$$

where $1 = \chi I$ and $x \otimes y$ is given by

$$(x \otimes y)(\alpha, \beta) = x(\alpha)\, y(\beta) \quad \forall \ \alpha, \beta \in I.$$

Similarly, $\hat{\pi}: L^1(\mathfrak{B}) \to L^1(\mathfrak{A})$ [54D] is given by

$$\hat{\pi}(x^\cdot) = (x \otimes 1)^\cdot \quad \forall \ x \in \mathfrak{L}^1(\nu).$$

On the other hand, $P_\pi: L^1(\mathfrak{A}) \to L^1(\mathfrak{B})$ is given by

$$\langle P_\pi u, v\rangle = \langle u, \hat{\pi} v\rangle \quad \forall \ u \in L^1(\mathfrak{A}), \quad v \in L^\infty(\mathfrak{B})$$

[54F], i.e.

$$\langle P_\pi x^\cdot, y^\cdot\rangle = \int x \times (y \otimes 1)\, d\mu \quad \forall \ x \in \mathfrak{L}^1(\mu), \quad y \in L^\infty(\mathrm{T})$$

[using the representation of the duality between $L^1(\mathfrak{A})$ and $L^\infty(\mathfrak{A})$ given in 63F]. Now Fubini's theorem [MUNROE, § 28, or WILLIAMSON, p. 63, § 4.2] tells us that, given $x \in \mathfrak{L}^1(\mu)$,

$$z(\alpha) = \int x(\alpha, \beta)\, \nu(d\beta) \quad \text{exists} \quad \nu\text{–p.p. } (\alpha),$$

and that

$$\int z\, d\nu = \int x\, d\mu.$$

So, applying this to $x \times (y \otimes 1)$,

$$\langle P_\pi x^\cdot, y^\cdot\rangle = \int x \times (y \otimes 1)\, d\mu = \int y(\alpha)\, z(\alpha)\, \nu(d\alpha) = \langle z^\cdot, y^\cdot\rangle.$$

As this is true for every $y \in L^\infty(\mathrm{T})$, $P_\pi x^\cdot = z^\cdot$.

Thus P_π 'averages' members of $L^1(\mathfrak{A})$ over vertical lines, in a fashion perfectly analogous to 5XC.

6XD Sequence spaces Suppose, in 6XAa, that $X = \mathbf{N}$. Then $M(\Sigma)$, which is isomorphic to $L^0(\Sigma, \mu)$, becomes $\mathbf{R}^{\mathbf{N}}$, and we may call its subspaces 'sequence spaces'.

A Riesz subspace E of $\mathbf{R}^{\mathbf{N}}$ is easily seen to be order-dense iff it includes s_0, the space of sequences with only finitely many non-zero terms. In this case we may use 65B to identify $E^\times$ with

$$\{y : x \times y \in l^1(\mathbf{N}) \quad \forall \ x \in E\},$$

identifying l^1 with $\mathfrak{L}^1 \cong L^1$. Observe that $(\mathbf{R}^{\mathbf{N}})^\times = s_0$ and $(s_0)^\times = \mathbf{R}^{\mathbf{N}}$. This result also includes the identifications of $(l^1)^\times$, $(l^\infty)^\times$ and $(c_0)^\times$ given in 2XA–2XC.

6XE Extended Fatou norms For any semi-finite measure space (X, Σ, μ), we may extend $\| \ \|_1$ and $\| \ \|_\infty$ to the whole of $L^0(\Sigma, \mu)$ by writing

$$\|u\|_1 = \infty \quad \text{for} \quad u \in L^0 \setminus L^1,$$
$$\|u\|_\infty = \infty \quad \text{for} \quad u \in L^0 \setminus L^\infty.$$

Because the norm topologies on L^1 and L^∞ are Fatou and Levi, it is now clear (reversing part of the argument of 65E) that $\| \ \|_1$ and $\| \ \|_\infty$ are extended Fatou norms in the sense of 65C, and also that they are associates in the sense of 65D [using 52G and 63G].

6XF L^p spaces $(1 < p < \infty)$ (a) Suppose that $p > 1$ in $\mathbf{R}$. Set $q = (p-1)^{-1} p$, so that $p^{-1} + q^{-1} = 1$. Let (X, Σ, μ) be any semi-finite measure space. If $x, y \in M(\Sigma)$,

$$x^{\cdot} = y^{\cdot} \Leftrightarrow x = y \text{ p.p.} \Rightarrow |x|^p = |y|^p \quad \text{p.p.} \Leftrightarrow (|x|^p)^{\cdot} = (|y|^p)^{\cdot}.$$

Thus we may define $|x^{\cdot}|^p$ to be equal to $(|x|^p)^{\cdot}$. Now we write

$$\|u\|_p = (\| \, |u|^p \|_1)^{1/p}$$

for each $u \in L^0(\Sigma, \mu)$, where $\| \ \|_1$ is permitted to take the value ∞, as in 6XE. Then $\| \ \|_p$ is an extended Fatou norm and its associate is $\| \ \|_q$.

Proof (i) Hölder's inequality If $\alpha, \beta \in \mathbf{R}$, then

$$p^{-1} \alpha^p + q^{-1} \beta^q \geqslant \alpha\beta$$

[HARDY, LITTLEWOOD & PÓLYA, § 2.5]. It follows that if $u, v \geqslant 0$ in L^0,

$$p^{-1} |u|^p + q^{-1} |v|^q \geqslant u \times v.$$

Consequently $\|u \times v\|_1 \leqslant 1$ whenever $\|u\|_p \leqslant 1$ and $\|v\|_q \leqslant 1$ (since $p^{-1} + q^{-1} = 1$). So $\|u \times v\|_1 \leqslant \|u\|_p \|v\|_q$ whenever $\|u\|_p$ and $\|v\|_q$ are finite, for

$$\|\alpha u\|_p = |\alpha| \, \|u\|_p \quad \forall \ \alpha \in \mathbf{R}, \quad u \in L^0.$$

(ii) It follows that

$$\|u\|_p = \sup \{\|u \times v\|_1 : \|v\|_q \leqslant 1\}$$

for every $u \in L^0$. **P** We have just seen that

$$\|u\|_p \geqslant \sup \{\|u \times v\|_1 : \|v\|_q \leqslant 1\}$$

for every u such that $\|u\|_p < \infty$, and therefore for every $u \in L^0$. If $\|u\|_p = 0$, then $|u|^p = 0$ so $u = 0$, and the equality is certainly satisfied.

If $0 < \|u\|_p < \infty$, set $v = |u|^{p/q}$. Then

$$\|v\|_q = (\| |v|^q \|_1)^{1/q} = (\| |u|^p \|_1)^{1/q} = (\|u\|_p)^{p/q} = \frac{1}{\beta}$$

say. So $\|\beta v\|_q = 1$, and

$$\|u \times \beta v\|_1 = \beta \| |u| \times |v| \|_1 = \beta \| |u|^p \|_1 = \|u\|_p,$$

using the fact that $1 + (p/q) = p$, so that $p - (p/q) = 1$. So in this case also the equality is satisfied. Finally, if $\|u\|_p = \infty$, we use the fact that (X, Σ, μ) is semi-finite to see that

$$\begin{aligned} \infty = \|u\|_p &= \sup\{\|w\|_p : w \in S(\mathfrak{A}^f),\ 0 \leqslant w \leqslant |u|\} \\ &= \sup\{\|w \times v\|_1 : w \in S,\ 0 \leqslant w \leqslant |u|,\ \|v\|_q \leqslant 1\} \\ &\leqslant \sup\{\|u \times v\|_1 : \|v\|_q \leqslant 1\} \leqslant \infty. \end{aligned}$$

(iii) It follows at once that $\| \ \|_p$ satisfies the conditions (i)–(iv) of 65C, and it satisfies (v) and (vi) because $\| \ \|_1$ does. Similarly $\| \ \|_q$ is an extended Fatou norm, and is of course the associate of $\| \ \|_p$.

(b) We normally write $\boldsymbol{L}^p$ for the Banach space

$$\{u : \|u\|_p < \infty\}$$

with the norm $\| \ \|_p$. As $\| \ \|_p$ satisfies the condition 65H(iv) (because $\| \ \|_1$ does), $(\boldsymbol{L}^p)' = (\boldsymbol{L}^p)^\sim = (\boldsymbol{L}^p)^\times$ may be identified with $\boldsymbol{L}^q$, at least when (X, Σ, μ) is Maharam. Thus, in this case, $\boldsymbol{L}^p$ is reflexive. *Actually, $\boldsymbol{L}^p$ always has the countable sup property (because $\boldsymbol{L}^1$ does), so is necessarily Dedekind complete, even when (X, Σ, μ) is not Maharam. We can see that the norm topology of $\boldsymbol{L}^p$ is Levi because the norm topology of $\boldsymbol{L}^1$ is Levi; thus $\boldsymbol{L}^p$ is always complete [see 25Na]. The argument of 65A shows that $\boldsymbol{L}^q$ can be thought of as an order-dense Riesz subspace of $(\boldsymbol{L}^p)^\times$, which by 17F is solid. Now, because $\boldsymbol{L}^q$ is a Banach space and the norm on $\boldsymbol{L}^p$ is that induced by the duality [part (ii) of the proof of (a) above], a $\mathfrak{T}_s(\boldsymbol{L}^p, \boldsymbol{L}^q)$-bounded set is $\| \ \|_p$-bounded; so $\boldsymbol{L}^p$ is Levi for $\mathfrak{T}_s(\boldsymbol{L}^p, \boldsymbol{L}^q)$. It follows from 33B that $\boldsymbol{L}^p$ can be identified with $(\boldsymbol{L}^q)^\times$. Similarly, $\boldsymbol{L}^q = (\boldsymbol{L}^p)^\times = (\boldsymbol{L}^p)'$, so again $\boldsymbol{L}^p$ is reflexive.

6XG $\boldsymbol{L}^2$ spaces If in 6XF we set $p = q = 2$, we get the outstanding special case

$$\boldsymbol{L}^2 = \{u : u \in \boldsymbol{L}^0,\ u \times u \in \boldsymbol{L}^1\}.$$

The identification between $\boldsymbol{L}^2$ and $(\boldsymbol{L}^2)'$ corresponds of course to an inner product, writing

$$(u|v) = \int u \times v \quad \forall\ u,\ \ v \in \boldsymbol{L}^2.$$

We know that L^2 is complete; therefore it is a real Hilbert space. In view of its great importance, it is worth noting that many of the arguments of 6XF can be simplified in this special case.

***6XH Orlicz spaces** Let us consider the properties that the unit ball U of a Banach function space L^ρ defined by an extended Fatou norm ρ must have. It is a subset of an L^0 space which is solid and convex [65C(i)–(iii)], and $\mathscr{I}U \subseteq U$ [65C(iv)]. If $u > 0$ in L^0, there is a $v \in U$ such that $0 < v \leqslant u$ [65C(v)], and if $\alpha u \in U$ for every $\alpha \in \mathbf{R}$, then $u = 0$ [65C(vi)]. Clearly these properties are necessary and sufficient for U to be $\{u : \rho(u) \leqslant 1\}$ for some extended Fatou norm ρ.

We may regard 6XF as defining $\| \ \|_p$ by

$$\{u \colon \|u\|_p \leqslant 1\} = \{u \colon \textstyle\int |u|^p \leqslant 1\}.$$

Now the function $\alpha \mapsto \alpha^p : \mathbf{R}^+ \to \mathbf{R}^+$ is a continuous monotonic function which, for $p \geqslant 1$, is *convex*, i.e.

$$(\gamma\alpha + (1-\gamma)\beta)^p \leqslant \gamma\alpha^p + (1-\gamma)\beta^p$$

whenever $\alpha, \beta \geqslant 0$ and $0 \leqslant \gamma \leqslant 1$. This is clearly the essential property to ensure that $\{u\colon \int |u|^p \leqslant 1\}$ is convex.

So now suppose that $\Phi\colon \mathbf{R}^+ \to [0, \infty]$ is any convex monotonic non-decreasing function satisfying the following:

(i) Φ is continuous on the left, and $\Phi(0) = 0$;

(ii) there is an $\alpha > 0$ such that $\Phi(\alpha) < \infty$;

(iii) Φ is not identically zero.

Let U be

$$\{u\colon \exists\ x \in \boldsymbol{M}(\Sigma),\quad x^{\cdot} = u,\quad \textstyle\int \Phi(|x(t)|)\,dt \leqslant 1\}.$$

(Because Φ is permitted to take the value ∞, we must be careful to choose x such that $\Phi(|x(t)|) < \infty$ for all t; if this is impossible, then $u \notin U$. Because Φ is monotonic, the function $t \mapsto \Phi(|x(t)|)$ will always be measurable, using the criterion of 62D.)

It is immediate from the definition, because Φ is convex and monotonic, that U is solid and convex. It is a little harder to see that $\mathscr{I}U \subseteq U$. But, for any $u \geqslant 0$ in U, we may define $\Phi u \in L^1$ by

$$\Phi u = (\Phi x)^{\cdot} \quad \text{where} \quad x^{\cdot} = u.$$

Now if $\varnothing \subset A \uparrow u_0$ in $(L^0)^+$ and $A \subseteq U$,

$$\{\Phi u : u \in A\}\uparrow$$

in the unit ball of L^1, and has a supremum v_0 say. Using the left-continuity of Φ, we can show that $\Phi u_0 \leqslant v_0$, so that $u_0 \in U$. Conditions

(ii) and (iii) on Φ now ensure the other properties of U (since, as always, the measure space is supposed semi-finite).

For a description of the way in which Φ and its 'complementary Young's function' can be used to generate an associated pair of Banach function spaces, I refer you now to ZAANEN, chapter 5, §§ 4–6.

Apart from the obvious cases of L^p, $1 \leqslant p < \infty$, many other spaces can be defined in this way. In particular, setting

$$\Phi(\alpha) = 0 \quad \text{for} \quad \alpha \leqslant 1, \quad \infty \text{ for } \alpha > 1,$$

we find that U is just the unit ball of L^∞.

***6XI L^p spaces ($0 < p < 1$)** Let (X, Σ, μ) be a semi-finite measure space, and suppose that $0 < p < 1$. For $u \in L^0(\Sigma, \mu)$, define

$$\|u\|_p = \| \, |u|^p \|_1,$$

where $|u|^p$ is defined as in 6XF. Since $(\alpha + \beta)^p \leqslant \alpha^p + \beta^p$ for all $\alpha, \beta \in \mathbf{R}^+$, $\|u+v\|_p \leqslant \|u\|_p + \|v\|_p$ for all u, $v \in L^0$; now it is easy to see that $\| \; \|_p : L^0 \to [0, \infty]$ satisfies all the conditions of 65C except (ii). Instead, we see that

$$\|\alpha u\|_p = |\alpha|^p \, \|u\|_p \quad \forall \; u \in L^0, \alpha \in \mathbf{R}.$$

However, this is enough to ensure that $\| \; \|_p$ is a Fatou pseudo-norm on $L^p = \{u : \|u\|_p < \infty\}$.

Just as in 6XF, L^p now has the countable sup property and is Dedekind complete, and its pseudo-norm topology is Levi; so L^p is an example of a complete metrizable topological Riesz space which is not (except in trivial cases) locally convex.

Further reading for Chapter 6 It has unfortunately been impossible to include more than a few of the concepts of abstract measure theory. Probably the best introduction to general modern measure theory is still HALMOS M.T. One of the most serious omissions above is the lack of any reference to product measures; HEWITT & STROMBERG [chapter VI] are very sound, though they restrict their attention to spaces of countable magnitude. An alternative look at the foundations, with some of the ideas of § 64, can be found in IONESCU TULCEA [chapter I]. Here also is a proof of the Lifting Theorem, one of the most remarkable results of pure measure theory. There are a great many important results in DUNFORD & SCHWARTZ; it is instructive to seek to refine their theorems and proofs in the light of the work above.

Since many of the most fruitful developments of measure theory have been stimulated by problems arising in the theory of probability, it is worth making an effort to cross the language barrier between the two subjects. A concise and energetic survey of measure theory from a probabilist's viewpoint is in MEYER. Here again it may be profitable to investigate whether the ideas of Chapters 4 and 5 above could be used to clarify some of the concepts.

7. Representation of linear functionals

Suppose that we are given a set X, a Riesz subspace $\boldsymbol{E}$ of $\mathbf{R}^X$, and an increasing linear functional $f: \boldsymbol{E} \to \mathbf{R}$. My object in this chapter is to discuss conditions under which f is an 'integral', that is, when there is a measure μ on X such that $\int x d\mu$ exists and is equal to fx for every $x \in \boldsymbol{E}$. A necessary, and nearly sufficient, condition is that f should be 'sequentially smooth' [71B–71G]. Further conditions on f and $\boldsymbol{E}$ lead, of course, to stronger results [§§ 72, 73].

The outlines of the theory are not hard to appreciate. Unfortunately, the technical refinements needed for the strongest results are complex, and in their most general forms they are difficult to grasp intuitively. Each extra scrap of information costs us a good deal of hard work. I shall try to summarize the theory in a way which will show its essential structure and maintain reasonable simplicity in the theorems, though the proofs will inevitably be lengthy. The general approach of the first two sections of this chapter is close to that of TOPSØE T.M.

71 Sequentially smooth functionals

The first representation theorem I give [71G] is the most natural, in the sense that the hypotheses and consequences are most directly related here. In order not to have to repeat them in the next section, I remove certain parts of the argument into separate lemmas. The first of these is a substantial result, giving a powerful method of constructing measure spaces.

71A Theorem Let X be a set, and $\mathscr{K} \subseteq \mathscr{P}X$ a sublattice containing $\varnothing$ and closed under countable intersections. Let $\lambda: \mathscr{K} \to \mathbf{R}^+$ be an increasing function such that:

(i) $\lambda\varnothing = 0$;

(ii) if $H, K \in \mathscr{K}$ and $H \cap K = \varnothing$,

$$\lambda(H \cup K) \geq \lambda H + \lambda K;$$

(iii) If $F, H \in \mathscr{K}$ and $H \subseteq F$,

$$\lambda F \leq \lambda H + \sup\{\lambda K : K \in \mathscr{K}, K \subseteq F \backslash H\};$$

(iv) if $\langle K_n\rangle_{n\in\mathbb{N}}$ is a decreasing sequence in $\mathscr{K}$ and $\bigcap_{n\in\mathbb{N}} K_n = \varnothing$, then $\inf_{n\in\mathbb{N}} \lambda K_n = 0$.

Then λ has an extension to a complete locally determined measure μ defined on a σ-algebra $\Sigma \supseteq \mathscr{K}$ such that

(α) $\mu E = \sup\{\lambda K : K \subseteq E,\ K \in \mathscr{K}\}$
for every $E \in \Sigma$;

(β) if $E \subseteq X$ and $E \cap K \in \Sigma$ for every $K \in \mathscr{K}$, then $E \in \Sigma$.

Proof Define $\lambda_*: \mathscr{P}X \to [0,\infty]$ by

$$\lambda_* A = \sup\{\lambda K : K \in \mathscr{K},\ K \subseteq A\}$$

for every $A \subseteq X$. Set

$$\Sigma = \{E : E \subseteq X,\ \lambda K \leqslant \lambda_*(K \cap E) + \lambda_*(K \setminus E) \quad \forall\ K \in \mathscr{K}\},$$

and let μ be the restriction of λ_* to Σ. The whole of the rest of the proof consists of a demonstration that μ is the required measure.

(**a**) Let us begin with two simple observations. If $A,\ B \subseteq X$ and $A \cap B = \varnothing$, then $\lambda_*(A \cup B) \geqslant \lambda_* A + \lambda_* B$.

$$\textbf{P}\ \lambda_* A + \lambda_* B = \sup\{\lambda H + \lambda K : H \subseteq A,\ K \subseteq B\}$$
$$\leqslant \sup\{\lambda(H \cup K) : H \subseteq A,\ K \subseteq B\}$$
$$\leqslant \lambda_*(A \cup B),$$

using condition (ii). **Q**

(**b**) If $E \in \Sigma$, then

$$\lambda_* A = \lambda_*(A \cap E) + \lambda_*(A \setminus E) \quad \forall\ A \subseteq X.$$

$$\textbf{P}\quad \lambda_* A = \sup\{\lambda K : K \in \mathscr{K},\ K \subseteq A\}$$
$$\leqslant \sup\{\lambda_*(K \cap E) + \lambda_*(K \setminus E) : K \in \mathscr{K},\ K \subseteq A\}$$
$$\leqslant \lambda_*(A \cap E) + \lambda_*(A \setminus E)$$

(because λ_* is obviously increasing)

$$\leqslant \lambda_* A,$$

by (a) above. **Q**

(**c**) Suppose that $E,\ F \in \Sigma$. Then $X \setminus E$ and $E \cup F \in \Sigma$. If $E \cap F = \varnothing$, then $\lambda_*(E \cup F) = \lambda_* E + \lambda_* F$.

P It is immediate from the definition of Σ that if $E \in \Sigma$ then $X \setminus E \in \Sigma$. Now let $K \in \mathscr{K}$. Then

$$\lambda K \leqslant \lambda_*(K \cap F) + \lambda_*(K \setminus F)$$

(because $F \in \Sigma$)

$$= \lambda_*(K \cap F) + \lambda_*((K \backslash F) \cap E) + \lambda_*((K \backslash F) \backslash E)$$

(because $E \in \Sigma$, using (b) above with $A = K \backslash F$)

$$= \lambda_*(K \cap (E \cup F)) + \lambda_*(K \backslash (E \cup F))$$

(because $F \in \Sigma$, using (b) above with $A = K \cap (E \cup F)$). As K is arbitrary, $E \cup F \in \Sigma$. Finally, if $E \cap F = \varnothing$, then, again using (b),

$$\lambda_*(E \cup F) = \lambda_*((E \cup F) \cap E) + \lambda_*((E \cup F) \backslash E) = \lambda_* E + \lambda_* F. \quad \mathbf{Q}$$

(**d**) So we see that Σ is a subalgebra of $\mathscr{P}X$. If $K \in \mathscr{K}$, then $\lambda_* K = \lambda K$, because λ is increasing. So if $H, F \in \mathscr{K}$,

$$\lambda H \leqslant \lambda(H \cap F) + \sup\{\lambda K : K \in \mathscr{K}, K \subseteq H \backslash F\}$$
$$= \lambda_*(H \cap F) + \lambda_*(H \backslash F).$$

As H is arbitrary, $F \in \Sigma$. Thus $\mathscr{K} \subseteq \Sigma$ and $\mu = \lambda_* = \lambda$ on $\mathscr{K}$.

(**e**) Now suppose that $\langle E_n \rangle_{n \in \mathbf{N}}$ is an increasing sequence in Σ and that $E = \bigcup_{n \in \mathbf{N}} E_n$. Then $E \in \Sigma$ and $\mu E = \lim_{n \to \infty} \mu E_n$.

P I shall rely heavily on the fact that (by (c) above)

$$\Sigma^f = \{F : F \in \Sigma, \lambda_* F < \infty\}$$

is a subring of $\mathscr{P}X$ and $\lambda_* : \Sigma^f \to \mathbf{R}$ is additive.

Let $H \in \mathscr{K}$ and $\epsilon > 0$. For each $n \in \mathbf{N}$, choose an $H_n \in \mathscr{K}$ such that $H_n \subseteq H \backslash E_n$ and $\lambda H_n \geqslant \lambda_*(H \backslash E_n) - 2^{-n}\epsilon$. Now we know that, for each $n \in \mathbf{N}$, $H_n \cup H_{n+1} \subseteq H \backslash E_n$; so

$$\lambda(H_n \cup H_{n+1}) \leqslant \lambda_*(H \backslash E_n) \leqslant \lambda H_n + 2^{-n}\epsilon;$$

and $$\lambda_*(H_{n+1} \backslash H_n) = \lambda(H_n \cup H_{n+1}) - \lambda H_n \leqslant 2^{-n}\epsilon.$$

So, if we set $K_n = \bigcap_{i \leqslant n} H_i$ for each $n \in \mathbf{N}$,

$$\lambda_*(H_{n+1} \backslash K_n) \leqslant \textstyle\sum_{i \leqslant n} \lambda_*(H_{i+1} \backslash H_i) \leqslant \sum_{i \leqslant n} 2^{-i}\epsilon \leqslant 2\epsilon,$$

and

$$\lambda K_{n+1} = \lambda H_{n+1} - \lambda_*(H_{n+1} \backslash K_n) \geqslant \lambda_*(H \backslash E_{n+1}) - 2^{-n-1}\epsilon - 2\epsilon$$
$$\geqslant \lambda_*(H \backslash E_{n+1}) - 3\epsilon.$$

Let $K = \bigcap_{n \in \mathbf{N}} K_n \subseteq H \backslash E$. We know that

$$\lambda_*(H \cap E) + \lambda_*(H \backslash E) \geqslant \lim_{n \to \infty} \lambda_*(H \cap E_n) + \lambda K.$$

To compute λK, consider $\lambda_*(H \backslash K)$. If $F \in \mathscr{K}$ and $F \subseteq H \backslash K$, then $\langle F \cap K_n \rangle_{n \in \mathbf{N}} \downarrow \varnothing$ and $\langle \lambda(F \cap K_n) \rangle_{n \in \mathbf{N}} \to 0$; this is where we use the condition (iv) of the theorem, and this is why the decreasing sequence $\langle K_n\rangle_{n\in\mathbf{N}}$ had to be found. Now we see that

$$\lambda F = \lim_{n\to\infty}\lambda(F\cap K_n)+\lim_{n\to\infty}\lambda_*(F\backslash K_n)$$

$$= \lim_{n\to\infty}\lambda(F\backslash K_n) \leqslant \lim_{n\to\infty}\lambda_*(H\backslash K_n).$$

As F is arbitrary, $\lambda_*(H\backslash K) \leqslant \lim_{n\to\infty}\lambda_*(H\backslash K_n)$, and

$$\lambda K \geqslant \lim_{n\to\infty}\lambda K_n \geqslant \lim_{n\to\infty}\lambda_*(H\backslash E_n)-3\epsilon \qquad \ldots(*)$$

So

$$\lambda_*(H\cap E)+\lambda_*(H\backslash E) \geqslant \lim_{n\to\infty}\lambda_*(H\cap E_n)+\lim_{n\to\infty}\lambda_*(H\backslash E_n)-3\epsilon$$

$$= \lambda H - 3\epsilon.$$

As ϵ is arbitrary,

$$\lambda_*(H\cap E)+\lambda_*(H\backslash E) \geqslant \lambda H;$$

as H is arbitrary, $E\in\Sigma$.

Moreover, if, above, $H\subseteq E$, then $K = \varnothing$; so by equation (*), $\lim_{n\to\infty}\lambda_*(H\backslash E_n)\leqslant 3\epsilon$. As ϵ is arbitrary, $\lim_{n\to\infty}\lambda_*(H\backslash E_n) = 0$, so

$$\lambda H = \lim_{n\to\infty}\lambda_*(H\cap E_n)+\lim_{n\to\infty}\lambda_*(H\backslash E_n)$$

$$\leqslant \lim_{n\to\infty}\lambda_* E_n.$$

Now, because H is arbitrary,

$$\lambda_* E \leqslant \lim_{n\to\infty}\lambda_* E_n \leqslant \lambda_* E,$$

as required. **Q**

(f) Thus (X,Σ,μ) is a measure space, and clearly it has the property (α). If $E\in\Sigma$ and $\mu E > 0$, there is a $K\in\mathscr{K}$ such that $K\subseteq E$ and $\lambda K > 0$; now $\mu K = \lambda K < \infty$. Thus (X,Σ,μ) is semi-finite. If $E\subseteq X$ and $E\cap K\in\Sigma$ for every $K\in\mathscr{K}$, then

$$\lambda_*(K\cap E)+\lambda_*(K\backslash E) = \mu(K\cap E)+\mu(K\backslash E) = \mu K = \lambda K$$

for every $K\in\mathscr{K}$, so $E\in\Sigma$; thus Σ has the property (β), and as $\mathscr{K}\subseteq\Sigma^f$, it follows at once that (X,Σ,μ) is locally determined. If $E\subseteq F\in\Sigma$ and $\mu F = 0$, then for any $K\in\mathscr{K}$,

$$\lambda_*(K\cap E)+\lambda_*(K\backslash E) \geqslant \lambda_*(K\backslash F) \geqslant \lambda K-\lambda_*(K\cap F) = \lambda K,$$

so $E\in\Sigma$. Thus (X,Σ,μ) is complete.

71B Definition Let X be a set and E a Riesz subspace of $\mathbf{R}^X$. An increasing linear functional $f\colon E\to\mathbf{R}$ is **sequentially smooth** if $\langle fx_n\rangle_{n\in\mathbf{N}}\to 0$ whenever $\langle x_n\rangle_{n\in\mathbf{N}}$ is a sequence in E such that $\langle x_n(t)\rangle_{n\in\mathbf{N}}\downarrow 0$ for every $t\in X$.

Remarks In §72 I shall introduce 'smooth' functionals.

Note that the property of being sequentially smooth is not intrinsic to the functional f and the space E; it depends on the embedding of E in $\mathbf{R}^X$. A sequentially order-continuous increasing linear functional on E must be sequentially smooth, but the converse is false. For instance, it is not hard to show that there is no non-trivial sequentially order-continuous increasing linear functional on $C([0, 1])$, but every increasing linear functional on $C([0, 1])$ is smooth.

Consequently we must, at several points below, distinguish between bounds in E and bounds in $\mathbf{R}^X$. I shall use the phrase '$\inf A = z$ in $\mathbf{R}^X$' to mean that $z(t) = \inf_{x \in A} x(t)$ for every $t \in X$; while '$\sup A = z$ in E' would mean that z was the least member of E which was an upper bound for A.

71C Definition Let X be a set and E a Riesz subspace of $\mathbf{R}^X$. I shall say that a functional $f: E \to \mathbf{R}$ is an **integral** if there is a measure μ on X such that $\int x\,d\mu$ exists and is equal to fx for every $x \in E$. In this case, I shall call f 'the integral with respect to μ'.

An integral is always linear, increasing and sequentially smooth [Lebesgue's theorem, 63Md]. To obtain a converse result, we need an extra condition.

71D Definition Let X be a set and E a Riesz subspace of $\mathbf{R}^X$. I shall call E **truncated** if $x \wedge \chi X \in E$ for every $x \in E$.

71E Lemma Let X be a set and E a truncated Riesz subspace of $\mathbf{R}^X$. Let $x \in E$. Then (a) $x \wedge \alpha\chi X \in E$ for every $\alpha > 0$ (b) if

$$H = \{t: x(t) \geqslant 1\},$$

then there is a sequence $\langle x_n \rangle_{n \in \mathbf{N}}$ in E such that $\langle x_n \rangle_{n \in \mathbf{N}} \downarrow \chi H$ in $\mathbf{R}^X$.

Proof (a) For $x \wedge \alpha\chi X = \alpha(\alpha^{-1}x \wedge \chi X)$. (**b**) Set

$$x_n = 2^n(x \wedge \chi X - x \wedge \alpha_n \chi X),$$

where $\alpha_n = 1 - 2^{-n}$.

71F Lemma Let (X, Σ, μ) be a measure space and E a truncated Riesz subspace of $\mathbf{R}^X$ such that, for every $x \in E$,

$$H_x = \{t: x(t) \geqslant 1\} \in \Sigma.$$

Suppose that f is a sequentially smooth increasing linear functional on $\boldsymbol{E}$ such that, for each $x \in \boldsymbol{E}$,

$$\mu H_x = \inf\{fy : y \in \boldsymbol{E}, y \geqslant \chi H_x\}.$$

Then f is the integral with respect to μ.

Proof (**a**) Let $x \in \boldsymbol{E}^+$, and for each $n, i \in \mathrm{N}$ set

$$F_{ni} = \{t : x(t) \geqslant 2^{-n}(i+1)\}$$
$$= \{t : 2^n(i+1)^{-1}x(t) \geqslant 1\} \in \Sigma.$$

Set $$y_{nr} = \textstyle\sum_{i \leqslant r} 2^{-n}\chi F_{ni};$$

then $y_{nr}(t) = 2^{-n}k$ if there is a $k \leqslant r$ such that $2^{-n}k \leqslant x(t) < 2^{-n}(k+1)$, and $2^{-n}(r+1)$ otherwise. Since

$$\chi F_{ni} \leqslant 2^n[x \wedge 2^{-n}(i+1)\,\chi X - x \wedge 2^{-n}i\chi X],$$
$$\mu F_{ni} \leqslant 2^n[f(x \wedge 2^{-n}(i+1)\,\chi X) - f(x \wedge 2^{-n}i\chi X)].$$

So $$\int y_{nr}\,d\mu = \textstyle\sum_{i \leqslant r} 2^{-n}\mu F_{ni} \leqslant f(x \wedge 2^{-n}(r+1)\,\chi X) \leqslant fx.$$

Put $y_n = \sup_{r \in \mathbf{N}} y_{nr}$ in $\mathbf{R}^X$; then by B. Levi's theorem [63Ma], $\int y_n\,d\mu$ exists and is not greater than fx. But

$$y_n(t) = 2^{-n}i \quad \text{whenever} \quad 2^{-n}i \leqslant x(t) < 2^{-n}(i+1).$$

So $\langle y_n \rangle_{n \in \mathbf{N}} \uparrow x$ in $\mathbf{R}^X$, and by B. Levi's theorem again,

$$\int x\,d\mu \leqslant fx.$$

(**b**) Again, let x be any member of $\boldsymbol{E}^+$, and let $\epsilon > 0$. Then

$$x_n = x \wedge 2^n\chi X - x \wedge 2^{-n}\chi X \in \boldsymbol{E} \quad \forall\ n \in \mathrm{N},$$

and $\langle x_n \rangle_{n \in \mathbf{N}} \uparrow x$ in $\mathbf{R}^X$, i.e. $\langle x - x_n \rangle_{n \in \mathbf{N}} \downarrow 0$ in $\mathbf{R}^X$. So there is an $n \in \mathrm{N}$ such that $fx \leqslant fx_n + \epsilon$, because f is sequentially smooth. Now set

$$H = \{t : x(t) \geqslant 2^{-n}\} \in \Sigma,$$

and set $y = 2^n\chi H$, so that $x_n \leqslant y$. Let $z \in \boldsymbol{E}$ be such that $z \geqslant \chi H$ and $fz \leqslant \mu H + 2^{-n}\epsilon$. Then

$$fx \leqslant \epsilon + fx_n = \epsilon + f(2^n z) - f(2^n z - x_n)$$
$$\leqslant \epsilon + 2^n fz - \int (2^n z - x_n)\,d\mu$$

(applying (a) above to $2^n z - x_n$)

$$\leqslant \epsilon + 2^n\mu H + \epsilon - \int (y - x_n)\,d\mu$$
$$= 2\epsilon + 2^n\mu H - \int y\,d\mu + \int x_n\,d\mu$$
$$= 2\epsilon + \int x_n\,d\mu$$
$$\leqslant 2\epsilon + \int x\,d\mu.$$

As ϵ is arbitrary, $fx \leqslant \int x\,d\mu$. Thus $fx = \int x\,d\mu$ for every $x \in \mathbf{E}^+$. Clearly now $fx = \int x\,d\mu$ for every $x \in \mathbf{E}$, i.e. f is the integral with respect to μ.

71G Theorem Let X be a set and $\mathbf{E}$ a truncated Riesz subspace of $\mathbf{R}^X$. Let $f: \mathbf{E} \to \mathbf{R}$ be a sequentially smooth increasing linear functional. Then there is a complete locally determined measure μ on X such that f is the integral with respect to μ.

Proof (**a**) Let $\mathscr{K}$ be the family of all those sets $K \subseteq X$ such that there exists a sequence $\langle x_n \rangle_{n \in \mathbf{N}}$ in $\mathbf{E}$ with $\chi K = \inf_{n \in \mathbf{N}} x_n$ in $\mathbf{R}^X$. Define $\lambda: \mathscr{K} \to \mathbf{R}^+$ by

$$\lambda K = \inf\{fx : x \in \mathbf{E}, x \geqslant \chi K\} \quad \forall\ K \in \mathscr{K}.$$

I propose to prove that $\mathscr{K}$ and λ satisfy the conditions of 71A.

(**b**) Obviously $\varnothing \in \mathscr{K}$, setting $x_n = 0$ for each $n \in \mathrm{N}$. If H, $K \in \mathscr{K}$, let $\langle x_n \rangle_{n \in \mathbf{N}}$ and $\langle y_n \rangle_{n \in \mathbf{N}}$ be sequences in $\mathbf{E}$ such that $\chi H = \inf_{n \in \mathbf{N}} x_n$ and $\chi K = \inf_{n \in \mathbf{N}} y_n$ in $\mathbf{R}^X$; then

$$\chi(H \cup K) = \inf_{n \in \mathbf{N}} x_n \vee y_n,$$

so $H \cup K \in \mathscr{K}$. If $\langle K_n \rangle_{n \in \mathbf{N}}$ is any sequence in $\mathscr{K}$, then choose for each $n \in \mathrm{N}$ a sequence $\langle x_{ni} \rangle_{i \in \mathbf{N}}$ in $\mathbf{E}$ such that $\chi K_n = \inf_{i \in \mathbf{N}} x_{ni}$; then

$$\chi(\textstyle\bigcap_{n \in \mathbf{N}} K_n) = \inf_{n,\, i \in \mathbf{N}} x_{ni},$$

so $\bigcap_{n \in \mathbf{N}} K_n \in \mathscr{K}$. Thus $\mathscr{K}$ is a sublattice of $\mathscr{P}X$, closed under countable intersections.

(**c**) Of course λ is increasing and $\lambda\varnothing = 0$. If H, $K \in \mathscr{K}$ and $H \cap K = \varnothing$, let $\langle x_n \rangle_{n \in \mathbf{N}}$ and $\langle y_n \rangle_{n \in \mathbf{N}}$ be sequences in $\mathbf{E}$ such that $\chi H = \inf_{n \in \mathbf{N}} x_n$ and $\chi K = \inf_{n \in \mathbf{N}} y_n$ in $\mathbf{R}^X$. Now let x be any member of $\mathbf{E}$ such that $x \geqslant \chi(H \cup K)$. For each $n \in \mathrm{N}$, set

$$z_n = x \wedge \inf_{i \leqslant n}(x_i \wedge y_i).$$

Then $\langle z_n \rangle_{n \in \mathbf{N}} \downarrow 0$ in $\mathbf{R}^X$, so $\langle fz_n \rangle_{n \in \mathbf{N}} \to 0$. But

$$\begin{aligned} x &\geqslant (x \wedge \inf_{i \leqslant n} x_i) \vee (x \wedge \inf_{i \leqslant n} y_i) \\ &= x \wedge \inf_{i \leqslant n} x_i + x \wedge \inf_{i \leqslant n} y_i - z_n, \end{aligned}$$

so

$$\begin{aligned} fx &\geqslant f(x \wedge \inf_{i \leqslant n} x_i) + f(x \wedge \inf_{i \leqslant n} y_i) - f(z_n) \\ &\geqslant \lambda H + \lambda K - fz_n; \end{aligned}$$

taking the limit as $n \to \infty$, $fx \geqslant \lambda H + \lambda K$. As x is arbitrary,

$$\lambda(H \cup K) \geqslant \lambda H + \lambda K,$$

as required by 71A(ii).

(d) Now suppose that $F, H \in \mathscr{K}$ and that $F \subseteq H$. Fix on a sequence $\langle x_n \rangle_{n\in\mathbf{N}}$ in $\boldsymbol{E}$ such that $\chi F = \inf_{n\in\mathbf{N}} x_n$ in $\mathbf{R}^X$. Let $\epsilon > 0$. Take an $x \in \boldsymbol{E}$ such that $x \geqslant \chi H$ and $fx \leqslant \lambda H + \epsilon$. Set

$$K_0 = \left\{t : t \in F,\, x(t) \leqslant \frac{1}{1+\epsilon}\right\} \subseteq F \backslash H.$$

Then
$$K_0 = \left\{t : t \in F,\, (x_n - x)(t) \geqslant \frac{\epsilon}{1+\epsilon} \quad \forall\; n \in \mathbf{N}\right\}$$
$$= F \cap \bigcap_{n\in\mathbf{N}} \{t : (\epsilon^{-1}+1)(x_n - x)(t) \geqslant 1\}.$$

But $\{t : (\epsilon^{-1}+1)(x_n - x)(t) \geqslant 1\} \in \mathscr{K}$ for every $n \in \mathbf{N}$, by 71Eb. So, because $\mathscr{K}$ is closed under countable intersections [(b) above], $K_0 \in \mathscr{K}$. Now suppose that z is any member of $\boldsymbol{E}$ such that $z \geqslant \chi K_0$. Then $z + (1+\epsilon)x \geqslant \chi F$, so

$$\lambda F \leqslant (1+\epsilon) fx + fz.$$

As z is arbitrary,

$$\begin{aligned} \lambda F &\leqslant (1+\epsilon) fx + \lambda K_0 \\ &\leqslant (1+\epsilon)(\epsilon + \lambda H) + \sup\{\lambda K : K \in \mathscr{K},\, K \subseteq F \backslash H\}. \end{aligned}$$

As ϵ is arbitrary,

$$\lambda F \leqslant \lambda H + \sup\{\lambda K : K \in \mathscr{K},\, K \subseteq F \backslash H\},$$

and 71A(iii) is satisfied.

(e) Finally, suppose that $\langle K_n \rangle_{n\in\mathbf{N}} \downarrow \varnothing$ in $\mathscr{K}$. For each $n \in \mathbf{N}$, choose a sequence $\langle x_{ni} \rangle_{i\in\mathbf{N}}$ in $\boldsymbol{E}$ such that $\chi K_n = \inf_{i\in\mathbf{N}} x_{ni}$ in $\mathbf{R}^X$. Set

$$y_n = \inf_{i,j\leqslant n} x_{ij} \quad \forall\; n \in \mathbf{N}.$$

Then $\langle y_n \rangle_{n\in\mathbf{N}} \downarrow$ and $\inf_{n\in\mathbf{N}} y_n = \inf_{n\in\mathbf{N}} \chi K_n = 0$ in $\mathbf{R}^X$. So $\langle f y_n \rangle_{n\in\mathbf{N}} \to 0$. But $y_n \geqslant \chi K_n$, so $\langle \lambda K_n \rangle_{n\in\mathbf{N}} \to 0$.

(f) Thus all the conditions of 71A are satisfied, and there is a complete locally determined measure μ on X extending λ. But now we find [using 71Eb again] that μ satisfies the conditions of 71F, as all the sets H_x actually belong to $\mathscr{K}$. So f is the integral with respect to μ.

71H Exercises (a) Show that under the conditions of 71A, there is only one extension of λ to a complete locally determined measure with the property (α).

(b) Suppose that X is a set and $\mathscr{K}$ a sublattice of $\mathscr{P}X$ containing $\varnothing$. Let $\lambda : \mathscr{K} \to \mathbf{R}^+$ be an increasing functional satisfying the conditions (i)–(iii) only of 71A. Then λ has an extension to an additive functional on the subring of $\mathscr{P}$X generated by $\mathscr{K}$.

(**c**) Let (X, T, ν) be any measure space. Let $\mathscr{K} = \mathrm{T}^f$ and let λ be the restriction of ν to T^f. Show that $\mathscr{K}$ and λ satisfy the conditions of 71A, and that the measure μ produced by 71A is precisely the complete measure on X obtained by applying 64Ja and 64Jb to (X, T, ν).

(**d**) Let X be a set, $\mathfrak{A}$ a subring of $\mathscr{P}X$, and $\mu: \mathfrak{A} \to \mathbf{R}$ an increasing additive functional. Show that μ has an extension to a measure on X iff, whenever $\langle A_n \rangle_{n \in \mathbf{N}}$ is a decreasing sequence in $\mathfrak{A}$ such that $\bigcap_{n \in \mathbf{N}} A_n = \varnothing$, then $\langle \mu A_n \rangle_{n \in \mathbf{N}} \to 0$. [Hint: *either* show that

$$\hat{\mu}: \boldsymbol{S}(\mathfrak{A}) \to \mathbf{R}$$

is sequentially smooth, using the method of 42Jc, and apply 71G, *or* apply 71A directly to a suitable $\mathscr{K} \supseteq \mathfrak{A}$.]

*(**e**) Let X be a set and $\boldsymbol{E}$ a Riesz subspace of $\mathbf{R}^X$ containing χX. Let Σ_0 be the smallest σ-subalgebra of $\mathscr{P}X$ such that $\boldsymbol{E} \subseteq \boldsymbol{M}(\Sigma_0)$, i.e. the σ-subalgebra of $\mathscr{P}X$ generated by the sets $\{t : x(t) > \alpha\}$ as x runs through $\boldsymbol{E}$ and α through $\mathbf{R}$. Let $f: \boldsymbol{E} \to \mathbf{R}$ be a sequentially smooth increasing linear functional. Show that there is a *unique* measure μ_0 with domain Σ_0 such that f is the integral with respect to μ_0. [Hint: define $\mathscr{K}$, Σ and μ as in 71G. Show that μ and μ_0 agree on $\mathscr{K}$, and therefore on Σ_0.]

Notes and comments It is a slightly surprising fact that, in 71G, we cannot dispense with the condition that $\boldsymbol{E}$ should be truncated; there is a counterexample in 7XB. Otherwise, it is clear that this theorem goes about as far as it can, since if we have any measure space (X, T, ν), we can set $\boldsymbol{E} = \mathfrak{L}^1(\nu)$, the space of ν-integrable functions on X, and that the theorem will then generate the 'complete locally determined extension' of ν described in 64Ja–b [cf. 71Hc above].

Readers who have encountered the Carathéodory 'outer measure' construction for measure spaces will recognize that 71A is based on a similar 'inner measure' method. Up to a point they are interchangeable (thus 71Hd can be used to tackle product measures, while an outer measure method will give 71G), but outer measures are probably more important for general measure theory. However, I shall not have occasion to use them in this book. The advantage of 71A is that it directly generates a locally determined measure space, which outer measures often fail to do.

71G can also be thought of as a way of extending a sequentially smooth functional to a whole $\mathfrak{L}^1$-space; the point about an $\mathfrak{L}^1$-space being that it is closed under countable suprema and infima, at least of sets which are order-bounded in itself. Indeed there exist proofs which set out to make this extension directly [see LOOMIS, § 12, pp. 29 *et seq.*]; this approach has the advantage, when E is not truncated, that equivalents of $\mathfrak{L}^1$ and L^1 can be constructed even though they may not correspond to any measure. It is a remarkable fact that the range space $\mathbf{R}$ in this result can be replaced by a Dedekind σ-complete Riesz space F iff F is weakly σ-distributive [WRIGHT E.T.].

72 Smooth functionals: quasi-Radon measure spaces

A natural adaption of 71B leads us to the concept of 'smooth' functional, in which the decreasing sequence of 71B is replaced by a directed set. In order to give a description of smooth functionals which will correspond to 71G, we need to consider a special type of measure space. Because its important applications are to functionals defined on spaces of continuous functions, I express the theory in terms of measures on topological spaces, the 'quasi-Radon' measures.

72A Definition **A quasi-Radon measure space** is a quadruple $(X, \mathfrak{T}, \Sigma, \mu)$, where (X, Σ, μ) is a measure space and $\mathfrak{T}$ is a topology on X such that:

(i) (X, Σ, μ) is complete and locally determined;

(ii) $\mathfrak{T} \subseteq \Sigma$ (i.e. every open set is measurable);

(iii) if $E \in \Sigma$ and $\mu E > 0$, there is a $G \in \mathfrak{T}$ such that $\mu G < \infty$ and $\mu(E \cap G) > 0$;

(iv) $\mu E = \sup\{\mu F : F \subseteq E, F \text{ closed}\}$ for every $E \in \Sigma$;

(v) if $\varnothing \subset \mathcal{G} \uparrow$ in $\mathfrak{T}$, then

$$\mu(\bigcup \mathcal{G}) = \sup\{\mu G : G \in \mathcal{G}\}.$$

Remark This definition probably strikes you as undesirably long and technical. I hope that by the end of this section it will be apparent that there are good reasons for each feature. The following theorem is a reason for many of them.

72B Theorem A quasi-Radon measure space $(X, \mathfrak{T}, \Sigma, \mu)$ is decomposable.

Proof I shall apply the criterion of 64I.

Let us call a set $F \in \Sigma$ *supporting* if $\mu(G \cap F) > 0$ whenever G is an open set meeting F. Let $\mathcal{X}$ be the collection of all disjoint families of supporting sets of finite measure. By Zorn's lemma, $\mathscr{X}$ has a maximal member $\mathscr{A}$.

? Now suppose, if possible, that

$$\sup\{F^\bullet : F \in \mathscr{A}\} \neq X^\bullet$$

in the measure algebra $\mathfrak{A}$ of (X, Σ, μ). Then there must be a non-zero $a \in \mathfrak{A}$ such that $a \cap F^\bullet = 0$ for every $F \in \mathscr{A}$; i.e. there is an $E \in \Sigma$ such that $\mu E > 0$ but $\mu(E \cap F) = 0$ for every $F \in \mathscr{A}$.

(i) Let $G_0 \in \mathfrak{T}$ be such that $\mu G_0 < \infty$ and $\mu(G_0 \cap E) > 0$ [72A(iii)]. Every $F \in \mathscr{A}$ is supporting, so

$$\begin{aligned}\mathscr{A}_0 &= \{F : F \in \mathscr{A}, F \cap G_0 \neq \varnothing\} \\ &= \{F : F \in \mathscr{A}, \mu(F \cap G_0) > 0\} \\ &= \bigcup_{n \in \mathbf{N}} \{F : F \in \mathscr{A}, \mu(F \cap G_0) \geqslant 2^{-n}\},\end{aligned}$$

which is countable, because $\mu G_0 < \infty$ and $\mathscr{A}$ is disjoint, so that

$$\{F : F \in \mathscr{A}, \mu(F \cap G_0) \geqslant 2^{-n}\}$$

cannot have more than $2^n \mu G_0$ members. So

$$E_1 = G_0 \cap E \backslash \bigcup \mathscr{A}_0 \in \Sigma;$$

also, because $\mu(E \cap F) = 0$ for every $F \in \mathscr{A}_0$, $\mu(E \cap \bigcup \mathscr{A}_0) = 0$, and $\mu E_1 = \mu(G_0 \cap E) > 0$. Now $E_1 \cap F = \varnothing$ for every $F \in \mathscr{A}$.

(ii) Let $\mathscr{G} = \{G : G \in \mathfrak{T}, G \subseteq G_0, \mu(E_1 \cap G) = 0\}$. Then $\mathscr{G} \uparrow$; let $H = \bigcup \mathscr{G}$. We know that

$$\mu H = \sup\{\mu G : G \in \mathscr{G}\} \leqslant \mu G_0 < \infty$$

[73A(v)]; so for any $\epsilon > 0$ there is a $G \in \mathscr{G}$ such that $\mu G \geqslant \mu H - \epsilon$. Now

$$\mu(H \cap E_1) \leqslant \mu(G \cap E_1) + \mu(H \backslash G) \leqslant \epsilon.$$

As ϵ is arbitrary, $\mu(H \cap E_1) = 0$. Set $F_0 = E_1 \backslash H$; then $\mu F_0 = \mu E_1 > 0$, $F_0 \cap F = \varnothing$ for every $F \in \mathscr{A}$, $F_0 \subseteq G_0$, and $F_0 \cap G = \varnothing$ for every $G \in \mathscr{G}$.

(iii) Now suppose that $G \in \mathfrak{T}$ and that G meets F_0. Then $G \cap G_0$ meets F_0, so $G \cap G_0 \notin \mathscr{G}$ and

$$0 < \mu(E_1 \cap G \cap G_0) \leqslant \mu(F_0 \cap G) + \mu(E_1 \backslash F_0) = \mu(F_0 \cap G).$$

As G is arbitrary, we see that F_0 is supporting.

(iv) It follows that F_0 is a supporting set of finite measure, not $\varnothing$, such that $F_0 \cap F = \varnothing$ for every $F \in \mathscr{A}$; which contradicts the supposed maximality of $\mathscr{A}$. **X**

Since (X, Σ, μ) is complete and locally determined [72A (i)], it is decomposable [64I].

72C Definition Let X be a set and $\boldsymbol{E}$ a Riesz subspace of $\mathbf{R}^X$. An increasing linear functional $f\colon \boldsymbol{E} \to \mathbf{R}$ is **smooth** if $\inf_{x\in A} fx = 0$ whenever A is a non-empty set in $\boldsymbol{E}$ such that $A \downarrow 0$ in $\mathbf{R}^X$ [i.e. A is directed downwards and $\inf_{x\in A} x(t) = 0$ for every $t \in X$].

72D Proposition Let $(X, \mathfrak{T}, \Sigma, \mu)$ be a quasi-Radon measure space. Let $\boldsymbol{E}$ be a Riesz subspace of $\boldsymbol{C}(X)$ such that $fx = \int x\,d\mu$ exists for every $x \in \boldsymbol{E}$. Then $f\colon \boldsymbol{E} \to \mathbf{R}$ is smooth.

Proof Suppose that $\varnothing \subset A \subseteq \boldsymbol{E}$ and that $A \downarrow 0$ in $\mathbf{R}^X$. Fix $x_0 \in A$. Let $\epsilon > 0$. Let $\delta > 0$ be such that

$$\int x_0 \wedge \delta\chi X \leqslant \epsilon,$$

and let $G_0 = \{t : x_0(t) > \delta\}$. Then G_0 is open and $\mu G_0 < \infty$. If $\mu G_0 = 0$, set $y = x_0$; then $fy = \int x_0 \leqslant \epsilon$. Otherwise, let $n \in \mathbf{N}$ be such that

$$\int x_0 \wedge 2^n\chi X \geqslant \int x_0 - \epsilon,$$

and set $\eta = 2^{-n}\epsilon$. Then for any $E \in \Sigma$ such that $\mu E \leqslant \eta$,

$$\begin{aligned}\int_E x_0 &\leqslant \int_E 2^n\chi X + \int (x_0 - 2^n\chi X)^+ \\ &\leqslant 2^n\mu E + \epsilon \leqslant 2\epsilon.\end{aligned}$$

For $x \in A$, set

$$H_x = \{t : t \in G_0,\ x(t) < (\mu G_0)^{-1}\epsilon\}.$$

Then $\{H_x : x \in A\} \uparrow G_0$, so by 73A(v) there is an $x \in A$ such that $\mu H_x \geqslant \mu G_0 - \eta$, i.e. $\mu(G_0 \backslash H_x) \leqslant \eta$. Now let $y \in A$ be such that $y \leqslant x \wedge x_0$. Then

$$\begin{aligned}y &= y \times \chi H_x + y \times \chi(G_0 \backslash H_x) + y \times \chi(X \backslash G_0) \\ &\leqslant \epsilon(\mu G_0)^{-1}\chi G_0 + x_0 \times \chi(G_0 \backslash H_x) + x_0 \wedge \delta\chi X.\end{aligned}$$

So in this case also $fy = \int y \leqslant \epsilon + 2\epsilon + \epsilon = 4\epsilon$. As ϵ is arbitrary, $\inf_{y\in A} fy = 0$; as A is arbitrary, f is smooth.

72E Theorem Let X be a set and $\boldsymbol{E}$ a truncated Riesz subspace of $\mathbf{R}^X$ [definition: 71D]. Let $f\colon \boldsymbol{E} \to \mathbf{R}$ be a smooth increasing linear functional. Let $\mathfrak{T}$ be the coarsest topology on X such that every member of $\boldsymbol{E}$ is continuous. Then there is a measure μ on X, quasi-Radon for $\mathfrak{T}$, such that f is the integral with respect to μ.

Proof Let $\mathscr{K}$ be the collection of all $\mathfrak{T}$-closed sets $K \subseteq X$ for which there is some $x \in \boldsymbol{E}$ with $x \geqslant \chi K$. Define $\lambda: \mathscr{K} \to \mathbf{R}^+$ by

$$\lambda K = \inf\{fx : x \in \boldsymbol{E},\, x \geqslant \chi K\} \quad \forall\ K \in \mathscr{K}.$$

Just as in 71G, I propose to show that $\mathscr{K}$ and λ satisfy the conditions of 71A.

(a) It will help to begin with a brief examination of $\mathfrak{T}$. For each $x \in \boldsymbol{E}$, let H_x be $\{t : x(t) > 0\}$, and set

$$Y = \bigcup\{H_x : x \in \boldsymbol{E}\} = \{t : \exists\ x \in \boldsymbol{E},\, x(t) \neq 0\}.$$

Since $H_x \cap H_y = H_{x \wedge y}$ for all $x, y \in \boldsymbol{E}$, $\{H_x : x \in \boldsymbol{E}\}$ is a base for a topology $\mathfrak{T}_0$ on Y. Let $\mathfrak{T}_1 = \{G : G \subseteq X,\, G \cap Y \in \mathfrak{T}_0\}$; then $\mathfrak{T}_1$ is a topology on X.

Suppose that $x \in \boldsymbol{E}^+$. For any $\alpha \geqslant 0$,

$$\{t : x(t) > \alpha\} = H_z \in \mathfrak{T}_0 \subseteq \mathfrak{T}_1,$$

where $z = x - x \wedge \alpha\chi X \in \boldsymbol{E}$ [using 71Ea]. Similarly,

$$\{t : t \in Y,\, x(t) < \alpha\} = \bigcup\{B_y : y \in \boldsymbol{E}\},$$

where $B_y = \{t : (y \wedge \alpha\chi X - x)(t) > 0\} \in \mathfrak{T}_0$ for each $y \in \boldsymbol{E}$. So

$$\{t : t \in Y,\, x(t) < \alpha\} \in \mathfrak{T}_0 \quad \text{and} \quad \{t : x(t) < \alpha\} \in \mathfrak{T}_1.$$

Thus every member of $\boldsymbol{E}^+$, and therefore every member of $\boldsymbol{E}$, is continuous for $\mathfrak{T}_1$. So $\mathfrak{T} \subseteq \mathfrak{T}_1$.

This shows that, if $G \in \mathfrak{T}$ and $t \in G \cap Y$, then there is an $x \in \boldsymbol{E}$ such that

$$t \in H_x = \{u : x(u) > 0\} \subseteq G.$$

(b) It follows that for any $K \in \mathscr{K}$,

$$A = \{x : x \in \boldsymbol{E},\, x \geqslant \chi K\} \downarrow \chi K \quad \text{in} \quad \mathbf{R}^X.$$

P Of course $A \downarrow$, and by the definition of $\mathscr{K}$, $A \neq \varnothing$. Fix $x_0 \in A$, and let t be any point of X. If $t \in K$, then $x_0 \wedge \chi X \in A$, so $\inf_{x \in A} x(t) = 1$. If $t \in X \setminus Y$, then $x_0(t) = 0$, so $\inf_{x \in A} x(t) = 0$. If $t \in Y \setminus K$, then by (a) above there is a $z \in \boldsymbol{E}$ such that

$$t \in \{u : z(u) > 0\} \subseteq X \setminus K.$$

Now $(x_0 - nz)^+ \in A$ for every $n \in \mathbf{N}$, so

$$\inf_{x \in A} x(t) \leqslant \inf_{n \in \mathbf{N}} (x_0 - nz)^+(t) = 0.$$

Thus in all cases $\inf_{x \in A} x(t) = (\chi K)(t)$, as required. **Q**

(c) Now we can proceed along the same lines as in 71G. Clearly, $\varnothing \in \mathscr{K}$, as $\varnothing$ is closed and $\chi\varnothing \leqslant 0 \in \boldsymbol{E}$. If $H, K \in \mathscr{K}$, and if $\chi H \leqslant x$ and $\chi K \leqslant y$ where x and y belong to $\boldsymbol{E}$, then $H \cup K$ is closed and

$$\chi(H \cup K) \leqslant x + y \in \boldsymbol{E};$$

so $H \cup K \in \mathscr{K}$. If $\langle K_n \rangle_{n\in\mathbf{N}}$ is any sequence in $\mathscr{K}$, let $x \in \boldsymbol{E}$ be such that $\chi K_0 \leqslant x$; then $K = \bigcap_{n\in\mathbf{N}} K_n$ is closed and $\chi K \leqslant x$, so $K \in \mathscr{K}$. Thus $\mathscr{K}$ is a sublattice of $\mathscr{P}X$, closed under countable intersections and containing $\varnothing$.

(**d**) Of course λ is increasing, and $\lambda\varnothing = 0$. If H and K belong to $\mathscr{K}$ and $H \cap K = \varnothing$, then by (b) above

$$A = \{x : x \in \boldsymbol{E},\, x \geqslant \chi H\} \downarrow \chi H \quad \text{in} \quad \mathbf{R}^X,$$

$$B = \{y : y \in \boldsymbol{E},\, y \geqslant \chi K\} \downarrow \chi K \quad \text{in} \quad \mathbf{R}^X.$$

So $\{x \wedge y : x \in A,\, y \in B\} \downarrow 0$ in $\mathbf{R}^X$. It follows, because f is smooth, that for any $\epsilon > 0$ there exist $x_1 \in A$ and $y_1 \in B$ such that $f(x_1 \wedge y_1) \leqslant \epsilon$. There is also a $z_1 \in \boldsymbol{E}$ such that $z_1 \geqslant \chi(H \cup K)$ and

$$\begin{aligned}\lambda(H \cup K) &\geqslant fz_1 - \epsilon \geqslant f(z_1 \wedge (x_1 \vee y_1)) - \epsilon \\ &= f(z_1 \wedge x_1) + f(z_1 \wedge y_1) - f(z_1 \wedge x_1 \wedge y_1) - \epsilon \\ &\geqslant \lambda H + \lambda K - 2\epsilon.\end{aligned}$$

As ϵ is arbitrary, $\lambda(H \cup K) \geqslant \lambda H + \lambda K$, and 71A(ii) is satisfied.

(**e**) Now suppose that $F, H \in \mathscr{K}$ and that $F \supseteq H$. Let $\epsilon > 0$. Take an $x \in \boldsymbol{E}$ such that $x \geqslant \chi H$ and $fx \leqslant \lambda H + \epsilon$. Let

$$K_0 = \left\{t : t \in F,\, x(t) \leqslant \frac{1}{1+\epsilon}\right\} \subseteq F \backslash H.$$

Then K_0 is closed and $K_0 \subseteq F$, so $K_0 \in \mathscr{K}$. Let $z \in \boldsymbol{E}$ be such that $z \geqslant \chi K_0$. Since $z + (1+\epsilon)x \geqslant \chi F$,

$$\lambda F \leqslant (1+\epsilon) fx + fz.$$

As z is arbitrary,

$$\begin{aligned}\lambda F &\leqslant (1+\epsilon) fx + \lambda K_0 \\ &\leqslant (1+\epsilon)(\lambda H + \epsilon) + \sup\{\lambda K : K \in \mathscr{K},\, K \subseteq F \backslash H\}.\end{aligned}$$

As ϵ is arbitrary,

$$\lambda F \leqslant \lambda H + \sup\{\lambda K : K \in \mathscr{K},\, K \subseteq F \backslash H\},$$

as required by 71A(iii).

(**f**) Finally, suppose that $\mathscr{A} \subseteq \mathscr{K}$ is non-empty, and that $\mathscr{A} \downarrow \varnothing$ in $\mathscr{P}X$ (i.e. $\mathscr{A}\downarrow$ and $\bigcap \mathscr{A} = \varnothing$). Then $\inf_{K\in\mathscr{A}} \lambda K = 0$. **P** Set

$$A = \{x : x \in \boldsymbol{E},\ \exists\ K \in \mathscr{A},\, x \geqslant \chi K\}.$$

Then $A\downarrow$ in $\boldsymbol{E}$, and

$$\inf A = \inf_{K\in\mathscr{A}} \chi K = 0$$

in $\mathbf{R}^X$ [using (b) above]. So

$$0 = \inf_{x \in A} fx = \inf_{K \in \mathscr{A}} \lambda K. \quad \mathbf{Q}$$

In particular, 71A(iv) is satisfied.

(**g**) Accordingly, there is a complete locally determined measure μ extending λ; and if Σ is the domain of μ,

(α) $\mu E = \sup\{\lambda K : K \in \mathscr{K}, K \subseteq E\}$ for every $E \in \Sigma$,

(β) If $E \subseteq X$ and $E \cap K \in \Sigma$ for every $K \in \mathscr{K}$, then $E \in \Sigma$.

From 71F we see again that f is the integral with respect to μ.

(**h**) We must now check that $(X, \mathfrak{T}, \Sigma, \mu)$ is quasi-Radon; let us go through the conditions of 72A in order.

(i) We know that (X, Σ, μ) is complete and locally determined.

(ii) If $G \in \mathfrak{T}$ and $K \in \mathscr{K}$, then $K \backslash G \in \mathscr{K} \subseteq \Sigma$; so $K \cap G \in \Sigma$. By (g) (β) above, $G \in \Sigma$.

(iii) If $E \in \Sigma$ and $\mu E > 0$, there is a $K \in \mathscr{K}$ such that $K \subseteq E$ and $\mu K > 0$. Now there is an $x \in \boldsymbol{E}$ such that $x \geqslant \chi K$. So if $G = \{t : x(t) > \frac{1}{2}\}$, $\mu G \leqslant 2fx < \infty$, while

$$\mu(G \cap E) \geqslant \mu(G \cap K) = \mu K > 0.$$

(iv) For every $E \in \Sigma$,

$$\mu E = \sup\{\mu K : K \in \mathscr{K}, K \subseteq E\}$$
$$\leqslant \sup\{\mu F : F \subseteq E, F \text{ closed}\} \leqslant \mu E,$$

using (g) (α).

(v) If $\varnothing \subset \mathscr{G} \uparrow G_0$ in $\mathfrak{T}$, let $K \in \mathscr{K}$ be such that $K \subseteq G_0$. Then $K \backslash G \in \mathscr{K}$ for every $G \in \mathscr{G}$, and $\{K \backslash G : G \in \mathscr{G}\} \downarrow \varnothing$. So by (f) above,

$$\inf\{\mu(K \backslash G) : G \in \mathscr{G}\} = \inf\{\lambda(K \backslash G) : G \in \mathscr{G}\} = 0.$$

So $$\mu K \leqslant \sup\{\mu(K \cap G) : G \in \mathscr{G}\} \leqslant \sup\{\mu G : G \in \mathscr{G}\}.$$

As K is arbitrary,

$$\mu G_0 \leqslant \sup\{\mu G : G \in \mathscr{G}\} \leqslant \mu G_0.$$

Thus the last condition of 73A is satisfied and the proof is complete.

***72F Lemma** Let $(X, \mathfrak{T}, \Sigma, \mu)$ be a quasi-Radon measure space.

(a) Let $\mathfrak{T}^f = \{G : G \in \mathfrak{T}, \mu G < \infty\}$. Then

(i) $\mu E = \sup\{\mu(E \cap G) : G \in \mathfrak{T}^f\}$ for every $E \in \Sigma$;

(ii) if $E \subseteq X$ and $E \cap G \in \Sigma$ for every $G \in \mathfrak{T}^f$, then $E \in \Sigma$.

(b) If ν is another quasi-Radon measure on $(X, \mathfrak{T})$, and if $\mu G = \nu G$ for every $G \in \mathfrak{T}$, then $\mu = \nu$.

Proof (a) (i) Let $\alpha = \sup\{\mu(G \cap E) : G \in \mathfrak{T}^f\}$. Then $\alpha \leqslant \mu E$, so if $\alpha = \infty$ then certainly $\alpha = \mu E$. On the other hand, if $\alpha < \infty$, we can choose a sequence $\langle G_n \rangle_{n \in \mathbf{N}}$ in $\mathfrak{T}^f$ such that

$$\mu(E \cap G_n) \geqslant \alpha - 2^{-n} \quad \forall\ n \in \mathbf{N}.$$

Now set $E_1 = E \backslash \bigcup_{n \in \mathbf{N}} G_n$. **?** If $\mu E_1 > 0$, there is a $G \in \mathfrak{T}^f$ such that $\mu(G \cap E_1) > 0$ [72A(iii)], and an $n \in \mathbf{N}$ such that $\mu(G \cap E_1) > 2^{-n}$. But now $G \cup G_n \in \mathfrak{T}^f$, so

$$\alpha \geqslant \mu(E \cap (G \cup G_n)) \geqslant \mu(G \cap E_1) + \mu(E \cap G_n) > 2^{-n} + \alpha - 2^{-n},$$

which is impossible. **X**

Thus $\mu E_1 = 0$, and

$$\mu E = \mu(E \cap \textstyle\bigcup_{n \in \mathbf{N}} G_n) = \lim_{n \to \infty} \mu(E \cap \bigcup_{i \leqslant n} G_i) \leqslant \alpha.$$

(ii) Now suppose that $E \subseteq X$ is such that $E \cap G \in \Sigma$ for every $G \in \mathfrak{T}^f$, and suppose that $F \in \Sigma^f$. Then by (i) there is a sequence $\langle G_n \rangle_{n \in \mathbf{N}}$ in $\mathfrak{T}^f$ such that $\sup_{n \in \mathbf{N}} \mu(F \cap G_n) = \mu F$, so that $\mu(F \backslash \bigcup_{n \in \mathbf{N}} G_n) = 0$. So

$$E \cap F \backslash \textstyle\bigcup_{n \in \mathbf{N}} G_n \in \Sigma \text{ because } (X, \Sigma, \mu) \text{ is complete, while}$$

$$E \cap F \cap \textstyle\bigcup_{n \in \mathbf{N}} G_n = \bigcup_{n \in \mathbf{N}} (E \cap G_n \cap F) \in \Sigma$$

by hypothesis. So $E \cap F \in \Sigma$. As F is arbitrary, $E \in \Sigma$, because (X, Σ, μ) is locally determined.

(b) The basic point we have to check is that μ and ν have the same domains of measurable sets. Let T be the domain of ν. Observe first that if $F \subseteq X$ is closed,

$$\begin{aligned} \mu F &= \sup\{\mu(F \cap G) : G \in \mathfrak{T}^f\}. \\ &= \sup\{\mu G - \mu(G \backslash F) : G \in \mathfrak{T}^f\} \\ &= \sup\{\nu G - \nu(G \backslash F) : G \in \mathfrak{T}^f\} \\ &= \nu F, \end{aligned}$$

using (a) (i) above and observing that

$$\mathfrak{T}^f = \{G : G \in \mathfrak{T},\ \mu G < \infty\} = \{G : G \in \mathfrak{T},\ \nu G < \infty\}.$$

So μ and ν agree on the closed sets. Now suppose that $E \in \mathrm{T}$. Let $G \in \mathfrak{T}^f$. By 72A(iv), applied to ν, there exist increasing sequences

$\langle F_n\rangle_{n\in\mathbf{N}}$ and $\langle H_n\rangle_{n\in\mathbf{N}}$ of closed subsets of $G\cap E$ and $G\backslash E$ respectively, such that

$$\sup_{n\in\mathbf{N}}\nu F_n = \nu(G\cap E), \quad \sup_{n\in\mathbf{N}}\nu H_n = \nu(G\backslash E).$$

Set $$F = \bigcup_{n\in\mathbf{N}}F_n \subseteq G\cap E, \quad H = \bigcup_{n\in\mathbf{N}}H_n \subseteq G\backslash E;$$

then F and H both belong to Σ, and

$$\begin{aligned}\mu(G\backslash(F\cup H)) &\leqslant \inf_{n\in\mathbf{N}}\mu(G\backslash(F_n\cup H_n))\\ &= \inf_{n\in\mathbf{N}}\nu(G\backslash(F_n\cup H_n))\\ &= \inf_{n\in\mathbf{N}}[\nu G-\nu F_n-\nu H_n]\\ &= \nu G-\nu(G\cap E)-\nu(G\backslash E)\\ &= 0.\end{aligned}$$

Now $(G\cap E)\backslash F\subseteq G\backslash(F\cup H)$; so, because (X,Σ,μ) is complete, $(G\cap E)\backslash F\in\Sigma$ and $G\cap E\in\Sigma$. As G is arbitrary, $E\in\Sigma$ [(a)(ii) above].

So we see that $\mathrm{T}\subseteq\Sigma$. Similarly, $\Sigma\subseteq\mathrm{T}$. Now it is obvious from 72A(iv) that μ and ν are equal, since we have seen that they agree on the closed sets.

72G Exercises (a) Let $(X,\mathfrak{T},\Sigma,\mu)$ be a quasi-Radon measure space, and $E\in\Sigma$. Then $\inf\{\mu G: G\in\mathfrak{T},\ G\supseteq E\}$ is either ∞ or μE.

(b) Let $(X,\mathfrak{T})$ be a topological space, and let μ and ν be two quasi-Radon measures on $(X,\mathfrak{T})$ which agree on the closed sets. Then they are equal.

(c) Suppose that, in 72E, $Y = X$, i.e. for every $t\in X$ there is an $x\in\boldsymbol{E}$ such that $x(t)\neq 0$. Then there is only one measure μ which is quasi-Radon for $(X,\mathfrak{T})$ and such that f is the integral with respect to μ.

(d) Let $(X,\mathfrak{T},\Sigma,\mu)$ be a quasi-Radon measure space and suppose that $\varnothing\subset\mathscr{G}\uparrow G_0$ in $\mathfrak{T}$. Then

$$\mu(E\cap G_0) = \sup\{\mu(E\cap G): G\in\mathscr{G}\}$$

for every $E\in\Sigma$.

Notes and comments The correspondence between 71G and 72E is made clear by part (b) of the proof of 72E; in 71G the family $\mathscr{K}$ consists of those sets K such that χK is the infimum of a sequence in $\boldsymbol{E}$, while in 72E we take sets K such that χK is the infimum of an arbitrary family in $\boldsymbol{E}$. The effect of introducing the topology is to reduce the rather obscure relationship between a measure and a space of functions on which it acts as a smooth functional to relatively straightforward relationships between both and the topology. For instance, putting 72E and 72B together, we have the result that a smooth functional is the integral with respect to a decomposable measure; it is not clear that there is any easy direct proof of this.

Quasi-Radon measure spaces have many remarkable properties; but the majority belong to pure measure theory and we shall have to pass them by. My aim has been to characterize smooth linear functionals in a way which will be an effective step towards the classical theorems of the next section. Note carefully the technical result 72Fb. A quasi-Radon measure is determined by its values on the open sets, and can therefore for some purposes be identified with its restriction to the Borel sets.

73 Radon measures and Riesz' theorem

Much the most important applications of the work of the last section are those in 73D and 73E below. In these results we find ourselves considering locally compact Hausdorff quasi-Radon measure spaces in which compact sets are of finite measure. These are the Radon measure spaces of twenty years ago; but in the last decade it has been shown that a rather wider class can be studied with the same methods. Again, I shall have to pass by the most interesting properties of these spaces, and concentrate on the results most closely related to those we have seen already.

73A Definition A **Radon measure space** is a quadruple $(X, \mathfrak{T}, \Sigma, \mu)$, where (X, Σ, μ) is a measure space and $\mathfrak{T}$ is a Hausdorff topology on X such that:

(i) (X, Σ, μ) is complete and locally determined;

(ii) $\mathfrak{T} \subseteq \Sigma$;

(iii) if $t \in X$, there is a $G \in \mathfrak{T}$ such that $t \in G$ and $\mu G < \infty$;

(iv) if $E \in \Sigma$,

$$\mu E = \sup\{\mu F : F \subseteq E,\ F \text{ compact}\}.$$

(Note that because $\mathfrak{T}$ is Hausdorff, every compact set is closed and therefore measurable.)

73B Because of the rather different forms of the definitions, I had better set out in detail the following result:

Proposition A Radon measure space $(X, \mathfrak{T}, \Sigma, \mu)$ is quasi-Radon.

Proof We know that 72A(i) and 72(Aii) are satisfied.

If $E \in \Sigma$ and $\mu E > 0$, there is a compact $F \subseteq E$ such that $\mu F > 0$ [73A (iv)]. We know that

$$\mathscr{G} = \{G : G \in \mathfrak{T}, \mu G < \infty\}\uparrow,$$

and $\bigcup \mathscr{G} = X$ by 73A (iii). So, as F is compact, there must be a $G \in \mathscr{G}$ such that $F \subseteq G$. Now $\mu G < \infty$ and $\mu(E \cap G) \geqslant \mu F > 0$. Thus 72A (iii) is satisfied.

72A(iv) is obvious from 73A(iv), because every compact set is closed.

If $\mathscr{G}$ is a non-empty set in $\mathfrak{T}$ and $\mathscr{G} \uparrow G_0$, then $\mu G_0 = \sup\{\mu F : F \subseteq G_0$, F compact$\}$. But if $F \subseteq G_0$ and F is compact, there must be a $G \in \mathscr{G}$ such that $G \supseteq F$. So

$$\mu G_0 \leqslant \sup\{\mu G : G \in \mathscr{G}\} \leqslant \mu G_0.$$

Thus 72A(v) is satisfied, and $(X, \mathfrak{T}, \Sigma, \mu)$ is indeed a quasi-Radon measure space.

73C In a locally compact space, we have something close to a converse result.

Lemma Let $(X, \mathfrak{T}, \Sigma, \mu)$ be a quasi-Radon measure space such that $\mathfrak{T}$ is locally compact and Hausdorff. Suppose moreover that $\mu F < \infty$ for every compact set F. Then $(X, \mathfrak{T}, \Sigma, \mu)$ is a Radon measure space.

Proof We know that 73A(i) and 73A(ii) are true.

If $t \in X$, let F be a compact neighbourhood of t, and let $G = \operatorname{int} F$. Then $\mu G \leqslant \mu F < \infty$. So 73A(iii) is satisfied.

Suppose that E is any measurable set, and that $\alpha < \mu E$. By 72A (iv), there is a closed set $F \subseteq E$ such that $\alpha < \mu F$. By 72Fa (i), there is an open set G such that $\mu G < \infty$ and $\alpha < \mu(G \cap F)$. Set

$$\mathscr{G} = \{H : H \in \mathfrak{T}, H \subseteq G, \bar{H} \text{ is compact}\}.$$

As $\mathfrak{T}$ is locally compact and Hausdorff, $\mathscr{G} \uparrow G$. By 72A(v), there is an $H \in \mathscr{G}$ such that

$$\mu(G \backslash H) = \mu G - \mu H < \mu(G \cap F) - \alpha.$$

Now $$\mu(\bar{H} \cap F) \geqslant \mu(H \cap F) \geqslant \mu(G \cap F) - \mu(G \backslash H) > \alpha.$$

But $\bar{H} \cap F$ is compact and $\bar{H} \cap F \subseteq F \subseteq E$. As α is arbitrary,

$$\mu E \leqslant \sup\{\mu K : K \subseteq E, K \text{ compact}\} \leqslant \mu E,$$

and 73A(iv) is satisfied.

73D Riesz' theorem: first form Let $(X, \mathfrak{T})$ be a locally compact Hausdorff topological space. Let $\boldsymbol{K} = \boldsymbol{K}(X)$ be the space of continuous functions on X with compact support [A2D]. Let $f: \boldsymbol{K} \to \mathbf{R}$ be an increasing linear functional. Then there exists a unique Radon measure μ on X such that f is the integral with respect to μ.

Proof (**a**) I show first that f is smooth. **P** Suppose that $\varnothing \subset A \subseteq \boldsymbol{K}$ and that $A \downarrow 0$ in $\mathbf{R}^X$. Fix $x_0 \in A$. Let $G = \{t : x_0(t) > 0\}$; then $\bar{G}$ is compact, so there is an $x_1 \in \boldsymbol{K}$ such that $x_1 \geqslant \chi\bar{G}$ in $\mathbf{R}^X$ [A2Eb].

Let $\epsilon > 0$. Let $\delta > 0$ be such that $\delta f(x_1) \leqslant \epsilon$. For each $x \in A$, set

$$H_x = \{t : x(t) < \delta\}.$$

Since $A \downarrow 0$ in $\mathbf{R}^X$, $\{H_x : x \in A\} \uparrow X$. Because $\bar{G}$ is compact, there is an $x \in A$ such that $H_x \supseteq \bar{G}$.

Now let $y \in A$ be such that $y \leqslant x \wedge x_0$. Then $y \leqslant \delta x_1$, so $fy \leqslant \delta f(x_1) \leqslant \epsilon$. As ϵ is arbitrary, $\inf_{y \in A} fy = 0$; as A is arbitrary, f is smooth. **Q**

(**b**) Because $\mathfrak{T}$ is completely regular, it is the coarsest topology on X such that every member of $\boldsymbol{K}$ is continuous [A2Ea]. And of course $\boldsymbol{K}$ is truncated. So by 72E there is a measure μ on X, quasi-Radon for $\mathfrak{T}$, such that f is the integral with respect to μ.

If $F \subseteq X$ is compact, there is an $x \in \boldsymbol{K}$ such that $x \geqslant \chi F$ [A2Eb]; so $\mu F \leqslant fx < \infty$. It follows by 73C that μ is a Radon measure.

(**c**) We see also that

$$\mu G = \sup\{fx : x \in \boldsymbol{K},\, x \leqslant \chi G\}$$

for every open set $G \subseteq X$. **P** We know that

$$\mu G = \sup\{\mu F : F \subseteq G,\, F \text{ compact}\}.$$

But if F is a compact subset of G, there is an $x \in \boldsymbol{K}$ such that

$$\chi F \leqslant x \leqslant \chi G$$

[A2Eb]. So $\mu F \leqslant \int x\, d\mu = fx$. Thus.

$$\mu G \leqslant \sup\{fx : x \in \boldsymbol{K},\, x \leqslant \chi G\} \leqslant \mu G. \quad \mathbf{Q}$$

Now, if ν is any other Radon measure on X such that $fx = \int x\, d\nu$ for every $x \in \boldsymbol{K}$, the same argument shows that

$$\nu G = \sup\{fx : x \in \boldsymbol{K},\, x \leqslant \chi G\} = \mu G$$

for every $G \in \mathfrak{T}$. So $\nu = \mu$ by 72Fb. Thus μ is unique.

73E Riesz' theorem: second form Let $(X, \mathfrak{T})$ be a locally compact Hausdorff topological space. Let C_0 be the space of continuous functions on X converging to 0 at infinity [A2C]. Let $f: C_0 \to \mathbf{R}$ be an increasing linear functional. Then there is a unique Radon measure μ on X such that f is the integral with respect to μ, and $\mu X < \infty$.

Proof Consider the restriction of f to $K = K(X)$. This is still an increasing linear functional, so by 73D there is a unique Radon measure μ on X such that $fx = \int x\,d\mu$ for every $x \in K$.

Now, because C_0 is complete for $\| \;\|_\infty$, f is continuous [25E], so

$$\|f\| = \sup\{|fx| : x \in C_0, \|x\|_\infty \leqslant 1\} < \infty.$$

But if $F \subseteq X$ is compact, there is an $x \in K$ such that $\chi F \leqslant x \leqslant \chi X$, and now

$$\mu F \leqslant fx \leqslant \|f\|.$$

So

$$\mu X = \sup\{\mu F : F \subseteq X, F \text{ compact}\} \leqslant \|f\| < \infty.$$

If now x is any continuous function from X to $\mathbf{R}$, x is certainly measurable, since every open set in X is measurable; so any bounded continuous function is integrable, because the integrable functions are solid in the measurable functions [63Ea], and the constant functions are integrable. Thus $\hat{\mu}(x) = \int x\,d\mu$ exists for every $x \in C_0$, and

$$|\textstyle\int x\,d\mu| \leqslant \int |x|\,d\mu \leqslant \int \|x\|_\infty \chi X\,d\mu = \|x\|_\infty \mu X.$$

So $\hat{\mu}: C_0 \to \mathbf{R}$ is a continuous linear functional, agreeing with f on K; as K is dense in C_0, $\hat{\mu}$ agrees with f on C_0, i.e. f is the integral with respect to μ.

***Remark** The original theorem of RIESZ referred to the special case $X = [0, 1]$ (in which case the two forms coincide) and showed that an arbitrary member of $C([0, 1])'$ could be expressed as a Riemann–Stieltjes integral.

***73F** It will be helpful later to be able to refer to an argument I have already used in 73D.

Lemma Let $(X, \mathfrak{T}, \Sigma, \mu)$ be a completely regular Radon measure space. Then

(a) for any compact set $K \subseteq X$,

$$\mu K = \inf\{\textstyle\int x\,d\mu : x \in C(X), \chi K \leqslant x \leqslant \chi X\};$$

(b) for any open set $G \subseteq X$,

$$\mu G = \sup\{\textstyle\int x\,d\mu : x \in C(X), 0 \leqslant x \leqslant \chi G\}.$$

Proof (**a**) Let $\mathcal{G} = \{G : G \in \mathfrak{T}, \mu G < \infty\}$. Then $\mathcal{G} \uparrow X$ by 73A (iii). So there must be a $G \in \mathcal{G}$ such that $G \supseteq K$.

Let $\epsilon > 0$. Let $F \subseteq G \backslash K$ be a compact set such that $\mu F \geqslant \mu(G \backslash K) - \epsilon$ [73A(iv)]. Set $H = G \backslash F$; then H is open, $H \supseteq K$, and $\mu H \leqslant \mu K + \epsilon$. Because $\mathfrak{T}$ is completely regular and K is compact, there is an $x \in \mathbf{C}(X)$ such that

$$\chi K \leqslant x \leqslant \chi H.$$

Now x is certainly measurable, so $\int x d\mu$ exists and

$$\int x d\mu \leqslant \mu H \leqslant \mu K + \epsilon.$$

As ϵ is arbitrary,

$$\mu K \geqslant \inf\{\int x d\mu : x \in \mathbf{C}, \chi K \leqslant x \leqslant \chi X\} \geqslant \mu K.$$

(**b**) If $\alpha < \mu G$, then there is a compact set $K \subseteq G$ such that $\mu K \geqslant \alpha$ [73A(iv)]. Let H be an open set of finite measure including K, as in (a) above. Let $x \in \mathbf{C}$ be such that $\chi K \leqslant x \leqslant \chi(G \cap H)$. Then $\int x d\mu$ exists and

$$\int x d\mu \geqslant \mu K \geqslant \alpha.$$

Thus $$\mu G \leqslant \sup\{\int x d\mu : x \in \mathbf{C}, 0 \leqslant x \leqslant \chi G\} \leqslant \mu G.$$

73G Exercises (**a**) Let $(X, \mathfrak{T})$ be a Hausdorff topological space. Let μ and ν be two Radon measures on $(X, \mathfrak{T})$ agreeing on the compact sets. Then they are equal.

(**b**) Let $(X, \mathfrak{T}, \Sigma, \mu)$ be a Radon measure space. Let $E \subseteq X$ be such that $E \cap F \in \Sigma$ for every compact set $F \subseteq X$. Then $E \in \Sigma$.

(**c**) Let X, $\mathfrak{T}$ and f: $\mathbf{K}(X) \to \mathbf{R}$ satisfy the hypotheses of 73D. Let $\mathcal{K}$ be the set of compact subsets of X, and define λ: $\mathcal{K} \to \mathbf{R}$ by

$$\lambda F = \inf\{fx : x \in \mathbf{K}, x \geqslant \chi F\} \quad \forall \ F \in \mathcal{K}.$$

Show that $\mathcal{K}$ and λ satisfy the conditions of 71A, and so prove 73D without using 72E.

(**d**) Let X, $\mathfrak{T}$ and f: $\mathbf{C}_0(X) \to \mathbf{R}$ satisfy the hypotheses of 73E. Show that f is smooth, and so prove 73E without using 73D.

(**e**) (Prokhorof's theorem) Let $(X, \mathfrak{T})$ be a completely regular Hausdorff topological space, and f: $\mathbf{C}_b(X) \to \mathbf{R}$ an increasing linear functional. Show that f is the integral with respect to a Radon measure on X iff it is **tight**, i.e.

for every $\epsilon > 0$ there is a compact set $F \subseteq X$
such that $fx \leqslant \epsilon$ whenever $x \leqslant \chi(X \backslash F)$.

[Hint: use 72E to find a quasi-Radon measure μ; now show that μ satisfies 73A(iv).]

*(**f**) Let (X,ρ) be a complete metric space. Show that every smooth increasing linear functional on $C_b(X)$ is tight.

*(**g**) Let (X,ρ) be a separable metric space. Show that every sequentially smooth increasing linear functional on $C_b(X)$ is smooth.

(**h**) Let $(X,\mathfrak{T})$ be a Hausdorff topological space and $\mathscr{K}$ the set of compact subsets of X. Let $\lambda\colon \mathscr{K} \to \mathbf{R}$ be an increasing functional such that (i) $\lambda\varnothing = 0$, (ii) $\lambda(H \cup K) = \lambda H + \lambda K$ whenever $H, K \in \mathscr{K}$ and $H \cap K = \varnothing$, (iii) if $F, H \in \mathscr{K}$ and $F \supseteq H$, then

$$\lambda F = \lambda H + \sup\{\lambda K : K \in \mathscr{K}, K \subseteq F\backslash H\};$$

(iv) for every $t \in X$ there is an open set G containing t such that $\sup\{\lambda K : K \in \mathscr{K}, K \subseteq G\} < \infty$. Then λ has a unique extension to a Radon measure on X.

(**i**) Let $(X,\mathfrak{T})$ be a Hausdorff topological space and $\mathfrak{B}$ the family of Borel subsets of X (i.e. $\mathfrak{B}$ is the σ-subalgebra of $\mathscr{P}X$ generated by $\mathfrak{T}$). Then a measure μ on X with domain $\mathfrak{B}$ can be extended to a Radon measure on X iff (i) it is **regular** (sometimes called **inner regular**) i.e.

$$\mu E = \sup\{\mu F : F \subseteq E, F \text{ compact}\} \quad \forall\ E \in \mathfrak{B};$$

(ii) it is locally finite, i.e. every point of X has a neighbourhood of finite measure. In this case the extension of μ is unique, and is precisely that obtained by applying the processes of 64Ja and 64Jb to $(X, \mathfrak{B}, \mu)$.

*(**j**) Using (f) and (g) above and 71He, or otherwise, show that if (X,ρ) is a complete separable metric space, then any measure μ defined on the Borel sets of X, such that $\mu X < \infty$, must be inner regular.

Notes and comments The theorems above have innumerable consequences in many branches of mathematics, and a classification of their applications would fill a book by itself; I shall make no attempt to describe them here.

Observe that the 'sequential' theorem 71G is inadequate for the applications above. If we try to apply 71G to the situation of 73D or 73E, we shall of course find a measure μ_0, since the increasing linear functional f is certainly sequentially smooth. But the domain Σ_0 of μ_0 may be too small. For we know only that functions in $\boldsymbol{K}(X)$ or $\boldsymbol{C}_0(X)$ are measurable; with a little ingenuity (using 71A(β)), we can prove that

$$\{t : x(t) > 0\} \in \Sigma_0$$

whenever $x \in C(X)$. But the σ-subalgebra generated by these sets, the σ-algebra of **Baire** sets, may fail to contain all compact sets [7XC]. Of course, we know that f corresponds to some Radon measure μ on X, and it is not hard to show that μ is the unique Radon measure extending μ_0. TOPSØE T.M. [§ 5] attempts to analyse the relationship between μ and μ_0.

For further information on the remarkable properties of Radon measures, I refer you to SCHWARTZ. *I ought to point out that his definition of Radon measure does not quite coincide with mine. His 'Radon measures' are my 'locally finite inner regular Borel measures', described in 73Gi. These need not be decomposable or even Maharam. But in view of the one-to-one correspondence described in 73Gi, the difference is only technical.

7X Examples for Chapter 7

The simplest example of the theorems of this chapter is of course the development of the Lebesgue integral from the Riemann integral [7XA]. 7XB shows that the conditions of 71G are all necessary for the result. 7XC shows how the procedures of 71G and 72E can produce different measures when applied to the same linear functional.

7XA Lebesgue measure on $\mathbf{R}^m$ There are various ways of defining a Riemann-type integral on continuous functions of compact support on $\mathbf{R}^m$; for instance, given $x \in \boldsymbol{K}(\mathbf{R}^m)$,

$$fx = \lim_{n\to\infty} \sum_{i_1=-\infty}^{\infty} \sum_{i_2=-\infty}^{\infty} \cdots \sum_{i_m=-\infty}^{\infty} (2^{-n})^m x(2^{-n}i_1, \ldots, 2^{-n}i_m)$$

exists, because x is uniformly continuous. Now $f: \boldsymbol{K}(\mathbf{R}^m) \to \mathbf{R}$ is an increasing linear functional, so by 73D there is a Radon measure μ on $\mathbf{R}^m$ such that $fx = \int x d\mu$ for every $x \in \boldsymbol{K}(\mathbf{R}^m)$. And of course μ is Lebesgue measure on $\mathbf{R}^m$.

In this case, because every compact set in $\mathbf{R}^m$ is a G_δ set, it does not matter whether we use 71G or 72E; the basic family $\mathscr{K}$ is the set of all compact subsets of $\mathbf{R}^m$ in either case.

7XB A non-truncated function space Let X be the unit interval $[0, 1]$, and let I be the σ-ideal of meagre subsets of X, i.e. the σ-ideal of $\mathscr{P}X$ generated by the nowhere dense closed sets. Let $\boldsymbol{E}$ be the set of all those functions $x: X \to \mathbf{R}$ such that there exists some $\alpha \in \mathbf{R}$ with

$$\{\beta : \beta \in X, x(\beta) \neq \alpha(1+\beta)\} \in I.$$

In this case, write $f(x) = \alpha$. Because $X \notin I$, f: $\boldsymbol{E} \to \mathbf{R}$ is well-defined. Then we see that $\boldsymbol{E}$ is an order-dense Riesz subspace of $\mathbf{R}^X$ and that f: $\boldsymbol{E} \to \mathbf{R}$ is a Riesz homomorphism. Moreover, because X is not a countable union of members of I, f is sequentially smooth.

However, f is not an integral. **P** **?** Suppose, if possible, that μ is a measure on X, with domain Σ, such that $\int x\,d\mu = fx$ for every $x \in \boldsymbol{E}$. Since the function e given by

$$e(\beta) = 1 + \beta \quad \forall\ \beta \in X$$

belongs to $\boldsymbol{E}$, $\mu X \leqslant f(e) = 1$; also, because e is measurable, every open set in $[0, 1]$ belongs to Σ. Let $\mathscr{G}$ be the algebra of regular open sets in X [4XF]. If $\langle G_n \rangle_{n \in \mathbf{N}} \downarrow \varnothing$ in $\mathscr{G}$, then $\operatorname{int}(\overline{\bigcap_{n \in \mathbf{N}} G_n})$ is a regular open set included in every G_n, so must be empty; thus $\bigcap_{n \in \mathbf{N}} G_n \in I$ and

$$0 = \mu(\bigcap_{n \in \mathbf{N}} G_n) = \lim_{n \to \infty} \mu G_n.$$

Thus if ν is the restriction of μ to $\mathscr{G}$, ν: $\mathscr{G} \to \mathbf{R}$ is an increasing countably additive functional, and $\nu X = \mu X \geqslant \frac{1}{2} f(e) = \frac{1}{2}$. But this is impossible, because X is separable and Hausdorff and without isolated points [4XFh]. **XQ**

So we see that the condition that $\boldsymbol{E}$ be truncated in 71G is essential. I do not know whether it is still needed in 72E.

7XC A Baire measure which is not a Borel measure Let ω_1 be the first uncountable ordinal, and let X be the set of all ordinals not greater than ω_1. Then X is a well-ordered set, therefore Dedekind complete, and it has greatest and least members, so in its order topology it is compact [1XC]. For $E \subseteq X$, ω_1 is a cluster point of E iff E is uncountable. If x: $X \to \mathbf{R}$ is a continuous function, then for every $n \in \mathbf{N}$

$$\{t : t \in X,\ |x(t) - x(\omega_1)| \geqslant 2^{-n}\}$$

must be countable, so $\{t : t \in X,\ x(t) \neq x(\omega_1)\}$ is countable. From this we see easily that the Baire sets of X are precisely the countable sets not containing ω_1, and their complements.

Now suppose we define f: $\boldsymbol{C}(X) \to \mathbf{R}$ by $fx = x(\omega_1)$ for every $x \in \boldsymbol{C}(X)$. If we follow the procedure of 71G, we get a measure μ_0 defined on the Baire sets by

$$\mu_0 E = 0 \quad \text{if } E \text{ is countable,} \quad 1 \quad \text{otherwise.}$$

In particular, $\{\omega_1\}$ is not measurable. While of course the procedure of 72E produces a measure μ defined on all subsets of X by

$$\mu E = 0 \quad \text{if} \quad \omega_1 \notin E, \quad \mu E = 1 \quad \text{if} \quad \omega_1 \in E.$$

μ is of course a Radon measure, the unique Radon measure extending μ_0, and is also produced by applying 73D.

*Actually, there is another Borel measure extending μ_0. The point is that, given any Borel set $E \subseteq X$, *either* there is an uncountable closed set F such that $F \backslash \{\omega_1\} \subseteq E$ *or* there is an uncountable closed set F such that $F \backslash \{\omega_1\} \subseteq X \backslash E$. (The proof proceeds by showing that the collection of sets E with this property is a σ-subalgebra of $\mathscr{P}X$.) In the first case set $\mu_1 E = 1$; otherwise set $\mu_1 E = 0$. Then μ_1 is a countably additive functional extending μ_0; it is a Borel measure which is not regular [73Gi].

Further reading for Chapter 7 A great deal of work has been done on measures on topological spaces. Once again, the best starting point is probably HALMOS M.T., chapter X [§§ 50-6]. A more elaborate treatment is in BOURBAKI VI, where Riesz' theorem is made the basis of the whole theory, and a 'measure' is regarded primarily as a linear functional on $\boldsymbol{K}(X)$. Many of the same ideas may be found in SCHWARTZ, who takes a line rather closer to that of §§72–3 above.

An important direction of generalization for the work of this chapter is towards a theory of vector-valued measures and integrals. Measures taking values in normed spaces are discussed by DUNFORD & SCHWARTZ, [chapter III]. Perhaps closer to the ideas of this book is the notion of a measure which takes values in an ordered linear space; the elementary techniques are exhaustively investigated by MCSHANE, while some more sophisticated results have been given recently in a series of papers by WRIGHT.

8. Weak compactness

Weakly compact sets have proved fascinating objects of study in the theory of linear topological spaces. In addition to the rich harvest of general theory, many results have been proved for special spaces, which are often topological Riesz spaces. It is my object here to collect these together, and to give some new theorems. As we shall be considering pairs of Riesz spaces in duality, this chapter will carry on from Chapter 3. At a number of points I shall call on the more advanced results in the general theory of linear topological spaces laid out in the appendix [§ A1].

81 Weak compactness in $E^\sim$

This section is a foundation for the rest. It will be helpful to have some results concerning weakly compact sets in $E^\sim$ safely established before we go on to weakly compact sets in $E^\times$. It turns out that the 'normal' topology on $E^\sim$, $\mathfrak{T}_{|s|}(E^\sim, E)$, the coarsest locally solid topology finer than $\mathfrak{T}_s(E^\sim, E)$, is useful for technical reasons. The results below are generally straightforward, with the exception of 81F(v) and 81H, where some relatively deep ideas come to the surface.

81A Definitions Let E and F be Riesz spaces. A bilinear functional $\langle\ ,\ \rangle: E\times F\to\mathbf{R}$ is **positive** if $\langle x,y\rangle\geqslant 0$ whenever $x\in E^+$ and $y\in F^+$; in this case we say that E and F are 'in positive duality'.

In this case, $\mathfrak{T}_{|s|}(F, E)$ is that linear space topology on F which has for basic neighbourhoods of 0 the sets

$$U_x=\{y:\langle x,|y|\rangle\leqslant 1\}$$

as x runs through E^+. This is locally solid and locally convex; it is the coarsest locally solid linear space topology finer than $\mathfrak{T}_s(F,E)$. It is defined by the Riesz seminorms of the form

$$y\mapsto\langle x,|y|\rangle$$

as x runs through E^+.

If F is a Riesz subspace of $E^\sim$, then $\mathfrak{T}_{|s|}(F, E)$ is the topology of uniform convergence on order-bounded sets of E [16Ea]; the same is true if E is a solid linear subspace (or any locally order-dense Riesz subspace) of $F^\sim$ [31D, 31Ea].

81B Proposition Let E and F be Riesz spaces in positive duality, and let E_1 be the image of E in F^* under the canonical map $x \mapsto \hat{x}$ given by

$$\hat{x}(y) = \langle x, y\rangle \quad \forall\ y \in F, x \in E.$$

Then $E_1 \subseteq F^\sim$ and the dual F' of F under $\mathfrak{T}_{|s|}(F, E)$ is precisely the solid hull of E_1 in $F^\sim$.

Proof Clearly, if $x \in E^+$, then $\hat{x} \in F^{\sim +}$. As $E = E^+ - E^+$, $E_1 \subseteq F^\sim$. It also follows that the solid hull of E_1 is

$$\{g : \exists\ x \in E^+, |g| \leqslant \hat{x}\} = G \quad \text{say}.$$

Suppose $g \in G$; then there is an $x \in E^+$ such that $|g| \leqslant \hat{x}$, and

$$|gy| \leqslant |g|\,(|y|) \leqslant \hat{x}(|y|) = \langle x, |y|\rangle \quad \forall\ y \in F,$$

so g is continuous. Conversely, if $g \in F'$, there is an $x \in E^+$ such that

$$|gy| \leqslant \langle x, |y|\rangle = \hat{x}(|y|) \quad \forall\ y \in F,$$

so $g \in F^\sim$ and $|g| \leqslant x$; thus $g \in G$.

81C Proposition Let E be any Riesz space. Then $\mathfrak{T}_{|s|}(E^\sim, E)$ is a complete Levi Hausdorff Lebesgue Fatou topology.

Proof 16Db gives us 'Fatou'; 23M gives us 'Levi'; 16Ec gives us 'Lebesgue'. It is Hausdorff because E separates the points of $E^\sim$. Now completeness follows either from Nakano's theorem [23K] or by applying Grothendieck's criterion [A1D], because $E^\sim$ is precisely the set of linear functionals on E which are bounded on the order-bounded sets, and $\mathfrak{T}_{|s|}(E^\sim, E)$ is the topology of uniform convergence on these.

81D Corollary Let E be any Riesz space. Then $\mathfrak{T}_{|s|}(E^\times, E)$ is a complete Levi Hausdorff Lebesgue Fatou topology.

Proof Observe that $\mathfrak{T}_{|s|}(E^\times, E)$ is just the topology on $E^\times$ induced by $\mathfrak{T}_{|s|}(E^\sim, E)$. Now $E^\times$ is a band in $E^\sim$ [16H]; its properties therefore follow at once from those of $E^\sim$ (recalling that by 22Ec $E^\times$ is closed).

81E Lemma Let E be a Riesz space and A a subset of $E^\sim$. Let C be the solid hull of A in $E^\sim$. Then, for any $x \in E$,

$$\sup_{f \in C} |fx| = \sup_{f \in A} |f|\,(|x|).$$

Proof For we see from 31D that, given any $f \in E^\sim$ and $x \in E$,

$$|f|(|x|) = \sup\{|gx| : |g| \leqslant |f|\}.$$

81F Now we can describe the convex relatively weakly compact sets in $E^\sim$. The only really surprising condition is (v).

Proposition Let E be a Riesz space and A a subset of $E^\sim$. Then these are equivalent:

(i) The solid convex hull of A in $E^\sim$ is relatively compact for $\mathfrak{T}_s(E^\sim, E)$.

(ii) The convex hull of A is relatively compact for $\mathfrak{T}_s(E^\sim, E)$.

(iii) The solid hull of A in $E^\sim$ is bounded for $\mathfrak{T}_s(E^\sim, E)$.

(iv) A is bounded for $\mathfrak{T}_{|s|}(E^\sim, E)$.

*(v) A is bounded for $\mathfrak{T}_s(E^\sim, E)$ and $\sup_{n \in \mathbb{N}} \sup_{f \in A} |fx_n| < \infty$ whenever $\langle x_n \rangle_{n \in \mathbb{N}}$ is a disjoint sequence in E^+ which is bounded above in E^+.

Proof (a) (i) $\Rightarrow$ (ii) and (i) $\Rightarrow$ (iii) $\Leftrightarrow$ (iv) $\Rightarrow$ (v) are elementary, in view of 81E. I shall prove that (ii) $\Rightarrow$ (iv) $\Rightarrow$ (i) and that (v) $\Rightarrow$ (iv).

(b) (ii) $\Rightarrow$ (iv) Let B be the $\mathfrak{T}_s(E^\sim, E)$-closed convex hull of A in $E^\sim$. Then B is convex and $\mathfrak{T}_s(E^\sim, E)$-compact. So, if $C \subseteq E$ is $\mathfrak{T}_s(E, E^\sim)$-bounded,

$$\sup\{|fx| : f \in B, x \in C\} < \infty$$

by the Banach–Mackey theorem [A1I]. In particular, consider the case $C = [-x, x]$, where $x \in E^+$; then

$$\infty > \sup\{|fy| : f \in B, y \in C\}$$
$$= \sup\{|f|(x) : f \in B\}$$

by 16Ea. As x is arbitrary, B, and therefore A, is $\mathfrak{T}_{|s|}(E^\sim, E)$-bounded.

(c) (iv) $\Rightarrow$ (i) For $x \in E$, set

$$\rho(x) = \sup_{f \in A} |f|(|x|) < \infty.$$

Then ρ is a Riesz seminorm on E, so induces a locally solid linear space topology $\mathfrak{T}$ on E. Let E' be the dual of E for $\mathfrak{T}$; then $E' \subseteq E^\sim$ [22D]. Let $U = \{x : \rho(x) \leqslant 1\}$; then U is a solid neighbourhood of 0 in E, so U^0 is a convex solid $\mathfrak{T}_s(E', E)$-compact set [22D]. But $A \subseteq U^0 \subseteq E' \subseteq E^\sim$, so the solid convex hull of A is included in U^0, and is relatively compact.

***(d) (v) $\Rightarrow$ (iv)** For $x \in E$. define

$$\rho(x) = \sup_{f \in A} |fx| < \infty.$$

Let F be $\{y : y \in E, \sup\{\rho(x) : |x| \leqslant |y|\} < \infty\}$.

Then F is a solid linear subspace of E. **P** Clearly F is solid. If y_1, y_2 belong to F, then

$$\sup\{\rho(x): |x| \leqslant |y_1+y_2|\} \leqslant \sup\{\rho(x): |x| \leqslant |y_1|+|y_2|\}$$
$$= \sup\{\rho(x_1+x_2): |x_1| \leqslant |y_1|, |x_2| \leqslant |y_2|\}$$

[using 14Jb]

$$\leqslant \sup\{\rho(x): |x| \leqslant |y_1|\}$$
$$+\sup\{\rho(x): |x| \leqslant |y_2|\} < \infty.$$

Thus F is closed under addition, and is therefore a solid linear subspace. **Q**

? Now suppose, if possible, that $F \neq E$. Then there is an $x_0 \in E^+ \backslash F$. Let E_0 be the solid linear subspace of E generated by x_0, and let $F_0 = F \cap E_0$, so that F_0 is a solid linear subspace of E_0 not containing x_0. We now find that there is a Riesz homomorphism $f: E_0 \to \mathbf{R}$ such that $fx = 0$ for every $x \in F_0$ but $fx_0 \neq 0$. **P** Let $\mathscr{X}$ be the set of all those solid linear subspaces of E_0 which include F_0 but do not contain x_0. By Zorn's lemma, $\mathscr{X}$ has a maximal member H say. Any solid linear subspace of E_0 properly including H must contain x_0, and therefore must be E_0 itself. Thus H is a maximal proper solid linear subspace of E_0, and the Riesz space quotient E_0/H has no non-trivial solid linear subspace. As $H \neq E_0$, $E_0/H \neq \{0\}$, and $E_0/H \cong \mathbf{R}$ [15G(v)]. Now let f be the composition of this isomorphism with the canonical map from E_0 to E_0/H; then f is a Riesz homomorphism and the kernel of f is H, which includes F_0 but does not contain x_0. **Q**

Let $fx_0 = \alpha$. Now construct sequences $\langle z_n\rangle_{n\in\mathbf{N}}$ and $\langle w_n\rangle_{n\in\mathbf{N}}$ as follows. Set $z_0 = x_0$. Having chosen z_n such that $0 \leqslant z_n \leqslant x_0$ and $fz_n = \alpha$, we know that $z_n \notin F_0$, so there is a u such that $|u| \leqslant z_n$ and

$$\rho(u) > 4n+\rho(z_n).$$

Set

$$v = u-\alpha^{-1}f(u)\,z_n.$$

Then $fv = 0$; also, as $|fu| = f(|u|) \leqslant f(z_n) = \alpha$,

$$\rho(v) \geqslant \rho(u)-\alpha^{-1}|fu|\,\rho(z_n) \geqslant \rho(u)-\rho(z_n) > 4n.$$

So $\max(\rho(v^+), \rho(v^-)) > 2n$; set $w = \frac{1}{2}v^+$ or $w = \frac{1}{2}v^-$ in such a way that $\rho(w) > n$. Again because $|\alpha^{-1}fu| \leqslant 1$, $-2z_n \leqslant v \leqslant 2z_n$, so $0 \leqslant w \leqslant z_n$; and $fw = 0$ because $fv = 0$.

Let $g \in A$ be such that $|g(w)| \geqslant n$, and let $\beta = |g|\,(z_n)$. Set

$$w_n = (w-\beta^{-1}z_n)^+.$$

Then $0 \leqslant w_n \leqslant w \leqslant z_n$, and

$$\rho(w_n) \geqslant |g(w_n)| = |g(w)-g(w \wedge \beta^{-1}z_n)| \geqslant |g(w)|-\beta^{-1}|g|\,(z_n) \geqslant n-1.$$

Set $z_{n+1} = (z_n - \beta w)^+$. Then

$$0 \leqslant z_{n+1} \leqslant z_n, \quad z_{n+1} \wedge w_n = 0 \quad \text{and} \quad f(z_{n+1}) = f(z_n) = \alpha.$$

Continue.

Now, because $\langle z_n \rangle_{n \in \mathbf{N}}$ is decreasing, $\langle w_n \rangle_{n \in \mathbf{N}}$ is a disjoint sequence in E^+, bounded above by x_0, and $\langle \rho(w_n) \rangle_{n \in \mathbf{N}} \to \infty$; which is impossible. **X**

Thus $F = E$, and

$$\begin{aligned} \sup_{f \in A} |f|(x) &= \sup\{|fy| : |y| \leqslant x, f \in A\} \\ &= \sup\{\rho(y) : |y| \leqslant x\} < \infty \end{aligned}$$

for every $x \in E^+$; i.e. A is $\mathfrak{T}_{|s|}(E^\sim, E)$-bounded.

81G Lemma Let E be a uniformly complete Archimedean Riesz space. If $A \subseteq E^\sim$ is bounded for $\mathfrak{T}_s(E^\sim, E)$, it is bounded for $\mathfrak{T}_{|s|}(E^\sim, E)$.

Proof Let $x \in E^+$. Let E_x be the solid linear subspace of E generated by x, and let $\| \ \|_x$ be the norm on E_x derived from its order unit x [25H]. Consider A as a set of functions on E_x. Since $A \subseteq E^\sim$, each member of A is bounded on the unit ball of E_x, i.e. is continuous for $\| \ \|_x$. Also, A is bounded for $\mathfrak{T}_s(E^\sim, E_x)$. By the uniform boundedness theorem (this is where we use the fact that E is uniformly complete, so that E_x is complete under $\| \ \|_x$), A is equicontinuous for $\| \ \|_x$, i.e.

$$\sup\{|fy| : f \in A,\ |y| \leqslant x\} < \infty.$$

But this is just

$$\sup\{|f|(x) : f \in A\} < \infty.$$

As x is arbitrary, A is $\mathfrak{T}_{|s|}(E^\sim, E)$-bounded, as stated.

***81H** The following lemma will be used later to give a criterion for compactness in the topology $\mathfrak{T}_s(E^\sim, E^{\sim\times})$ under suitable conditions.

Lemma Let E be an Archimedean Riesz space and F a Riesz subspace of $E^\sim$. Let $A \subseteq F$ be a non-empty $\mathfrak{T}_s(F, E)$-bounded set such that, whenever $\langle x_n \rangle_{n \in \mathbf{N}}$ is a disjoint sequence in E^+ which is bounded above,

$$\langle \sup_{f \in A} |fx_n| \rangle_{n \in \mathbf{N}} \to 0.$$

Then for every $x_0 \in E^+$ and $\epsilon > 0$, there is an $f_0 \in F^+$ such that

$$(|f| - |f| \wedge f_0)(x_0) \leqslant \epsilon \quad \forall\ f \in A.$$

Proof (a) By 81F(v), A is bounded for $\mathfrak{T}_{|\mathfrak{s}|}(\boldsymbol{E}^{\sim}, \boldsymbol{E})$. Define ρ: $\boldsymbol{E} \to \mathbf{R}^+$ by

$$\rho(x) = \sup\{|f|(|x|) : f \in A\};$$

then ρ is a Riesz seminorm on $\boldsymbol{E}$. Let $\mathfrak{T}$ be the topology on $\boldsymbol{E}$ defined by ρ. Then $\mathfrak{T}$ satisfies the conditions of 24H. **P** Of course order-bounded sets are bounded, as $\mathfrak{T}$ is locally solid. Now suppose that $\langle x_n \rangle_{n\in\mathbf{N}}$ is a disjoint sequence in $\boldsymbol{E}^+$, bounded above by x. Choose, for each $n \in \mathbf{N}$, an $f_n \in A$ such that $|f_n|(|x_n|) \geqslant \rho(x_n) - 2^{-n}$, and a $y_n \in \boldsymbol{E}$ such that $|y_n| \leqslant x_n$ and $|f_n y_n| \geqslant \rho(x_n) - 2.2^{-n}$. Then $\langle y_n^+ \rangle_{n\in\mathbf{N}}$ and $\langle y_n^- \rangle_{n\in\mathbf{N}}$ are both disjoint sequences in $\boldsymbol{E}^+$, bounded above by x, so

$$\begin{aligned}\rho(x_n) &\leqslant |f_n y_n| + 2.2^{-n} \\ &\leqslant \sup_{f\in A}|f y_n^+| + \sup_{f\in A}|f y_n^-| + 2.2^{-n} \\ &\to 0 \quad \text{as} \quad n \to \infty.\end{aligned}$$

Thus $\langle x_n \rangle_{n\in\mathbf{N}} \to 0$, as required. **Q**

(**b**) **?** Now suppose, if possible, that the result is false. Then there is an $x_0 \in \boldsymbol{E}^+$ and an $\epsilon > 0$ such that, for every $g \in \boldsymbol{F}^+$,

$$\sup_{f\in A}(|f| - g)^+(x_0) = \sup_{f\in A}(|f| - |f| \wedge g)(x_0) > 3\epsilon.$$

So we may choose a sequence $\langle f_n \rangle_{n\in\mathbf{N}}$ in A such that

$$(|f_n| - 2^n \textstyle\sum_{i<n} |f_i|)^+(x_0) > 3\epsilon \quad \forall\ n \in \mathbf{N}.$$

Next choose, for each $n \in \mathbf{N}$, an element y_n of $\boldsymbol{E}$ such that $0 \leqslant y_n \leqslant x_0$ and

$$(|f_n| - 2^n \textstyle\sum_{i<n} |f_i|)(y_n) \geqslant 3\epsilon.$$

It follows that

$$|f_i|(y_n) \leqslant 2^{-n} |f_n|(y_n) \leqslant 2^{-n}\alpha \quad \forall\ i < n,$$

where

$$\alpha = \sup\{|f|(x_0) : f \in A\} < \infty;$$

at the same time, $|f_n|(y_n) \geqslant 3\epsilon$.

Consider now, for $m, n \in \mathbf{N}$,

$$z_{mn} = \sup_{m \leqslant i \leqslant m+n} y_i.$$

For each $m \in \mathbf{N}$, $\langle z_{mn} \rangle_{n\in\mathbf{N}}$ is an increasing sequence, bounded above by x_0. So by 24H it is Cauchy for $\mathfrak{T}$, and there is an $r \in \mathbf{N}$ such that

$$\rho(z_{mn} - z_{mr}) \leqslant 2^{-m}\epsilon \quad \forall\ n \geqslant r.$$

Set $w_m = z_{mr}$. Now if $m \leqslant i \leqslant m+r$, $\rho(y_i - w_m)^+ = 0$; while if $i > m+r$,

$$\rho(y_i - w_m)^+ \leqslant \rho(z_{mi} - w_m) \leqslant 2^{-m}\epsilon.$$

Thus $\rho(y_i - w_m)^+ \leqslant 2^{-m}\epsilon$ for every $i \geqslant m$.

Now set $v_m = \inf_{i \leqslant m} w_i$ for each $m \in \mathbf{N}$. Since $\langle v_m \rangle_{m\in\mathbf{N}}$ is a decreasing sequence, bounded below, it is also Cauchy for $\mathfrak{T}$, and

$$\lim_{n\to\infty} \rho(v_n - v_{n+1}) = 0.$$

However, examine $|f_m|(v_m - v_{m+1})$, for any $m \in \mathbf{N}$. We know that

$$|f_m|(y_m - v_m)^+ \leqslant \sum_{i\leqslant m} |f_m|(y_m - w_i)^+ \leqslant \sum_{i\leqslant m} \rho(y_m - w_i)^+$$
$$\leqslant \sum_{i\leqslant m} 2^{-i}\epsilon \leqslant 2\epsilon.$$

So

$$|f_m|(v_m) \geqslant |f_m|(y_m \wedge v_m) = |f_m|(y_m) - |f_m|(y_m - v_m)^+$$
$$\geqslant |f_m y_m| - 2\epsilon \geqslant \epsilon.$$

On the other hand,

$$|f_m|(v_{m+1}) \leqslant |f_m|(w_{m+1}) \leqslant \sum_{i\geqslant m+1} |f_m|(y_i)$$
$$\leqslant \sum_{i\geqslant m+1} 2^{-i}\alpha = 2^{-m}\alpha.$$

So $\qquad \rho(v_m - v_{m+1}) \geqslant |f_m|(v_m - v_{m+1}) \geqslant \epsilon - 2^{-m}\alpha \quad \forall\ m \in \mathbf{N},$

and $\langle \rho(v_m - v_{m+1})\rangle_{m\in\mathbf{N}} \nrightarrow 0$. **X**

This contradiction shows that the lemma is true.

81I Exercises (a) Let E be an L-space. Then $\mathfrak{T}_{|s|}(E, E')$ is the norm topology on E.

(**b**) Let E and F be Riesz spaces in positive separating duality. Then $\mathfrak{T}_{|s|}(F, E)$ is the topology on F induced by its embedding as a Riesz subspace of the product of a suitable family of L-spaces.

*(**c**) Let E and F be Riesz spaces in positive duality. Suppose that $\mathfrak{T}_{|s|}(F, E)$ is Hausdorff and complete. Show that it is Lebesgue and Levi, and that F is perfect. [Hint: 26B.]

(**d**) Let E be a perfect Riesz space. Then E is complete for $\mathfrak{T}_b(E, E^\times)$, which is Fatou and Levi.

(**e**) Let E be a perfect Riesz space and $\mathfrak{T}$ a compatible complete metrizable linear space topology on E. Then $\mathfrak{T} = \mathfrak{T}_b(E, E^\times)$.

(**f**) Let E be any Riesz space. Then the Mackey topology $\mathfrak{T}_k(E, E^\sim)$ is locally solid, and is the finest locally convex topology on E for which order-bounded sets are bounded.

(**g**) Let E be a uniformly complete Archimedean Riesz space. Then $\mathfrak{T}_k(E, E^\sim) = \mathfrak{T}_b(E, E^\sim)$.

*(**h**) Let E be a uniformly complete Archimedean Riesz space. Then $E^\sim$ is complete for $\mathfrak{T}_k(E^\sim, E)$.

Notes and comments From 81Ia and 81Ib we can see that these $\mathfrak{T}_{|S|}$-topologies are a simple generalization of L-space topologies. It is from this point of view that they are treated in KÖTHE [§ 30.4 *et seq.*]. However, we shall be interested in them merely because they give efficient characterizations of the convex weakly compact sets in $\boldsymbol{E}^{\sim}$, in 81E(iv) and 81G.

In part (d) of the proof of 81F we saw that, given a Riesz space $\boldsymbol{E}_0$ with an order-unit x_0, and a proper solid subspace $\boldsymbol{F}_0$ of $\boldsymbol{E}_0$, there is a Riesz homomorphism f: $\boldsymbol{E}_0 \to \mathbf{R}$ such that $fy = 0$ for every $y \in \boldsymbol{F}_0$ but $fx_0 \neq 0$. This can be regarded as a first step towards the representation theorem for Archimedean Riesz spaces with order units, which asserts that any such space is isomorphic to a norm-dense Riesz subspace of $C(X)$ for some compact space X [LUXEMBURG & ZAANEN R.S., Theorem 45.3; see also the note concerning M-spaces with units at the end of § 26 above].

81F(v) is a striking result. Here it is used only in the proof of 81H, which has important applications in §§ 82 and 83.

82 Weak compactness in $\boldsymbol{E}^{\times}$

In this section I run through the principal characteristics of those sets in $\boldsymbol{E}^{\times}$ which are compact for $\mathfrak{T}_s(\boldsymbol{E}^{\times}, \boldsymbol{E})$. The most powerful results obtain when $\boldsymbol{E}$ is uniformly complete or Dedekind σ-complete; but it seems worth while to express the basic theorem 82E in terms of arbitrary Archimedean Riesz spaces. From these theorems we can deduce, as simple corollaries, some results concerning locally convex Lebesgue topologies.

I have starred the paragraphs 82A–82D because, although they are interesting in themselves, they will not be used again.

***82A Lemma** Let $\boldsymbol{E}$ be an Archimedean Riesz space. Let $\langle x_n \rangle_{n \in \mathbf{N}}$ be a decreasing sequence in $\boldsymbol{E}^+$. Then

$$A = \{f : f \in \boldsymbol{E}^{\times}, \langle |f|(x_n) \rangle_{n \in \mathbf{N}} \to 0\}$$

is closed for $\mathfrak{T}_s(\boldsymbol{E}^{\times}, \boldsymbol{E})$.

Proof Let

$$C = \{y : y \in \boldsymbol{E}, 0 \leqslant y \leqslant x_n \quad \forall \; n \in \mathbf{N}\}.$$

Then, just as in 15C,

$$\{x_n - y : n \in \mathbf{N}, y \in C\} \downarrow 0$$

in E. So, for every $f \in E^\times$,

$$\inf\{|f|(x_n - y) : n \in \mathbf{N}, y \in C\} = 0.$$

Consequently,

$$A = \{f : f \in E^\times, |f|(y) = 0 \quad \forall\ y \in C\}$$
$$= \{f : f \in E^\times, fz = 0 \text{ whenever } |z| \leqslant y \in C\},$$

which of course is $\mathfrak{T}_s(E^\times, E)$-closed.

***82B Proposition** Let E be an Archimedean Riesz space. Suppose that $A \subseteq E^\times$ is relatively countably compact (i.e. every sequence in A has a cluster point in $E^\times$) for $\mathfrak{T}_s(E^\times, E)$. Then A is relatively compact for $\mathfrak{T}_s(E^\times, E)$.

Proof (**a**) Certainly A is bounded for $\mathfrak{T}_s(E^\times, E)$, so A is relatively compact in E^* for $\mathfrak{T}_s(E^*, E)$. Let f be any member of the closure $\bar{A}$ of A in E^*. My aim is to show that $f \in E^\times$, using the criterion of 32D.

(**b**) Suppose that $\varnothing \subset B \downarrow 0$ in E. If $x_0 \in B$, then

$$\{x : x \in B, x \leqslant x_0\} \downarrow 0,$$

so for any $g \in E^\times$ and $\epsilon > 0$ there is an $x \in B$ such that $x \leqslant x_0$ and $|g|(x) \leqslant \epsilon$. It is therefore possible to choose inductively sequences $\langle g_n \rangle_{n \in \mathbf{N}}$ and $\langle y_n \rangle_{n \in \mathbf{N}}$ such that:

y_0 is an arbitrary member of B;

$g_n \in A$ and $|g_n y_i - f y_i| \leqslant 2^{-n} \quad \forall\ i \leqslant n$

(choosing g_n, using the fact that f is in the weak closure of A);

$y_{n+1} \in B, \quad y_{n+1} \leqslant y_n \quad \text{and} \quad (\sum_{i \leqslant n} |g_i|)(y_{n+1}) \leqslant 2^{-n}$

(choosing y_{n+1}, using the facts that $B \downarrow 0$ and $\sum_{i \leqslant n} |g_i| \in E^\times$).

Now let g be any cluster point of $\langle g_n \rangle_{n \in \mathbf{N}}$ in $E^\times$; this is where we use the hypothesis that A is relatively countably compact. Then

$$g y_i = \lim_{n \to \infty} g_n y_i = f y_i \quad \forall\ i \in \mathbf{N}.$$

Since $\qquad D = \{h : h \in E^\times, \lim_{i \to \infty} |h|(y_i) = 0\}$

is $\mathfrak{T}_s(E^\times, E)$-closed [82A], and contains every g_n, it contains g. But now

$$\inf_{y \in B} |fy| \leqslant \inf_{i \in \mathbf{N}} |f y_i| = \inf_{i \in \mathbf{N}} |g y_i|$$
$$\leqslant \inf_{i \in \mathbf{N}} |g|(y_i) = 0.$$

As B is arbitrary, $f \in \boldsymbol{E}^\times$ by 32D. As f is arbitrary, $\bar{A} \subseteq \boldsymbol{E}^\times$; but $\bar{A}$ is $\mathfrak{T}_s(\boldsymbol{E}^*, \boldsymbol{E})$-compact, i.e. $\mathfrak{T}_s(\boldsymbol{E}^\times, \boldsymbol{E})$-compact, and A is therefore relatively $\mathfrak{T}_s(\boldsymbol{E}^\times, \boldsymbol{E})$-compact.

***82C Proposition** Let $\boldsymbol{E}$ be an Archimedean Riesz space, and A a subset of $\boldsymbol{E}^\times$. Then these are equivalent:

(i) The convex hull of A is relatively compact for $\mathfrak{T}_s(\boldsymbol{E}^\times, \boldsymbol{E})$;

(ii) A is relatively compact for $\mathfrak{T}_s(\boldsymbol{E}^\times, \boldsymbol{E})$ and bounded for $\mathfrak{T}_{|s|}(\boldsymbol{E}^\times, \boldsymbol{E})$.

Proof (a) (i) $\Rightarrow$ (ii) follows from 81F (ii) $\Rightarrow$ (iv).

(b) (ii) $\Rightarrow$ (i) By 82B, it is enough to show that every sequence in the convex hull $\Gamma(A)$ of A has a cluster point in $\boldsymbol{E}^\times$. But given a sequence $\langle g_n \rangle_{n \in \mathbf{N}}$ in $\Gamma(A)$, there must be a sequence $\langle f_n \rangle_{n \in \mathbf{N}}$ in A such that $g_n \in \Gamma(A_0)$ for every $n \in \mathrm{N}$, where $A_0 = \{f_n : n \in \mathrm{N}\}$. Since A_0 is $\mathfrak{T}_{|s|}(\boldsymbol{E}^\times, \boldsymbol{E})$-bounded, its $\mathfrak{T}_s(\boldsymbol{E}^\sim, \boldsymbol{E})$-closed convex hull C is $\mathfrak{T}_s(\boldsymbol{E}^\sim, \boldsymbol{E})$-compact [81F (iv) $\Rightarrow$ (ii)]. Once again, I shall use 32D to show that $C \subseteq \boldsymbol{E}^\times$.

Suppose that $\varnothing \subset B \downarrow 0$ in $\boldsymbol{E}$. As in 82B above, we may choose a decreasing sequence $\langle y_n \rangle_{n \in \mathbf{N}}$ in B such that

$$(\textstyle\sum_{i \leqslant n} |f_i|)(y_n) \leqslant 2^{-n} \quad \forall\ n \in \mathrm{N}.$$

Define $T: \boldsymbol{E}^\sim \to \boldsymbol{l}^\infty(\mathrm{N})$ by

$$Tf = \langle f y_n \rangle_{n \in \mathbf{N}} \quad \forall\ f \in \boldsymbol{E}^\sim.$$

Then T is linear and is continuous for $\mathfrak{T}_s(\boldsymbol{E}^\sim, \boldsymbol{E})$ and $\mathfrak{T}_s(\boldsymbol{l}^\infty, \boldsymbol{s}_0)$, where $\boldsymbol{s}_0$ is the space of those sequences in $\mathbf{R}$ which have only finitely many non-zero terms.

Consider $$D = \{f : f \in \boldsymbol{E}^\times, \langle |f|(y_n) \rangle_{n \in \mathbf{N}} \to 0\}.$$

Then D is $\mathfrak{T}_s(\boldsymbol{E}^\times, \boldsymbol{E})$-closed by 82A, and $D \supseteq A_0$, so $D \supseteq \bar{A}_0$, which is $\mathfrak{T}_s(\boldsymbol{E}^\times, \boldsymbol{E})$-compact because A is supposed relatively $\mathfrak{T}_s(\boldsymbol{E}^\times, \boldsymbol{E})$-compact. Also $T[D] \subseteq \boldsymbol{c}_0(\mathrm{N})$. So $T[\bar{A}_0]$ is a $\mathfrak{T}_s(\boldsymbol{c}_0, \boldsymbol{s}_0)$-compact subset of $\boldsymbol{c}_0$.

But $T[\bar{A}_0]$ is $\|\ \|_\infty$-bounded, because $\bar{A}_0$ is $\mathfrak{T}_{|s|}(\boldsymbol{E}^\times, \boldsymbol{E})$-bounded, and $\{y_n : n \in \mathrm{N}\}$ is order-bounded in $\boldsymbol{E}$. So $\mathfrak{T}_s(\boldsymbol{c}_0, \boldsymbol{s}_0)$ agrees with $\mathfrak{T}_s(\boldsymbol{c}_0, \boldsymbol{l}^1)$ on $T[\bar{A}_0]$, because $\boldsymbol{s}_0$ is $\|\ \|_1$-dense in $\boldsymbol{l}^1 = \boldsymbol{l}^1(\mathrm{N})$, and $T[\bar{A}_0]$ is $\mathfrak{T}_s(\boldsymbol{c}_0, \boldsymbol{l}^1)$-compact. Now by Krein's theorem its $\mathfrak{T}_s(\boldsymbol{c}_0, \boldsymbol{l}^1)$-closed convex hull K is $\mathfrak{T}_s(\boldsymbol{c}_0, \boldsymbol{l}^1)$-compact, therefore $\mathfrak{T}_s(\boldsymbol{l}^\infty, \boldsymbol{s}_0)$-closed. So $T^{-1}[K]$ is a convex $\mathfrak{T}_s(\boldsymbol{E}^\sim, \boldsymbol{E})$-closed set, and includes C.

Thus $T[C] \subseteq K \subseteq \boldsymbol{c}_0$. So if $g \in C$,

$$\inf_{y \in B} |gy| \leqslant \inf_{n \in \mathbf{N}} |g y_n| = \inf_{n \in \mathbf{N}} |(Tg)(n)| = 0.$$

As B is arbitrary, $g \in E^\times$. Thus $C \subseteq E^\times$. But $\langle g_n \rangle_{n \in \mathbf{N}}$ has a cluster point in C, so has a cluster point in $E^\times$; by 81B, this shows that A is relatively compact for $\mathfrak{T}_s(E^\times, E)$.

***82D Corollary** If E is a uniformly complete Archimedean Riesz space and $A \subseteq E^\times$ is relatively compact for $\mathfrak{T}_s(E^\times, E)$, so is $\Gamma(A)$.

Proof For by 81G any $\mathfrak{T}_s(E^\times, E)$-bounded set is $\mathfrak{T}_{|s|}(E^\times, E)$-bounded.

82E Theorem Let E be an Archimedean Riesz space, and A a non-empty subset of $E^\times$. Then these are equivalent:

(i) The solid convex hull of A is relatively compact for $\mathfrak{T}_s(E^\times, E)$.

(ii) $\{|f| : f \in A\}$ is relatively countably compact for $\mathfrak{T}_s(E^\times, E)$.

(iii) A is bounded for $\mathfrak{T}_s(E^\times, E)$ and, whenever $\varnothing \subset B \downarrow 0$ in E,

$$\inf_{x \in B} \sup_{f \in A} |fx| = 0.$$

(iv) For any $x \in E^+$ and $\epsilon > 0$, there is an $f_0 \in E^{\times +}$ such that

$$(|f| - |f| \wedge f_0)(x) \leqslant \epsilon \quad \forall\ f \in A.$$

(v) A is bounded for $\mathfrak{T}_s(E^\times, E)$ and, whenever $\langle x_n \rangle_{n \in \mathbf{N}}$ is a disjoint sequence in E^+ which is bounded above in E,

$$\langle \sup_{f \in A} |fx_n| \rangle_{n \in \mathbf{N}} \to 0.$$

Proof (a) (i) $\Rightarrow$ (ii) is trivial. I shall prove that

$$\text{(ii)} \Rightarrow \text{(v)} \Rightarrow \text{(iv)} \Rightarrow \text{(iii)} \Rightarrow \text{(i)}.$$

(b) (ii) $\Rightarrow$ (v) Suppose that A satisfies (ii). In this case,

$$\sup_{f \in A} |fx| \leqslant \sup_{f \in A} |f|\,(|x|) < \infty \quad \forall\ x \in E,$$

so A is certainly bounded for $\mathfrak{T}_s(E^\times, E)$. **?** Suppose, if possible, that $\langle x_n \rangle_{n \in \mathbf{N}}$ is a disjoint sequence in E^+, bounded above, such that $\langle \sup_{f \in A} |fx_n| \rangle_{n \in \mathbf{N}} \not\to 0$. In this case, there is an $\epsilon > 0$ and a strictly increasing sequence $\langle n(k) \rangle_{k \in \mathbf{N}}$ in $\mathbf{N}$ such that

$$\sup_{f \in A} |fx_{n(k)}| > \epsilon \quad \forall\ k \in \mathbf{N},$$

and now we may choose a sequence $\langle f_k \rangle_{k \in \mathbf{N}}$ in A such that

$$|f_k|\,(x_{n(k)}) \geqslant |f_k(x_{n(k)})| \geqslant \epsilon \quad \forall\ k \in \mathbf{N}.$$

Let f be a cluster point of $\langle |f_k| \rangle_{k \in \mathbf{N}}$ in E. Set

$$B = \{\sup_{i \leqslant k} x_{n(i)} : k \in \mathbf{N}\} = \{\textstyle\sum_{i \leqslant k} x_{n(i)} : k \in \mathbf{N}\},$$

and let C be the set of upper bounds of B in E. Then $B\uparrow$ and $C\downarrow$, so $C-B\downarrow$, and $C-B\downarrow 0$ [15C]. So there is a $z\in C-B$ such that $fz<\epsilon$, i.e. there is a $y\in C$ and a $k\in \mathbf{N}$ such that

$$f(y-\sum_{i\leqslant k}x_{n(i)})<\epsilon.$$

However, if $j>k$, $x_{n(j)}+\sum_{i\leqslant k}x_{n(i)}\leqslant y$, so

$$|f_j|\,(y-\sum_{i\leqslant k}x_{n(i)})\geqslant |f_j|\,(x_{n(j)})\geqslant \epsilon.$$

As f is a cluster point of $\langle |f_j|\rangle_{j\in\mathbf{N}}$, $f(y-\sum_{i\leqslant k}x_{n(i)})\geqslant\epsilon$; contradicting the choice of y and k. **X**

(c) **(v)** ⇒ **(iv)** is 81H.

(d) **(iv)** ⇒ **(iii)** Suppose that A satisfies the condition (iv).

(α) A is bounded. **P** For every $x\in E$, there is an $f_0\in E^{\times+}$ such that $(|f|-|f|\wedge f_0)(|x|)\leqslant 1$ for every $f\in A$; but now

$$\sup_{f\in A}|fx|\leqslant 1+f_0(|x|)<\infty. \quad \mathbf{Q}$$

(β) Given that $\varnothing\subset B\downarrow 0$ in E, let $\epsilon>0$ and $x_0\in B$. Then there is an $f_0\in E^{\times+}$ such that $(|f|-|f|\wedge f_0)(x_0)\leqslant\epsilon$ for every $f\in A$, and an $x\in B$ such that $x\leqslant x_0$ and $f_0x\leqslant\epsilon$. But now

$$\begin{aligned}|fx|\leqslant |f|\,(x)&\leqslant (|f|-|f|\wedge f_0)(x)+f_0x\\&\leqslant (|f|-|f|\wedge f_0)(x_0)+\epsilon\leqslant 2\epsilon\end{aligned}$$

for every $f\in A$. As ϵ is arbitrary, A satisfies (iii).

(e) **(iii)** ⇒ **(i)** Suppose A satisfies the condition (iii). Let C be the solid convex hull of A in $E^\times$.

(α) C also satisfies the condition (iii). **P** If $x\in E^+$ and $\alpha\in\mathbf{R}^+$,

$$\{f:f\in E^\times,\ |f|\,(x)\leqslant\alpha\}=\{f:f\in E^\times,\ |y|\leqslant x\Rightarrow |fy|\leqslant\alpha\}$$

is a solid convex set in $E^\times$; so

$$\sup_{f\in C}|fx|=\sup\{|fy|:f\in A,\ |y|\leqslant x\}.$$

Now define ρ: $E\to\mathbf{R}^+$ by $\rho(x)=\sup_{f\in A}|fx|$ for every $x\in E$. Then ρ is a seminorm; let $\mathfrak{T}$ be the linear space topology it generates on E. The condition (iii) on A asserts that $\mathfrak{T}$ is Lebesgue. Accordingly order-bounded sets are bounded [24Db]; so, for any $x\in E^+$,

$$\sup_{f\in C}|fx|=\sup\{\rho(y):|y|\leqslant x\}<\infty,$$

and C is $\mathfrak{T}_s(E^\times,E)$-bounded. Also, if $\varnothing\subset B\downarrow 0$ in E,

$$\inf_{x\in B}\sup_{f\in C}|fx|=\inf_{x\in B}\sup_{|y|\leqslant x}\rho(y)=0$$

by 24Bc. **Q**

(β) C is relatively compact for $\mathfrak{T}_s(E^\times, E)$. **P** As C is bounded, it is relatively compact in E^* for $\mathfrak{T}_s(E^*, E)$. Let ϕ belong to the closure $\bar{C}$ of C in E^*. Suppose that $\varnothing \subset B \downarrow 0$ in E. Then

$$\inf_{x\in B}|\phi x| \leqslant \inf_{x\in B}\sup_{f\in C}|fx| = 0,$$

so $\phi \in E^\times$ by 32D. As ϕ is arbitrary, $\bar{C} \subseteq E^\times$; so C is relatively compact for $\mathfrak{T}_s(E^\times, E)$. **Q**

82F Corollary Let E be an Archimedean Riesz space. Then the Mackey topology $\mathfrak{T}_k(E, E^\times)$ is Lebesgue iff it is locally solid.

Proof (a) Suppose that $\mathfrak{T} = \mathfrak{T}_k(E, E^\times)$ is Lebesgue. Let U be a neighbourhood of 0 for $\mathfrak{T}$. Then $U^0 \subseteq E^\times$ has the property (iii) of 82E. So its solid convex hull C is relatively $\mathfrak{T}_s(E^\times, E)$-compact, and C^0 is a neighbourhood of 0 for $\mathfrak{T}$. Now C^0 is a solid neighbourhood of 0 for $\mathfrak{T}$, and is included in U. Thus $\mathfrak{T}$ is locally solid.

(b) The converse follows from 24G.

82G Theorem Let E be a Dedekind σ-complete Riesz space, and A a non-empty subset of $E^\times$. Then the following are equivalent:

(i) A is relatively countably compact for $\mathfrak{T}_s(E^\times, E)$.

(ii) The solid convex hull of A is relatively compact for $\mathfrak{T}_s(E^\times, E)$.

(iii) A is bounded for $\mathfrak{T}_s(E^\times, E)$ and, whenever $\langle x_n\rangle_{n\in\mathbf{N}} \downarrow 0$ in E,

$$\langle \sup_{f\in A}|fx_n|\rangle_{n\in\mathbf{N}} \to 0.$$

Proof (a) (ii) ⇒ (i) is trivial; and (ii) ⇒ (iii) is a special case of 82E(i) ⇒ (iii). I shall prove that (i) ⇒ (ii) and that (iii) ⇒ (ii), in both cases using the criterion 82E(v).

(b) (i) ⇒ (ii) Suppose that $\langle x_n\rangle_{n\in\mathbf{N}}$ is a disjoint sequence in E^+ which is bounded above. Then, because E is Dedekind σ-complete, $x = \sup_{n\in\mathbf{N}} x_n$ exists in E, and now

$$x = \sup_{n\in\mathbf{N}}\textstyle\sum_{i<n} x_i = \sum_{i\in\mathbf{N}} x_i,$$

the sum being for $\mathfrak{T}_s(E, E^\times)$. Similarly, any subsequence of $\langle x_n\rangle_{n\in\mathbf{N}}$ is summable for $\mathfrak{T}_s(E, E^\times)$. So by A1Hb,

$$\langle \sup_{f\in A}|fx_n|\rangle_{n\in\mathbf{N}} \to 0,$$

and we see from 82E(v) ⇒ (i) that the solid convex hull of A is relatively compact.

(c) (iii) $\Rightarrow$ (ii) Again suppose that $\langle x_n \rangle_{n \in \mathbf{N}}$ is a disjoint sequence in $\boldsymbol{E}$ which is bounded above. Let $x = \sup_{n \in \mathbf{N}} x_n$. Then

$$\langle x - \textstyle\sum_{i<n} x_i \rangle_{n \in \mathbf{N}} \downarrow 0$$

in $\boldsymbol{E}$, so $\quad \alpha_n = \sup_{f \in A} |f(x - \sum_{i<n} x_i)| \to 0 \quad \text{as} \quad n \to \infty,$

and $\quad \sup_{f \in A} |fx_n| \leqslant |\alpha_{n+1}| + |\alpha_n| \to 0 \quad \text{as} \quad n \to \infty.$

As $\langle x_n \rangle_{n \in \mathbf{N}}$ is arbitrary, the solid convex hull of A is relatively compact.

82H Corollary If $\boldsymbol{E}$ is a Dedekind σ-complete Riesz space, any locally convex linear space topology $\mathfrak{T}$ on $\boldsymbol{E}$ for which $\boldsymbol{E}' \subseteq \boldsymbol{E}^\times$ is Lebesgue. In particular, the Mackey topology $\mathfrak{T}_k(\boldsymbol{E}, \boldsymbol{E}^\times)$ is Lebesgue and therefore [by 82F] locally solid.

Proof If U is a closed absolutely convex neighbourhood of 0 for $\mathfrak{T}$, then U^0 is $\mathfrak{T}_s(\boldsymbol{E}', \boldsymbol{E})$-compact, and therefore $\mathfrak{T}_s(\boldsymbol{E}^\times, \boldsymbol{E})$-compact. So by 82G its convex solid hull is relatively $\mathfrak{T}_s(\boldsymbol{E}^\times, \boldsymbol{E})$-compact. Now we see from 82E(iii) that if $\varnothing \subset B \downarrow 0$ in $\boldsymbol{E}$, B meets $U^{00} = U$. As B and U are arbitrary, $\mathfrak{T}$ is Lebesgue.

82I Corollary Let $\boldsymbol{E}$ be an Archimedean Riesz space. Let $\boldsymbol{F}$ be the solid linear subspace of $\boldsymbol{E}^{\times\times}$ generated by the canonical image of $\boldsymbol{E}$ in $\boldsymbol{E}^{\times\times}$ [32B]. Then a subset A of $\boldsymbol{E}^\times$ is relatively compact for $\mathfrak{T}_s(\boldsymbol{E}^\times, \boldsymbol{F})$ iff the solid convex hull of A is relatively compact for $\mathfrak{T}_s(\boldsymbol{E}^\times, \boldsymbol{E})$.

Proof The point is that $\mathfrak{T}_s(\boldsymbol{E}^\times, \boldsymbol{F})$, being finer than $\mathfrak{T}_s(\boldsymbol{E}^\times, \boldsymbol{E})$, is Levi. So it follows from 33B that the canonical map from $\boldsymbol{E}^\times$ to $\boldsymbol{F}^\times$ is one-to-one and onto, and that we may identify $\boldsymbol{E}^\times$ with $\boldsymbol{F}^\times$.

Now if $A \subseteq \boldsymbol{E}^\times = \boldsymbol{F}^\times$ is relatively compact for $\mathfrak{T}_s(\boldsymbol{E}^\times, \boldsymbol{F})$, so is its solid convex hull, by 82G (i) $\Rightarrow$ (ii), because $\boldsymbol{F}$ is certainly Dedekind σ-complete. As $\mathfrak{T}_s(\boldsymbol{E}^\times, \boldsymbol{E})$ is coarser than $\mathfrak{T}_s(\boldsymbol{E}^\times, \boldsymbol{F})$, the solid convex hull of A is relatively compact for $\mathfrak{T}_s(\boldsymbol{E}^\times, \boldsymbol{E})$.

Conversely, suppose that the solid convex hull of A is relatively $\mathfrak{T}_s(\boldsymbol{E}^\times, \boldsymbol{E})$-compact. Let $\phi \in \boldsymbol{F}^+$. Then by the definition of $\boldsymbol{F}$ there is an $x \in \boldsymbol{E}^+$ such that $\hat{x} \geqslant \phi$. Now, given $\epsilon > 0$, there is an $f_0 \in \boldsymbol{E}^{\times +}$ such that

$$\epsilon \geqslant (|f| - |f| \wedge f_0)(x) \geqslant \langle |f| - |f| \wedge f_0, \phi \rangle$$

for every $f \in A$, using 82E (i) $\Rightarrow$ (iv). Now, using 82E(iv) $\Rightarrow$ (i), we find that A is relatively compact for $\mathfrak{T}_s(\boldsymbol{E}^\times, \boldsymbol{F})$.

82J Corollary Let $\boldsymbol{E}$ be a Dedekind σ-complete Riesz space. Then $\boldsymbol{E}^\times$ is sequentially complete for $\mathfrak{T}_s(\boldsymbol{E}^\times, \boldsymbol{E})$.

Proof Let $\langle f_n\rangle_{n\in\mathbf{N}}$ be a sequence in $E^\times$ which is Cauchy for $\mathfrak{T}_s(E^\times, E)$. I shall use 82E(v) to show that $\{f_n : n\in\mathbf{N}\}$ is relatively compact in $E^\times$.

Suppose that $\langle x_n\rangle_{n\in\mathbf{N}}$ is a disjoint sequence in E^+ which is bounded above. **?** Suppose, if possible, that

$$\langle \sup_{m\in\mathbf{N}} |f_m x_n|\rangle_{n\in\mathbf{N}} \not\to 0.$$

In this case, there is an $\epsilon > 0$ and a subsequence $\langle y_k\rangle_{k\in\mathbf{N}}$ of $\langle x_n\rangle_{n\in\mathbf{N}}$ such that

$$\sup_{m\in\mathbf{N}} |f_m y_k| > 2\epsilon \quad \forall\ k\in\mathbf{N}.$$

Construct a sequence $\langle g_r\rangle_{r\in\mathbf{N}}$ as follows. We know that $\langle y_k\rangle_{k\in\mathbf{N}} \to 0$ for $\mathfrak{T}_s(E, E^\times)$. So, given $r\in\mathbf{N}$, there is a $k(r)$ such that $k(r) \geqslant r$ and $|f_i(y_{k(r)})| \leqslant \epsilon$ for every $i \leqslant r$. Now let $m(r)$ be such that $|f_{m(r)}(y_{k(r)})| \geqslant 2\epsilon$. Set $g_r = f_{m(r)} - f_r$.

As certainly $m(r) > r$ for every $r\in\mathbf{N}$, $\langle g_r\rangle_{r\in\mathbf{N}} \to 0$ for $\mathfrak{T}_s(E^\times, E)$. So $\langle g_r\rangle_{r\in\mathbf{N}}$ is relatively compact for $\mathfrak{T}_s(E^\times, E)$; by 82G(i) $\Rightarrow$ (ii) and 82E(i) $\Rightarrow$ (v),

$$\langle \sup_{r\in\mathbf{N}} |g_r(y_j)|\rangle_{j\in\mathbf{N}} \to 0.$$

But $k(j) \geqslant j$ and $|g_j(y_{k(j)})| \geqslant \epsilon$ for every $j\in\mathbf{N}$; which entails a contradiction. **X**

As $\langle x_n\rangle_{n\in\mathbf{N}}$ is arbitrary, $\langle f_n\rangle_{n\in\mathbf{N}}$ is relatively compact for $\mathfrak{T}_s(E^\times, E)$, by 82E(v) $\Rightarrow$ (i). So it has a cluster point $f\in E^\times$. But now, of course, $\langle f_n\rangle_{n\in\mathbf{N}} \to f$ for $\mathfrak{T}_s(E^\times, E)$. Thus every Cauchy sequence has a limit, and $E^\times$ is sequentially complete for $\mathfrak{T}_s(E^\times, E)$.

Remark This result is true for all bands in $E^\sim$; see SCHAEFER W.C., where there are further remarks on sequential convergence in $E^\sim$.

82K Corollary Let E be a perfect Riesz space. (a) If $A \subseteq E$ is relatively compact for $\mathfrak{T}_s(E, E^\times)$, so is the solid convex hull of A. (b) E is sequentially complete for $\mathfrak{T}_s(E, E^\times)$.

Proof For E may be regarded as $(E^\times)^\times$, and $E^\times$ is certainly Dedekind complete; so the result is immediate from 82G and 82J.

82L Exercises **(a)** Let E be a Dedekind complete Riesz space. Then $[x, y]$ is compact for $\mathfrak{T}_s(E, E^\times)$ for every x, y in E.

(b) Let E be an Archimedean Riesz space and A a subset of $E^\times$. Then the solid convex hull of A is relatively compact for $\mathfrak{T}_s(E^\times, E)$ iff, whenever $\langle x_n\rangle_{n\in\mathbf{N}}$ is an order-bounded increasing sequence in E and $\langle f_n\rangle_{n\in\mathbf{N}}$ is a sequence in A, $\langle f_n(x_{n+1} - x_n)\rangle_{n\in\mathbf{N}} \to 0$.

*(c) Find an L-space E and a sequence $\langle x_n \rangle_{n \in \mathbf{N}}$ in E^+ which converges for $\mathfrak{T}_s(E, E^\times)$ but has no order-bounded subsequence.

Notes and comments The proofs of the main theorems 82E and 82G above are relatively elementary because substantial parts of the argument have been removed to §§81 and A1. For a more direct approach to these results, see FREMLIN A.K.S.I.

We may think of 82E(iv) and 82E(v) as being extremes, in that 82E(v) is the easiest of the five conditions to establish [see 82G, 82J], while 82E(iv) is the easiest to draw conclusions from [see 82I, 82Lb]. Thus it is unsurprising that the argument required to pass from (v) to (iv), using 24H, 82F(v) and 82H, is the hardest part of the work. It is perhaps worth observing that useful simplifications are possible if we allow ourselves to use the fact that $A \subseteq E^\times$.

For Dedekind σ-complete spaces, the conditions in 82E and 82G form a very powerful list of equivalences. Their most natural applications are to perfect Riesz spaces, in which they can be used to discuss weak compactness for $\mathfrak{T}_s(E, E^\times)$, as in 82K. This result has long been known for l^p and L^p spaces [6XF]; the generalization here is due essentially to DIEUDONNÉ.

*The propositions 82B–82D are immediately reminiscent of the theorems of Eberlein and Krein. It is therefore significant that the example 8XC shows that they cannot be deduced from these theorems. Other examples in § 8X show how the results proved here for uniformly complete or Dedekind σ-complete spaces cannot be extended to the next wider class. But I do not know whether 82B–82F are true for non-Archimedean spaces.

83 Weak compactness in L-spaces

The theorems of the last two sections have particularly direct applications to problems involving the identification of weakly compact sets in L-spaces. Since L-spaces are perfect [33F], we can immediately apply 82E and 82G to obtain various criteria, as in 83A. Moreover, the most important L-spaces are the L^1-spaces, and the identification of these with $L^\#$-spaces in § 52 sets them up for investigation by the methods developed above, as in 83F. The same arguments enable us to cope with L-spaces which are defined as the duals of M-spaces [see 83B, 83J]. One of our criteria for weak compactness in L^1-spaces [83F] can be generalized to characterize weak compactness in other perfect function spaces [83I].

83A Lemma Let E be an L-space and A a subset of E. Then these are equivalent:

(i) A is relatively compact for $\mathfrak{T}_s(E, E')$;

(ii) the solid convex hull of A is relatively compact for $\mathfrak{T}_s(E, E')$;

(iii) for every $\epsilon > 0$ there is an $x_0 \in E^+$ such that

$$\| |x| - |x| \wedge x_0 \| \leqslant \epsilon \quad \forall\ x \in A.$$

Proof Recall that an L-space is perfect [33F], so that E may be identified with $E^{\times\times} = E'^{\times}$ [26C]. Now E' is certainly Dedekind complete, so (i) and (ii) are equivalent by 82G. Next, if the solid convex hull of A is relatively compact, and if $\epsilon > 0$, then by 82E there is an $x_0 \in E^+$ such that

$$\| |x| - |x| \wedge x_0 \| = \int (|x| - |x| \wedge x_0) \leqslant \epsilon \quad \forall\ x \in A,$$

since $\int \in E'$ [26C]. On the other hand, if this condition is satisfied, and if $f \in E'$ and $\epsilon > 0$, then there is an $x_0 \in E^+$ such that

$$|f|\,(|x| - |x| \wedge x_0) \leqslant \|f\|\, \| |x| - |x| \wedge x_0 \| \leqslant \epsilon \quad \forall\ x \in A,$$

so A is relatively compact by 82E again.

83B Theorem Let G be a Riesz space with a Fatou norm $\| \ \|$ such that (α) $\|x \vee y\| = \max(\|x\|, \|y\|)$ whenever $x, y \geqslant 0$ in G (β) for any $x \in G^+$ and $\alpha > 0$, $\sup\{y : 0 \leqslant y \leqslant x, \|y\| \leqslant \alpha\}$ exists in G. Let A be a non-empty subset of G'. Then A is relatively compact for $\mathfrak{T}_s(G', G'')$ iff (i) it is bounded for $\mathfrak{T}_s(G', G)$ (ii) whenever $\langle x_n \rangle_{n \in \mathbf{N}}$ is a norm-bounded disjoint sequence in G^+,

$$\langle \sup_{f \in A} |f x_n| \rangle_{n \in \mathbf{N}} \to 0.$$

Proof We know that G' is an L-space [26D], so we may use the criterion of 83A.

(a) If A is relatively compact for $\mathfrak{T}_s(G', G'')$, then of course it is bounded for $\mathfrak{T}_s(G', G)$. Now suppose that $\langle x_n \rangle_{n \in \mathbf{N}}$ is a norm-bounded disjoint sequence in G^+; let $\alpha = \sup_{n \in \mathbf{N}} \|x_n\|$. Let $\epsilon > 0$. Then there is an $f_0 \in G'^+$ such that

$$\| |f| - |f| \wedge f_0 \| \leqslant \epsilon \quad \forall\ f \in A$$

[83A]. So, for any $n \in \mathbf{N}$,

$$\begin{aligned} \textstyle\sum_{i \leqslant n} f_0 x_i &= f_0(\textstyle\sum_{i \leqslant n} x_i) = f_0(\sup_{i \leqslant n} x_i) \leqslant \|f_0\| \| \sup_{i \leqslant n} x_i \| \\ &= \|f_0\| \sup_{i \leqslant n} \|x_i\| \leqslant \alpha \|f_0\|. \end{aligned}$$

So $\langle f_0 x_n \rangle_{n\in\mathbf{N}} \to 0$ and there is an $n_0 \in \mathbf{N}$ such that $f_0 x_n \leqslant \epsilon$ for every $n \geqslant n_0$. Now, for any $n \geqslant n_0$ and $f \in A$,

$$|fx_n| \leqslant |f|(x_n) \leqslant (|f| - |f| \wedge f_0)(x_n) + f_0 x_n$$
$$\leqslant \| |f| - |f| \wedge f_0 \| \|x_n\| + \epsilon \leqslant \alpha\epsilon + \epsilon.$$

So $\sup_{f\in A} |fx_n| \leqslant \epsilon(1+\alpha)$ for every $n \geqslant n_0$. As ϵ is arbitrary,

$$\langle \sup_{f\in A} |fx_n| \rangle_{n\in\mathbf{N}} \to 0,$$

as required.

(b) Conversely, suppose that A satisfies the conditions. Let $\epsilon > 0$.

(α) There is a $w_0 \in G^+$ such that

$$|f(w - w \wedge w_0)| \leqslant 2\epsilon$$

whenever $f \in A$, $w \in G^+$ and $\|w\| \leqslant 1$. **P ?** Suppose otherwise. Choose sequences $\langle x_n \rangle_{n\in\mathbf{N}}$ and $\langle y_n \rangle_{n\in\mathbf{N}}$ in G^+ and $\langle f_n \rangle_{n\in\mathbf{N}}$ in A inductively, as follows. $x_0 = 0$. Given x_n, let $y_n \in G^+$ and $f_n \in A$ be such that $\|y_n\| \leqslant 1$ and

$$|f_n(y_n - y_n \wedge x_n)| \geqslant 2\epsilon.$$

Set $x_{n+1} = x_n \vee \epsilon^{-1} \|f_n\| y_n$. Continue.

Now, for each $n \in \mathbf{N}$, set $\alpha_n = \epsilon \|f_n\|^{-1}$, and let

$$B_n = \{w : 0 \leqslant w \leqslant y_n, \|w\| \leqslant \alpha_n\}.$$

By the hypothesis (α), $B_n \uparrow$; by the hypothesis (β), $z_n = \sup B_n$ exists, and as $\| \;\|$ is a Fatou norm, $\|z_n\| \leqslant \alpha_n$. Of course $0 \leqslant z_n \leqslant y_n$. Set $u_n = (y_n - z_n - x_n)^+$. Then

$$\|u_n - (y_n - x_n)^+\| \leqslant \|z_n\| \leqslant \alpha_n,$$

so
$$|f_n u_n| \geqslant |f_n(y_n - x_n)^+| - \alpha_n \|f_n\| \geqslant 2\epsilon - \epsilon = \epsilon.$$

But examine $u_m \wedge u_n$, where $m < n$. Actually,

$$u_m \wedge \alpha_m u_n \leqslant (y_m - z_m) \wedge \alpha_m (y_n - x_n)^+$$
$$\leqslant (y_m - y_m \wedge \alpha_m y_n) \wedge \alpha_m (y_n - \alpha_m^{-1} y_m)^+$$

(because $0 \leqslant y_m \wedge \alpha_m y_n \leqslant y_m$ and $\|y_m \wedge \alpha_m y_n\| \leqslant \alpha_m \|y_n\| \leqslant \alpha_m$, so $y_m \wedge \alpha_m y_n \leqslant z_m$; while $x_n \geqslant x_{m+1} \geqslant \alpha_m^{-1} y_m$)

$$= (y_m - \alpha_m y_n)^+ \wedge (\alpha_m y_n - y_m)^+$$
$$= 0.$$

So by 14Kb $u_m \wedge u_n = 0$.

Thus $\langle u_n\rangle_{n\in\mathbf{N}}$ is a disjoint sequence in G^+, and $\|u_n\| \leqslant \|y_n\| \leqslant 1$ for every $n\in\mathbf{N}$. But

$$\sup_{f\in A}|fu_n| \geqslant |f_n u_n| \geqslant \epsilon \quad \forall\ n\in\mathbf{N},$$

contradicting the hypothesis (ii). **XQ**

(**β**) We now find that

$$|f|\,(|w|-|w|\wedge w_0) \leqslant 4\epsilon$$

whenever $f\in A$ and $\|w\| \leqslant 1$ in G. **P** Suppose that

$$0 \leqslant u \leqslant |w|-|w|\wedge w_0.$$

Set
$$v = u+|w|\wedge w_0;$$

then $|w|\wedge w_0 \leqslant v \leqslant |w|$, so $\|v\| \leqslant 1$ and

$$|w|\wedge w_0 \leqslant v\wedge w_0 \leqslant |w|\wedge w_0.$$

Thus $v-v\wedge w_0 = v-|w|\wedge w_0 = u$. Now

$$|fu| = |f(v-v\wedge w_0)| \leqslant 2\epsilon,$$

by part (α) above. As u is arbitrary,

$$f^+(|w|-|w|\wedge w_0) \leqslant 2\epsilon.$$

Similarly, $f^-(|w|-|w|\wedge w_0) \leqslant 2\epsilon$, so

$$|f|\,(|w|-|w|\wedge w_0) \leqslant 4\epsilon. \quad \mathbf{Q}$$

(**γ**) Observe next that A certainly satisfies the conditions of 81H, with $F = G'$, $E = G$. So there is an $f_0\in G'^+$ such that

$$(|f|-|f|\wedge f_0)(w_0) \leqslant \epsilon \quad \forall\ f\in A.$$

Now suppose that $f\in A$ and $\|w\| \leqslant 1$. Then

$$\begin{aligned}|(|f|-|f|\wedge f_0)(w)| &\leqslant (|f|-|f|\wedge f_0)(|w|)\\ &\leqslant (|f|-|f|\wedge f_0)(|w|-|w|\wedge w_0)\\ &\qquad +(|f|-|f|\wedge f_0)(w_0)\\ &\leqslant |f|\,(|w|-|w|\wedge w_0)+\epsilon\\ &\leqslant 4\epsilon+\epsilon = 5\epsilon.\end{aligned}$$

As w is arbitrary,

$$\|\,|f|-|f|\wedge f_0\| \leqslant 5\epsilon \quad \forall\ f\in A.$$

(**δ**) As ϵ is arbitrary, A is relatively compact by the criterion of 83A.

83C Theorem Let $\mathfrak{A}$ be a Boolean ring, and let A be a non-empty subset of $\boldsymbol{E} = \boldsymbol{S}(\mathfrak{A})' = \boldsymbol{L}^\infty(\mathfrak{A})' = \boldsymbol{L}^\infty(\mathfrak{A})^\sim$. Then A is relatively compact for $\mathfrak{T}_s(\boldsymbol{E}, \boldsymbol{E}')$ iff (i) $\sup_{f\in A}|f(\chi a)| < \infty$ for every $a \in \mathfrak{A}$, (ii) whenever $\langle a_n\rangle_{n\in\mathbf{N}}$ is a disjoint sequence in $\mathfrak{A}$,

$$\langle \sup_{f\in A}|f(\chi a_n)|\rangle_{n\in\mathbf{N}} \to 0.$$

Proof (a) Recall that by 25G $\boldsymbol{L}^{\infty\prime} = \boldsymbol{L}^{\infty\sim}$. Because $\boldsymbol{S}$ is a norm-dense subspace of $\boldsymbol{L}^\infty$, we may identify $\boldsymbol{S}'$ with $\boldsymbol{L}^{\infty\prime}$ as normed linear space; because the positive cone of $\boldsymbol{L}^\infty$ is just the norm-closure of the positive cone of $\boldsymbol{S}$, the positive cones of $\boldsymbol{S}'$ and $\boldsymbol{L}^{\infty\prime}$ coincide, so we may identify $\boldsymbol{S}'$ and $\boldsymbol{L}^{\infty\prime}$ as Riesz spaces.

(b) The space $\boldsymbol{S}$, with the norm $\| \ \|_\infty$, satisfies the conditions of 83B. **P** (α) is obvious, as $\boldsymbol{S}$ is defined to be a Riesz subspace of an $\boldsymbol{l}^\infty$ space. If $x \in \boldsymbol{S}^+$ and $\alpha > 0$, then we know by 42Eb that x is expressible as $\sum_{i<n}\alpha_i\chi a_i$, where $\langle a_i\rangle_{i<n}$ is disjoint in $\mathfrak{A}$ and $\alpha_i \geqslant 0$ for each $i < n$. Now it is easy to see that

$$z = \sum_{i<n}(\alpha_i \wedge \alpha)\,\chi a_i = \sup\{y : 0 \leqslant y \leqslant x,\ \|y\|_\infty \leqslant \alpha\}.$$

Because $\|z\|_\infty \leqslant \alpha$, we may conclude at the same time that $\| \ \|_\infty$ is a Fatou norm on $\boldsymbol{S}$ [42Re; cf. 43C]. **Q**

(c) Now we can use 83B. The condition (i) above is of course the same as saying that A is $\mathfrak{T}_s(\boldsymbol{S}', \boldsymbol{S})$-bounded, which is condition (i) of 83B. Similarly, condition (ii) above corresponds to condition (ii) of 83B. **P** For if $\langle a_n\rangle_{n\in\mathbf{N}}$ is disjoint in $\mathfrak{A}$, then $\langle \chi a_n\rangle_{n\in\mathbf{N}}$ is disjoint and norm-bounded in $\boldsymbol{S}^+$, so, if A is relatively compact, then

$$\langle \sup_{f\in A}|f(\chi a_n)|\rangle_{n\in\mathbf{N}} \to 0.$$

On the other hand, suppose that condition (ii) of this theorem is satisfied, and that $\langle x_n\rangle_{n\in\mathbf{N}}$ is a disjoint sequence in $\boldsymbol{S}^+$ such that $\gamma = \sup_{n\in\mathbf{N}}\|x_n\|_\infty < \infty$. For any n (unless $x_n = 0$), x_n can be expressed as $\sum_{j<m}\beta_j\chi b_j$, where $\sum_{j<m}\beta_j = \|x_n\|_\infty \leqslant \gamma$, and each $\beta_j > 0$ [42Ec], so that

$$\sup_{f\in A}|fx_n| \leqslant \gamma \max_{j<m}\sup_{f\in A}|f(\chi b_j)|.$$

Now let a_n be equal to one of the b_j and such that

$$\sup_{f\in A}|fx_n| \leqslant \gamma \sup_{f\in A}|f(\chi a_n)|;$$

we know that there is an $\alpha_n > 0$ such that $\alpha_n\chi a_n \leqslant x_n$. (If $x_n = 0$, set $a_n = 0$, $\alpha_n = 1$; then the same statements are valid.) Since, for $m \neq n$,

$$\alpha_n\chi a_n \wedge \alpha_m\chi a_m \leqslant x_n \wedge x_m = 0,$$

$\langle a_n \rangle_{n \in \mathbf{N}}$ is a disjoint sequence in $\mathfrak{A}$. So $\langle \sup_{f \in A} |f(\chi a_n)| \rangle_{n \in \mathbf{N}} \to 0$ and $\langle \sup_{f \in A} |f x_n| \rangle_{n \in \mathbf{N}} \to 0$, as required. **Q** Thus the conditions above are satisfied iff those of 83B are satisfied, i.e. iff A is relatively compact.

83D Corollary Let $\mathfrak{A}$ be a Boolean ring, and A a non-empty subset of $L^\# = L^\#(\mathfrak{A})$. Then A is relatively compact for $\mathfrak{T}_s(L^\#, L^{\#\prime})$ iff the following two conditions hold: (i) $\sup_{f \in A} |f(\chi a)| < \infty$ for every $a \in \mathfrak{A}$; (ii) $\langle \sup_{f \in A} |f(\chi a_n)| \rangle_{n \in \mathbf{N}} \to 0$ whenever $\langle a_n \rangle_{n \in \mathbf{N}}$ is a disjoint sequence in $\mathfrak{A}$.

Proof The point is that $L^\# = L^\infty(\mathfrak{A})^\times$ is a band in $E = L^\infty(\mathfrak{A})^\sim$. Consequently it is closed for the norm on E [22Ec], and is therefore closed for the weak topology $\mathfrak{T}_s(E, E')$. So a subset A of $L^\#$ is relatively compact for $\mathfrak{T}_s(L^\#, E')$, which coincides with $\mathfrak{T}_s(L^\#, L^{\#\prime})$, iff it is relatively compact for $\mathfrak{T}_s(E, E')$. The result now follows at once from 83C.

83E Corollary Let $\mathfrak{A}$ be a Boolean algebra, and A a subset of $L^\# = L^\#(\mathfrak{A})$; let C be the solid hull of A in $L^\#$. Then these are equivalent:

(i) A is relatively compact for $\mathfrak{T}_s(L^\#, L^{\#\prime})$;

(ii) C is relatively compact for $\mathfrak{T}_s(L^\#, L^\infty)$;

(iii) C is relatively compact for $\mathfrak{T}_s(L^\#, S)$, where $S = S(\mathfrak{A})$, $L^\infty = L^\infty(\mathfrak{A})$.

Proof (a) If A is relatively compact for $\mathfrak{T}_s(L^\#, L^{\#\prime})$ so is C [83A], so C is relatively compact for $\mathfrak{T}_s(L^\#, L^\infty)$. Thus (i) $\Rightarrow$ (ii). Of course (ii) $\Rightarrow$ (iii).

(b) (iii) $\Rightarrow$ (i) Recall that in 44Bb $L^\#$ is identified with the space of bounded completely additive functionals on $\mathfrak{A}$, while in 42L $S^\times$ is identified with the space of locally bounded completely additive functionals. But if $\mathfrak{A}$ has a 1, these are the same. So we may think of $L^\#$ as $S^\times$.

Now if C is relatively compact for $\mathfrak{T}_s(S^\times, S)$, then from 82E we see that for any $\epsilon > 0$ there is an $f_0 \in S^{\times +}$ such that

$$\| |f| - |f| \wedge f_0 \| = (|f| - |f| \wedge f_0)(\chi 1) \leqslant \epsilon \quad \forall\ f \in A.$$

So A is relatively compact for $\mathfrak{T}_s(L^\#, L^{\#\prime})$ by 83A. [Cf. 82I.]

83F Proposition Let $(\mathfrak{A}, \mu)$ be a semi-finite measure ring and A a non-empty subset of $L^1 = L^1(\mathfrak{A})$. Then A is relatively compact for

$\mathfrak{T}_s(L^1, L^{1\prime})$ iff (i) $\sup_{u\in A}|\langle u, \chi a\rangle| < \infty$ for every $a \in \mathfrak{A}$ (ii) for every $\epsilon > 0$ there is an $a \in \mathfrak{A}^f$ and a $\delta > 0$ such that

$$\sup_{u\in A}|\langle u, \chi b\rangle| \leqslant \epsilon \quad \text{whenever} \quad b \in \mathfrak{A},\ \mu(a \cap b) \leqslant \delta.$$

Proof Recall that we think of L^1 as identified with $L^{\#}_{-}(\mathfrak{A})$ [52E].

(a) If the conditions hold, then let $\langle a_n\rangle_{n\in\mathbf{N}}$ be any disjoint sequence in $\mathfrak{A}$. Given $\epsilon > 0$, find $a \in \mathfrak{A}^f$ and $\delta > 0$ as in (ii). Since $\langle a_n\rangle_{n\in\mathbf{N}}$ is disjoint, $\sum_{n\in\mathbf{N}}\mu(a \cap a_n) \leqslant \mu a < \infty$. So there is an n_0 such that $\mu(a \cap a_n) \leqslant \delta$ for all $n \geqslant n_0$. Now

$$\sup_{u\in A}|\langle u, \chi a_n\rangle| \leqslant \epsilon \quad \forall\ n \geqslant n_0.$$

As ϵ is arbitrary,

$$\lim_{n\to\infty}\sup_{u\in A}|\langle u, \chi a_n\rangle| = 0,$$

and A is relatively $\mathfrak{T}_s(L^1, L^{1\prime})$-compact by the criterion in 83D.

(b) Conversely, if A is relatively $\mathfrak{T}_s(L^1, L^{1\prime})$-compact, then certainly $\sup_{u\in A}|\langle u, \chi a\rangle| < \infty$ for every $a \in \mathfrak{A}$. Also, given $\epsilon > 0$, there is a $u_0 \in L^{1+}$ such that

$$\|\,|u| - |u| \wedge u_0\|_1 \leqslant \epsilon \quad \forall\ u \in A$$

[83A]. Now there is an $x_0 \in S(\mathfrak{A}^f)^+$ such that $\|x_0 - u_0\|_1 \leqslant \epsilon$, as S is dense in L^1, and there is an $a \in \mathfrak{A}^f$ and an $\alpha > 0$ such that $x_0 \leqslant \alpha\chi a$ [42Ec]. Accordingly

$$\|(|u| - \alpha\chi a)^+\|_1 \leqslant \|(|u| - u_0)^+\|_1 + \|(u_0 - \alpha\chi a)^+\|_1 \leqslant 2\epsilon$$

for every $u \in A$. Now set $\delta = \epsilon/\alpha$. If $b \in \mathfrak{A}$ and $\mu(a \cap b) \leqslant \delta$, then

$$\begin{aligned} |\langle u, \chi b\rangle| &\leqslant \langle |u|, \chi b\rangle \leqslant \langle (|u| - \alpha\chi a)^+, \chi b\rangle + \langle \alpha\chi a, \chi b\rangle \\ &\leqslant \|(|u| - \alpha\chi a)^+\|_1 + \alpha\mu(a \cap b) \leqslant 2\epsilon + \epsilon = 3\epsilon, \end{aligned}$$

for every $u \in A$. As ϵ is arbitrary, this establishes the second condition.

83G Lemma Let $\mathfrak{A}$ be a Dedekind σ-complete Boolean algebra. If $A \subseteq L^{\#}_{-}(\mathfrak{A})$ is bounded for $\mathfrak{T}_s(L^{\#}, S)$, it is bounded for the norm of $L^{\#}$.

Proof I shall use 81F, though of course more direct approaches are possible. Let C be the convex hull of A in $L^{\#}_{-}$. As in 83E, regard $L^{\#}$ as identified with $S^\times$. Since C is $\mathfrak{T}_s(S^\times, S)$-bounded, it is certainly relatively $\mathfrak{T}_s(S^*, S)$-compact. Let f_0 belong to the closure of C in S^*.

? If $f_0 \notin S^\sim$, then $f\chi\colon \mathfrak{A} \to \mathbf{R}$ is not locally bounded [42H], so by 42I there is a disjoint sequence $\langle b_n\rangle_{n\in\mathbf{N}} = \langle a_n \backslash a_{n+1}\rangle_{n\in\mathbf{N}}$ such that $f_0(\chi b_n) \geqslant n+1$ for every $n \in \mathbf{N}$. Now however

$$\sum_{i\in\mathbf{N}}\chi b_{n(i)} = \chi(\sup_{i\in\mathbf{N}} b_{n(i)})$$

exists for $\mathfrak{T}_s(S, L^{\#})$ for every strictly increasing sequence $\langle n(i)\rangle_{i\in\mathbf{N}}$. So by A1Hd $\{\chi b_n : n\in\mathbf{N}\}$ is $\mathfrak{T}_b(S, L^{\#})$-bounded and

$$\infty = \sup_{n\in\mathbf{N}} f_0(\chi b_n) \leqslant \sup_{n\in\mathbf{N}} \sup_{f\in C} |f(\chi b_n)| < \infty,$$

which is impossible. **X**

Thus $f_0 \in S^{\sim}$. As f_0 is arbitrary, C is relatively $\mathfrak{T}_s(S^{\sim}, S)$-compact, so by 81F A is $\mathfrak{T}_{|s|}(S^{\sim}, S)$-bounded, i.e.

$$\sup_{f\in A} \|f\| = \sup_{f\in A} |f|(\chi 1) < \infty.$$

83H Proposition Let $\mathfrak{A}$ be a Dedekind σ-complete Boolean algebra and A a subset of $L^{\#} = L^{\#}(\mathfrak{A})$. Then these are equivalent:

(i) A is relatively compact for $\mathfrak{T}_s(L^{\#}, L^{\#\prime})$;

(ii) A is relatively compact for $\mathfrak{T}_s(L^{\#}, L^{\infty})$;

(iii) A is relatively compact for $\mathfrak{T}_s(L^{\#}, S)$, where $S = S(\mathfrak{A})$, $L^{\infty} = L^{\infty}(\mathfrak{A})$.

Proof (a) **(i) $\Leftrightarrow$ (ii)** Of course (i) $\Rightarrow$ (ii). Conversely, if A is relatively compact for $\mathfrak{T}_s(L^{\#}, L^{\infty})$, then by 82G the solid hull of A is relatively compact for $\mathfrak{T}_s(L^{\#}, L^{\infty})$, since by 43Db L^{∞} is Dedekind σ-complete. Now it follows from 83E that A is relatively $\mathfrak{T}_s(L^{\#}, L^{\#\prime})$-compact.

(b) **(ii) $\Leftrightarrow$ (iii)** Of course (ii) $\Rightarrow$ (iii). On the other hand, if A is relatively compact for $\mathfrak{T}_s(L^{\#}, S)$, then by 83G its $\mathfrak{T}_s(L^{\#}, S)$-closure $\bar{A}$ is norm-bounded. So $\mathfrak{T}_s(L^{\#}, S)$ and $\mathfrak{T}_s(L^{\#}, L^{\infty})$ agree on $\bar{A}$, which is therefore $\mathfrak{T}_s(L^{\#}, L^{\infty})$-compact.

83I The same approach as in 83F gives a general criterion for weak compactness in Banach function spaces.

Theorem Let (X, Σ, μ) be a Maharam measure space and E an order-dense solid linear subspace of $L^0 = L^0(\Sigma, \mu)$. Let

$$F = \{w : w\in L^0,\ w\times u\in L^1 \quad \forall\ u\in E\},$$

so that F can be identified with $E^{\times}$ as in 65A. Let A be a non-empty subset of F. Then A is relatively compact for $\mathfrak{T}_s(F, E)$ iff it is bounded and, for every $\epsilon > 0$ and $u\in E^+$, there is an $H\in\Sigma^f$ and a $\delta > 0$ such that

$$\int_K |u\times w| \leqslant \epsilon \quad \forall\ w\in A$$

whenever $K\in\Sigma$ and $\mu(H\cap K) \leqslant \delta$.

(*Notation*: If $K\in\Sigma$ and $v\in L^1$, I write $\int_K v = \int v\times\chi K^{\cdot}$; i.e. $\int_K x^{\cdot} = \int_K x$ for any $x\in\mathfrak{L}^1(\mu)$.)

Proof (a) E, being solid in L^0, is Dedekind complete; so if A is relatively compact, the solid hull of A is relatively compact [82G], and for any $u \in E^+$ and $\epsilon > 0$ there is a $w_0 \in F^+$ such that

$$\langle |w| - |w| \wedge w_0, u\rangle \leqslant \epsilon \quad \forall \; w \in A.$$

Now $w_0 \times u \in L^1$, so there is a $v_0 \in S(\mathfrak{A}^f)^+$ such that $\|v_0 - w_0 \times u\|_1 \leqslant \epsilon$, and now there is an $a \in \mathfrak{A}^f$ and an $\alpha > 0$ such that $v_0 \leqslant \alpha\chi a$. Let $H \in \Sigma^f$ be such that $H^{\cdot} = a$, and let $\delta = \epsilon/\alpha$. Now suppose that $K \in \Sigma$ and that $\mu(H \cap K) \leqslant \delta$. Then for any $w \in A$,

$$\begin{aligned}\int_K |u \times w| &\leqslant \int_K u \times w_0 + \int u \times (|w| - |w| \wedge w_0)\\ &\leqslant \int_K v_0 + \|v_0 - u \times w_0\|_1 + \epsilon\\ &\leqslant \int_K \alpha\chi a + 2\epsilon\\ &= \alpha\mu(H \cap K) + 2\epsilon \leqslant 3\epsilon.\end{aligned}$$

Thus the condition is satisfied.

(**b**) On the other hand, suppose that the condition is satisfied. Suppose that $\langle u_n\rangle_{n\in\mathbf{N}}$ is any disjoint sequence in E^+, bounded above by u. Let $\epsilon > 0$. Let $H \in \Sigma^f$ and $\delta > 0$ be such that $\int_K |u \times w| \leqslant \epsilon$ whenever $w \in A$, $K \in \Sigma$ and $\mu(H \cap K) \leqslant \delta$. For each $n \in \mathbf{N}$, choose an $x_n \in M(\Sigma)^+$ such that $x_n^{\cdot} = u_n$, and let $K_n = \{t : x_n(t) > 0\}$. Then

$$\mu(K_m \cap K_n) = 0 \quad \text{whenever} \quad m \neq n,$$

so

$$\textstyle\sum_{m\in\mathbf{N}} \mu(K_m \cap H) \leqslant \mu H < \infty,$$

and there is an n_0 such that $\mu(K_n \cap H) \leqslant \delta$ for every $n \geqslant n_0$. Now, for $n \geqslant n_0$,

$$|\langle u_n, w\rangle| \leqslant \langle u \times \chi K_n^{\cdot}, |w|\rangle = \int_{K_n} |u \times w| \leqslant \epsilon,$$

for every $w \in A$. Thus

$$\lim_{n\to\infty} \sup_{w\in A} |\langle u_n, w\rangle| = 0.$$

By 82E, A is relatively compact for $\mathfrak{T}_s(F, E)$.

83J Proposition (GROTHENDIECK A.L. [théorème 2, p. 146]) Let X be a locally compact Hausdorff topological space. Let $K = K(X)$, normed by $\| \; \|_\infty$, and $E = K(X)' = C_0(X)'$ [A2D]. For $g \in E^+$, let μ_g be the corresponding Radon measure on X [73E]. Let A be a non-empty subset of E, and let $B = \{|f| : f \in A\}$. Then these are equivalent:

(i) A is relatively compact for $\mathfrak{T}_s(E, E')$;

(ii) A is bounded for $\mathfrak{T}_s(\boldsymbol{E},\boldsymbol{K})$ and, whenever $\langle x_n\rangle_{n\in\mathbf{N}}$ is a norm-bounded disjoint sequence in $\boldsymbol{K}^+$,

$$\langle \sup_{f\in A}|fx_n|\rangle_{n\in\mathbf{N}}\to 0;$$

(iii) A is bounded for $\mathfrak{T}_s(\boldsymbol{E},\boldsymbol{K})$ and, whenever $\langle G_n\rangle_{n\in\mathbf{N}}$ is a disjoint sequence of open sets in X and $\langle g(n)\rangle_{n\in\mathbf{N}}$ is a sequence in B,

$$\langle \mu_{g(n)}G_n\rangle_{n\in\mathbf{N}}\to 0;$$

(iv) A is bounded for $\mathfrak{T}_s(\boldsymbol{E},\boldsymbol{K})$ and, whenever $\langle F_n\rangle_{n\in\mathbf{N}}$ is a disjoint sequence of compact sets in X and $\langle g(n)\rangle_{n\in\mathbf{N}}$ is a sequence in B,

$$\langle \mu_{g(n)}F_n\rangle_{n\in\mathbf{N}}\to 0.$$

Proof (a) Observe that $\boldsymbol{K}$ and $\|\ \|_\infty$ obviously satisfy the conditions of 83B. So (ii) $\Rightarrow$ (i) is immediate.

(b) (i) $\Rightarrow$ (iv) Of course A is bounded for $\mathfrak{T}_s(\boldsymbol{E},\boldsymbol{K})$. Let $\langle F_n\rangle_{n\in\mathbf{N}}$ be a disjoint sequence of compact sets in X, let $\langle g(n)\rangle_{n\in\mathbf{N}}$ be a sequence in B, and let $\epsilon>0$. By 83A there is an $h\in\boldsymbol{E}^+$ such that

$$\|\,|f|-|f|\wedge h\|\leqslant\epsilon\quad\forall\ f\in A.$$

Now for any compact set $F\subseteq X$, there certainly exists an $x\in\boldsymbol{K}$ such that $\chi F\leqslant x$, as X is locally compact [A2Eb]. So we see from 73F that for any $f\in\boldsymbol{E}^+$

$$\mu_fF=\inf\{fx:x\in\boldsymbol{K},\ \chi F\leqslant x\leqslant\chi X\}.$$

Now if $f\in B$,

$$\mu_fF=\inf\{fx:x\in\boldsymbol{K},\ \chi F\leqslant x\leqslant\chi X\}$$
$$\leqslant\qquad\epsilon+\inf\{(f\wedge h)(x):x\in\boldsymbol{K},\ \chi F\leqslant x\leqslant\chi X\}$$

[since $\|f-f\wedge h\|\leqslant\epsilon$, so $fx\leqslant\epsilon+(f\wedge h)x$ whenever $0\leqslant x\leqslant\chi X$]

$$\leqslant\epsilon+\inf\{hx:x\in\boldsymbol{K},\ \chi F\leqslant x\leqslant\chi X\}$$
$$=\epsilon+\mu_hF.$$

But we can see that, because $\langle F_n\rangle_{n\in\mathbf{N}}$ is disjoint,

$$\textstyle\sum_{n\in\mathbf{N}}\mu_hF_n\leqslant\mu_hX<\infty\quad[73\text{E}].$$

So there is an n_0 such that $\mu_hF_n\leqslant\epsilon$ for every $n\geqslant n_0$. Now however

$$\mu_{g(n)}F_n\leqslant\epsilon+\mu_hF_n\leqslant2\epsilon\quad\forall\ n\geqslant n_0.$$

As ϵ is arbitrary,

$$\langle \mu_{g(n)}F_n\rangle_{n\in\mathbf{N}}\to 0.$$

(c) (iv) $\Rightarrow$ (iii) For each $n\in\mathbf{N}$, choose a compact set $F_n\subseteq G_n$ such that $\mu_{g(n)}F_n\geqslant\mu_{g(n)}G_n-2^{-n}$; this is possible because $\mu_{g(n)}$ is a Radon measure and $\mu_{g(n)}G_n \leqslant \mu_{g(n)}X < \infty$. Now $\langle F_n\rangle_{n\in\mathbf{N}}$ is disjoint, so $\langle \mu_{g(n)}F_n\rangle_{n\in\mathbf{N}} \to 0$, and it follows at once that $\langle \mu_{g(n)}G_n\rangle_{n\in\mathbf{N}} \to 0$.

(d) **(iii)** $\Rightarrow$ **(ii)** If $\langle x_n\rangle_{n\in\mathbf{N}}$ is disjoint and norm-bounded in $\boldsymbol{K}^+$, set $\alpha = \sup_{n\in\mathbf{N}} \|x_n\|_\infty$. For each $n \in \mathrm{N}$, let $G_n = \{t : x_n(t) > 0\}$; then $\langle G_n\rangle_{n\in\mathbf{N}}$ is a disjoint sequence of open sets in X. For each $n \in \mathrm{N}$, choose an $f_n \in A$ such that

$$|f_n x_n| \geqslant \sup_{f\in A} |fx_n| - 2^{-n},$$

and set $g(n) = |f_n| \in B$. Then

$$\sup_{f\in A} |fx_n| \leqslant |f_n x_n| + 2^{-n} \leqslant |f_n|(x_n) + 2^{-n}$$
$$\leqslant \alpha\mu_{g(n)}G_n + 2^{-n}.$$

But $\langle \mu_{g(n)}G_n\rangle_{n\in\mathbf{N}} \to 0$ by hypothesis, so $\langle \sup_{f\in A} |fx_n|\rangle_{n\in\mathbf{N}} \to 0$.

***83K Sequentially order-continuous linear functionals and countably additive functionals** Let $\boldsymbol{E}$ be any Archimedean Riesz space. Let $\boldsymbol{E}_c^\sim$ be the space of linear functionals $f: \boldsymbol{E} \to \mathbf{R}$ such that $\langle fx_n\rangle_{n\in\mathbf{N}} \to 0$ whenever $\langle x_n\rangle_{n\in\mathbf{N}} \downarrow 0$ in $\boldsymbol{E}$. In the results below, which can be taken as a set of exercises, I show how the methods of this book may be used to establish many properties of $\boldsymbol{E}_c^\sim$, along the same lines as those already developed for $\boldsymbol{E}^\times$.

(a) Let us call a linear space topology $\mathfrak{T}$ on $\boldsymbol{E}$ **sequentially Lebesgue** if $\langle x_n\rangle_{n\in\mathbf{N}} \to 0$ whenever $\langle x_n\rangle_{n\in\mathbf{N}} \downarrow 0$ in $\boldsymbol{E}$. (i) In this case, if $\langle x_n\rangle_{n\in\mathbf{N}} \downarrow 0$ and U is any neighbourhood of 0, there is an $n \in \mathrm{N}$ such that $[0, x_n] \subseteq U$ [cf. 24Bc]. (ii) Because $\boldsymbol{E}$ is Archimedean, order-bounded sets are bounded [cf. 24Db].

(b) (i) $\boldsymbol{E}_c^\sim \subseteq \boldsymbol{E}^\sim$ [use (a) (ii) above]; (ii) $\boldsymbol{E}_c^\sim$ is a Riesz subspace of $\boldsymbol{E}^\sim$ [use (a) (i)]; (iii) $\boldsymbol{E}_c^\sim$ is a solid linear subspace of $\boldsymbol{E}^\sim$; (iv) $\boldsymbol{E}_c^\sim$ is a band in $\boldsymbol{E}^\sim$ [cf. 16H]; (v) $\boldsymbol{E}_c^\sim$ is the set of those linear functionals on $\boldsymbol{E}$ which are expressible as the difference of two sequentially order-continuous increasing linear functionals. [See LUXEMBURG & ZAANEN B.F.S., note VI, § 20.]

(c) A subset of $\boldsymbol{E}_c^\sim$ which is relatively countably compact for $\mathfrak{T}_s(\boldsymbol{E}_c^\sim, \boldsymbol{E})$ is relatively compact for $\mathfrak{T}_s(\boldsymbol{E}_c^\sim, \boldsymbol{E})$ [cf. 82B].

(d) If $A \subseteq \boldsymbol{E}_c^\sim$, then $\Gamma(A)$ is relatively compact for $\mathfrak{T}_s(\boldsymbol{E}_c^\sim, \boldsymbol{E})$ iff A is relatively $\mathfrak{T}_s(\boldsymbol{E}_c^\sim, \boldsymbol{E})$-compact and $\mathfrak{T}_{|s|}(\boldsymbol{E}_c^\sim, \boldsymbol{E})$-bounded [cf. 82C].

(e) If A is a non-empty subset of $\boldsymbol{E}_c^\sim$, then these are equivalent: (i) the solid convex hull of A is relatively compact for $\mathfrak{T}_s(\boldsymbol{E}_c^\sim, \boldsymbol{E})$;

(ii) $\{|f| : f \in A\}$ is relatively countably compact for $\mathfrak{T}_s(E_c^\sim, E)$; (iii) A is bounded for $\mathfrak{T}_s(E_c^\sim, E)$ and, whenever $\langle x_n \rangle_{n\in\mathbf{N}} \downarrow 0$ in E,

$$\langle \sup_{f\in A} |fx_n| \rangle_{n\in\mathbf{N}} \to 0$$

[cf. 82E; see also 8XC].

(f) If E is Dedekind σ-complete and A is a non-empty subset of $E_c^\sim$, then these are equivalent: (i) A is relatively countably compact for $\mathfrak{T}_s(E_c^\sim, E)$; (ii) the solid convex hull of A is relatively compact for $\mathfrak{T}_s(E_c^\sim, E)$; (iii) A is bounded for $\mathfrak{T}_s(E_c^\sim, E)$ and, whenever $\langle x_n \rangle_{n\in\mathbf{N}}$ is a disjoint sequence in E^+ which is bounded above in E,

$$\langle \sup_{f\in A} |fx_n| \rangle_{n\in\mathbf{N}} \to 0;$$

(iv) for any $x \in E^+$ and $\epsilon > 0$, there is an $f_0 \in E_c^{\sim +}$ such that

$$(|f| - |f| \wedge f_0)(x) \leqslant \epsilon$$

for every $f \in A$ [cf. 82G].

(g) If E is Dedekind σ-complete, then $E_c^\sim$ is sequentially complete for $\mathfrak{T}_s(E_c^\sim, E)$ [cf. 82J].

(h) If $\mathfrak{A}$ is a Boolean algebra, then $S(\mathfrak{A})_c^\sim$ may be identified with the space of bounded countably additive functionals on $\mathfrak{A}$, and every member of $S(\mathfrak{A})_c^\sim$ extends uniquely to a member of $L^\infty(\mathfrak{A})_c^\sim$, so that $S(\mathfrak{A})_c^\sim$ can be identified with $L^\infty(\mathfrak{A})_c^\sim$. With its natural norm, $S(\mathfrak{A})_c^\sim$ is an L-space [cf. 44B, 42P, 83E proof, part (b)].

(i) Let $\mathfrak{A}$ be a Dedekind σ-complete Boolean algebra, let $S = S(\mathfrak{A})$, and let $H = S_c^\sim = L^\infty(\mathfrak{A})_c^\sim$, the space of all countably additive functionals on $\mathfrak{A}$ [(h) above, 42Q]. (i) If $A \subseteq H$ is bounded for $\mathfrak{T}_s(H, S)$, it is bounded for the norm of H [cf. 83G]. (ii) If $A \subseteq H$ is relatively compact for $\mathfrak{T}_s(H, S)$, it is relatively compact for $\mathfrak{T}_s(H, H')$ [cf. 83H].

***(j)** Let (X, Σ, μ) be the measure space consisting of $\mathbf{R}$ with Lebesgue measure. Then $\mathfrak{L}^1(\mu)_c^\sim$ is isomorphic, as Riesz space, to $s_0 \oplus L^\infty(\mathfrak{A})$, where s_0 is the space of functions on X zero except at finitely many points, and $\mathfrak{A}$ is the measure algebra of (X, Σ, μ). While $\mathfrak{L}^1(\mu)^\times \simeq s_0$.

83L Exercises **(a)** Let X be any set, and let s be the space of functions in $\mathbf{R}^X$ taking only finitely many values [cf. 4XCa]. Let A be a subset of $l^1 = l^1(X)$. Show that A is relatively countably compact for $\mathfrak{T}_s(l^1, s)$ iff A is relatively compact for the norm on $l^1(X)$. [Hint: use 83H and 83A.]

***(b)** Use the argument of KÖTHE, 22.4.2–3, to prove (a) above when $X = \mathbf{N}$. Now use this to prove Orlicz' theorem [A1H].

*(**c**) Let $(G, \| \ \|)$ be a normed Riesz space satisfying the conditions of 83B. Show that, whenever $A \subseteq G^+$ is non-empty and bounded above,

$$\sup_{x \in A} \|x\| = \inf_{y \in B} \|y\|,$$

where B is the set of upper bounds of A. Show that Theorem 83B applies to any Riesz space with a Riesz norm satisfying this condition. [See NAKANO C.A.C.R.]

*(**d**) Let $(G, \| \ \|)$ be a normed Riesz satisfying the conditions of 83B. (i) Show that the linear space $G \oplus \mathbf{R}$ can be ordered by saying that

$$x + \alpha e = (x, \alpha) \geqslant 0 \Leftrightarrow \|x^-\| \leqslant \alpha,$$

to produce an Archimedean Riesz space G_1 in which e is an order unit, and that the norm on G is now that induced by the norm $\| \ \|_e$ on G_1. (ii) Show that if X is the set of all Riesz homomorphisms $f: G \to \mathbf{R}$ such that $\|f\| = 1$, then G is isomorphic, as normed Riesz space, to a truncated Riesz subspace of $l^\infty(X)$.

Notes and comments The rather curious conditions of the basic theorem 83B are an attempt to find a common abstraction of the spaces $S(\mathfrak{A})$ and $K(X)$, when $\mathfrak{A}$ does not have a 1 or X is not compact, respectively. Another is offered in 83Lc. Just as in Chapter 7, but apparently for different reasons, we find ourselves dealing with truncated Riesz subspaces of M-spaces with units [see 83Ld]. In effect, 83B is a version of 81H in a situation in which we want to discuss disjoint sequences in G^+ which are order-bounded, not in G itself, but in G''. It would also be possible to prove 83B by adapting the argument of 81F/81H.

It is now easy, using the technical device in part (c) of the proof of 83C, to establish the important remaining criteria that have been given for weak compactness in L-spaces [83F, 83H, 83J]. Note that 83H can be proved by rather more direct arguments, since the ring $\mathfrak{A}$ is here an algebra, and 83B tells us no more than 81H does. Indeed, it can be derived from the work of § 82, since we are examining the dual pair $(S^\times, S)$ rather than $(S^\sim, S)$. Note that for l^1-spaces we have an equivalence between weak and norm compactness [83La]. KÖTHE [§ 30.6] generalizes this to other perfect sequence spaces. Another generalization of 83H is given by MOORE & REBER.

83J, however, seems to lie deeper. The substantial link in the chain is (ii) $\Rightarrow$ (i); this is due essentially to GROTHENDIECK A.L. and is the progenitor of 83B, 81H and even 24H. A number of authors have regarded this theorem as a criterion for weak compactness in spaces of regular measures, divorcing them from spaces of continuous functions, and thereby obtaining generalizations to spaces X which are not locally compact; see, for instance, GÄNSSLER.

83I is a version of a result due to DIEUDONNÉ, who points out that a set A in $F \cong E^\times$ is relatively weakly compact iff $\{w \times u : w \in A\}$ is relatively weakly compact in L^1 for every $u \in E^+$; this is because F is complete for $\mathfrak{T}_{|s|}(F, E)$ [81D], and therefore can be thought of as a closed subspace of a power of L^1 [cf. 81Ib].

*In 83K I try to show how the same arguments that I have used to study $E^\times$ can be applied to $E_c^\sim$. The point about $E_c^\sim$ is that if E is actually a subspace of $\mathbf{R}^X$, then $E_c^\sim$ may be more interesting than $E^\times$ [see 83Kj]. In view of the persistent appearance of sequences in § 82, it is not surprising that many of these results have forms applying to $E_c^\sim$.

8X Examples for Chapter 8

The following three examples show how relatively familiar spaces afford examples of several phenomena relevant to the results of § 82. In 8XA we have a Riesz space E which is not uniformly complete; consequently there can be a set $A \subseteq E^\times$ such that A is relatively compact for $\mathfrak{T}_s(E^\times, E)$, but its convex hull is not. In 8XB we have a Riesz space E which is uniformly complete but not Dedekind σ-complete, so there can be a set $A \subseteq E$ such that A is relatively compact for $\mathfrak{T}_s(E^\times, E)$ but its solid hull is not. Finally, in 8XC, we have a space E such that the topology $\mathfrak{T}_k(E^\times, E)$ is not complete; this shows that 82B and 82C cannot be deduced directly from Eberlein's and Krein's theorems. *At the same time we find that weakly compact sets in $E_c^\sim$ behave a little differently from those in $E^\times$.

8XA Let E be the subspace of $l^\infty(\mathbf{N})$ consisting of sequences which are eventually constant. Then E is a Riesz subspace of $l^\infty(\mathbf{N})$; it has an order unit $e = \chi\mathbf{N}$ and $\| \;\|_e = \| \;\|_\infty$ on E. E is isomorphic to $S(\mathfrak{A})$ where

$$\mathfrak{A} = \{I : I \subseteq \mathbf{N}, I \text{ is finite or } \mathbf{N}\backslash I \text{ is finite}\}.$$

We may identify $E^\times$ with $l^1(\mathbf{N})$ [use 65B or 4XCb].

Now let us define $\langle f_n \rangle_{n\in\mathbf{N}}$ in $E^\times$ by writing

$$f_n x = 2^{n+1}(x(n) - x(n+1)) \quad \forall \; x \in E, n \in \mathbf{N}.$$

Then $\langle f_n \rangle_{n\in\mathbf{N}} \to 0$ for $\mathfrak{T}_s(E^\times, E)$, so $A = \{f_n : n \in \mathbf{N}\}$ is relatively compact for $\mathfrak{T}_s(E^\times, E)$. But if, for each $n \in \mathbf{N}$, $g_n = \sum_{i<n} 2^{-(i+1)} f_i$, then $g_n x = x(0) - x(n)$ for every $x \in E$, so $\langle g_n \rangle_{n\in\mathbf{N}}$ has no cluster point in $E^\times$;

thus the convex hull of A is not relatively compact for $\mathfrak{T}_s(\boldsymbol{E}^\times, \boldsymbol{E})$. We note that, of course, A is not bounded for $\mathfrak{T}_{|s|}(\boldsymbol{E}^\times, \boldsymbol{E})$ [82C]. This can happen only because $\boldsymbol{E}$ is not uniformly complete [81G].

8XB Let $\boldsymbol{E}$ be the subspace of $\boldsymbol{l}^\infty(\mathrm{N})$ consisting of convergent sequences. Then $\boldsymbol{E}$ is a $\| \ \|_\infty$-closed Riesz subspace of $\boldsymbol{l}^\infty(\mathrm{N})$; with $\| \ \|_\infty$ it is an M-space with unit, isomorphic to $\boldsymbol{C}(Y)$, where Y is the one-point compactification of the discrete space N.

$\boldsymbol{E}^\times$ can be identified with $\boldsymbol{l}^1 = \boldsymbol{l}^1(\mathrm{N})$ (because $\boldsymbol{c}_0$ is order-dense in $\boldsymbol{E}$, and $(\boldsymbol{c}_0)^\times \cong \boldsymbol{l}^1$; or use 65B).

Define a sequence $\langle f_n \rangle_{n\in\mathrm{N}}$ in $\boldsymbol{E}^\times$ by

$$f_n x = x(2n) - x(2n+1) \quad \forall \ x \in \boldsymbol{E},\ n \in \mathrm{N}.$$

Set $A = \{f_n : n \in \mathrm{N}\}$. Then the convex hull of A is relatively compact for $\mathfrak{T}_s(\boldsymbol{E}^\times, \boldsymbol{E})$. **P** We can see that $\langle f_n \rangle_{n\in\mathrm{N}} \to 0$ for $\mathfrak{T}_s(\boldsymbol{E}^\times, \boldsymbol{E})$. Also, $\sum_{n\in\mathrm{N}} \alpha_n f_n$ exists for $\mathfrak{T}_s(\boldsymbol{E}^\times, \boldsymbol{E})$ for every sequence $\langle \alpha_n \rangle_{n\in\mathrm{N}}$ in $\boldsymbol{l}^1$. So if we define $T: \boldsymbol{l}^1 \to \boldsymbol{E}$ by

$$Tz = \sum_{n\in\mathrm{N}} z(n) f_n \quad \forall \ z \in \boldsymbol{l}^1,$$

T is a linear map continuous for $\mathfrak{T}_s(\boldsymbol{l}^1, \boldsymbol{c}_0)$ and $\mathfrak{T}_s(\boldsymbol{E}^\times, \boldsymbol{E})$. But the unit ball B of $\boldsymbol{l}^1$ is compact for $\mathfrak{T}_s(\boldsymbol{l}^1, \boldsymbol{c}_0)$, so $T[B]$ is compact for $\mathfrak{T}_s(\boldsymbol{E}^\times, \boldsymbol{E})$; now clearly $T[B]$ is the closed convex hull of A. **Q** Thus we see that A is equicontinuous for $\mathfrak{T}_k(\boldsymbol{E}, \boldsymbol{E}^\times)$.

Now, for each $n \in \mathrm{N}$, define $x_n \in \boldsymbol{E}$ by

$$x_n(i) = 0 \quad \text{if} \quad i \leqslant 2n, \quad x_n(i) = 1 \quad \text{if} \quad i > 2n.$$

Then $\langle x_n \rangle_{n\in\mathrm{N}} \downarrow 0$ in $\boldsymbol{E}$ but $f_n x_n = -1$ for every $n \in \mathrm{N}$. Thus $\mathfrak{T}_k(\boldsymbol{E}, \boldsymbol{E}^\times)$ is not Lebesgue. We can also see that, for each $n \in \mathrm{N}$, f_n^+ is given by $f_n^+(x) = x(2n)$ for every $x \in \boldsymbol{E}$; so we see that $\langle f_n^+ \rangle_{n\in\mathrm{N}}$ has no cluster point in $\boldsymbol{E}^\times$; thus the solid hull of A is not relatively compact for $\mathfrak{T}_s(\boldsymbol{E}^\times, \boldsymbol{E})$, and $\mathfrak{T}_k(\boldsymbol{E}, \boldsymbol{E}^\times)$ is not locally solid, therefore not Lebesgue [82F]. We have here a linear space topology for which $\boldsymbol{E}' \subseteq \boldsymbol{E}^\times$ but which is not Lebesgue.

We can see also that A satisfies the condition (i) of 82G, but neither of the others.

8XC Let X be an uncountable set and let $\boldsymbol{E}$ be the set of functions $x: X \to \mathbf{R}$ which are 'convergent at infinity', i.e. for which there exists a real number $f_0 x$ such that, for every $\epsilon > 0$,

$$\{t : |x(t) - f_0 x| \geqslant \epsilon\}$$

is finite. Then $\boldsymbol{E}$ is a $\| \ \|_\infty$-closed Riesz subspace of $\boldsymbol{l}^\infty(X)$; it is an M-space with unit, isomorphic to $\boldsymbol{C}(Y)$, where Y is the one-point compactification of the discrete space X.

If $\langle x_n\rangle_{n\in\mathbf{N}} \downarrow 0$ in $\boldsymbol{E}$, then $\langle \|x_n\|_\infty\rangle_{n\in\mathbf{N}} \to 0$. **P** Let $\epsilon > 0$. For each $n \in \mathbf{N}$, $\{t : x_n(t) \neq f_0 x_n\}$ is countable, so there is a $t \in X$ such that $x_n(t) = f_0 x_n$ for every $n \in \mathbf{N}$. Thus $\langle f_0 x_n\rangle_{n\in\mathbf{N}} \downarrow 0$ and there is an $m \in \mathbf{N}$ such that $f_0 x_m < \epsilon$. Now $X_0 = \{t : x_m(t) > \epsilon\}$ is finite, so there is an $n \geqslant m$ such that $x_n(t) \leqslant \epsilon$ for every $t \in X_0$. Clearly $\|x_i\|_\infty \leqslant \epsilon$ for every $i \geqslant n$. As ϵ is arbitrary, $\langle \|x_n\|_\infty\rangle_{n\in\mathbf{N}} \to 0$. **Q**

As in 8XB above, $\boldsymbol{E}^\times$ can be identified with $\boldsymbol{l}^1(X)$. The function $f_0 : \boldsymbol{E} \to \mathbf{R}$ is a Riesz homomorphism, and $\boldsymbol{E}^\sim = \boldsymbol{E}' = \boldsymbol{E}^\times \oplus \operatorname{lin}\{f_0\}$, since $\boldsymbol{E} \cong \boldsymbol{c}_0(X) \oplus \mathbf{R}$ as Banach space.

If $A = \{f : f \in \boldsymbol{E}^\times, \|f\| \leqslant 1\}$, then A satisfies (iii) of 82H but neither (i) nor (ii), since f_0 is in the closure of A for $\mathfrak{T}_s(\boldsymbol{E}^\sim, \boldsymbol{E})$. A is closed for $\mathfrak{T}_s(\boldsymbol{E}^\times, \boldsymbol{E})$, and for every $x \in \boldsymbol{E}$ there is a $g_0 \in A$ such that

$$g_0 x = \|x\|_\infty = \sup_{g\in A} gx,$$

since x must attain its bounds on X. But A is not compact for $\mathfrak{T}_s(\boldsymbol{E}^\times, \boldsymbol{E})$. By James' theorem [AlF], $\boldsymbol{E}^\times$ cannot be complete for $\mathfrak{T}_k(\boldsymbol{E}^\times, \boldsymbol{E})$.

*Observe that $\boldsymbol{E}^\sim = \boldsymbol{E}_c^\sim$, since the norm topology on $\boldsymbol{E}$ is 'sequentially Lebesgue' [83Ka]. The unit ball of $\boldsymbol{E}_c^\sim$ is therefore compact for $\mathfrak{T}_s(\boldsymbol{E}_c^\sim, \boldsymbol{E})$, but it does not satisfy either (iii) or (iv) of 83Kf; thus these cannot be added to the three conditions of 83Ke.

Further reading for Chapter 8 The earliest version of 82E/82G in an abstract context which I have been able to trace seems to have been in NAKANO M.S.O.L.S.; in the abridged version [NAKANO L.L., theorems 28.7–28.10] I find (a) a proof of 82G(i) $\Leftrightarrow$ 82G(iii) when $\boldsymbol{E}$ is a perfect Riesz space and $\boldsymbol{E}^\times$ has the countable sup property (b) a proof that, if $\boldsymbol{E}$ is perfect, then 82E(iii) is a necessary and sufficient condition for relative weak compactness. AMEMIYA O.T.L.S. proves 82G for Dedekind complete spaces $\boldsymbol{E}$, and gives some further references.

Extensions of these ideas may be found in FREMLIN A.K.S.I. and MOORE & REBER. Many of the important Riesz spaces (notably the $\boldsymbol{L}^p$ and Orlicz spaces [6XF–6XH]) have symmetry properties which mean that we can say much more about their weakly compact sets; see GARLING S.S.S. and FREMLIN S.S.; also GARLING I.O. and PHUONG-CÁC for yet further elaborations of these ideas.

For extensions of Grothendieck's theorem [83J], see GROTHENDIECK A.L., SEEVER, WELLS, TOPSØE C.S.M., GÄNSSLER, and SCHAEFER W.C.

Appendix

A1 Linear topological spaces

This book has been written on the assumption that readers will be acquainted with the elementary theory of linear topological spaces. But it may be helpful if I set out explicitly some of the results I use, though the following list is far from complete. In general I give references rather than proofs.

As in the rest of the book, all linear spaces are over the field **R** of real numbers.

A1A Notation Let E and F be real linear spaces in duality, i.e. endowed with a bilinear functional $\langle\ ,\ \rangle: E \times F \to \mathbf{R}$. The fundamental topologies on E associated with this duality I denote as follows: $\mathfrak{T}_s(E, F)$ is the **weak** topology, the topology of uniform convergence on finite subsets of F; $\mathfrak{T}_k(E, F)$ is the **Mackey** topology, the topology of uniform convergence on convex $\mathfrak{T}_s(F, E)$-compact sets; $\mathfrak{T}_b(E, F)$ is the **strong** topology, the topology of uniform convergence on $\mathfrak{T}_s(F, E)$-bounded sets.

A1B Mackey's theorem Let E and F be real linear spaces in duality, such that E separates the points of F. Let $\mathfrak{T}$ be a locally convex linear space topology on E. Then the dual of E for $\mathfrak{T}$ can be identified with F iff $\mathfrak{T}_s(E, F) \subseteq \mathfrak{T} \subseteq \mathfrak{T}_k(E, F)$.

Proof See BOURBAKI V, chapter IV, §2, no. 3, Theorem 2; KELLEY & NAMIOKA 18.8; KÖTHE 21.4.3; ROBERTSON & ROBERTSON, chapter III, §7, theorem 7; or SCHAEFER T.V.S. IV, 3.2.

A1C Closed graph theorem **(a)** Let E and F be complete metrizable linear topological spaces. Let $T: E \to F$ be a linear map such that the graph of T is closed in $E \times F$. Then T is continuous.

(b) Let E be a linear space, and $\mathfrak{S}$ and $\mathfrak{T}$ complete metrizable linear space topologies on E such that $\mathfrak{S} \subseteq \mathfrak{T}$. Then $\mathfrak{S} = \mathfrak{T}$.

Proof BOURBAKI V, Chapter I, §3, no. 3; KELLEY & NAMIOKA §11; SCHAEFER T.V.S., chapter III, §2; KÖTHE §15.12.

Remark I have stated this theorem without any assumption of local convexity, and this is the form in which I shall use it. But the reader who prefers to restrict himself to locally convex spaces will have no difficulty in seeing where to interpolate this hypothesis.

A1D Grothendieck's criterion Let $\boldsymbol{E}$ and $\boldsymbol{F}$ be real linear spaces in duality. Let $\mathfrak{M}$ be a family of subsets of $\boldsymbol{F}$, directed upwards and covering $\boldsymbol{F}$, such that each member of $\mathfrak{M}$ is balanced, convex and $\mathfrak{T}_s(\boldsymbol{F}, \boldsymbol{E})$-bounded. Let $\mathfrak{T}$ be the topology on $\boldsymbol{E}$ of uniform convergence on members of $\mathfrak{M}$. Then $\boldsymbol{E}$ is complete under $\mathfrak{T}$ iff every $g \in \boldsymbol{F}^*$ which is $\mathfrak{T}_s(\boldsymbol{F}, \boldsymbol{E})$-continuous on each member of $\mathfrak{M}$ is represented by a point of $\boldsymbol{E}$.

Proof KELLEY & NAMIOKA 16.9; KÖTHE 21.9.2; ROBERTSON & ROBERTSON, chapter VI §1, Cor. 1 to Th. 3 (p. 107); SCHAEFER T.V.S., IV.6.2, Cor. 2.

Remark I think that we need only the easier half of this result; that $\boldsymbol{E}$ is complete if the criterion is satisfied.

A1E Corollary Let $\boldsymbol{E}$ be a Banach space. Then its dual $\boldsymbol{E}'$ is complete for the Mackey topology $\mathfrak{T}_k(\boldsymbol{E}', \boldsymbol{E})$.

Proof $\mathfrak{T}_k(\boldsymbol{E}', \boldsymbol{E})$ is the topology of uniform convergence on the family $\mathfrak{M}$ of balanced convex weakly compact sets in $\boldsymbol{E}$. So suppose that $g \in \boldsymbol{E}^*$ is weakly continuous on each member of $\mathfrak{M}$. Let $\langle x_n \rangle_{n \in \mathbf{N}} \to 0$ for the strong (norm) topology on $\boldsymbol{E}$. Then $A = \{0\} \cup \{x_n : n \in \mathrm{N}\}$ is compact for the strong topology; because $\boldsymbol{E}$ is complete in the strong topology, the closed balanced convex hull B of A is compact. But now $B \in \mathfrak{M}$, so g is weakly continuous on B. As certainly $\langle x_n \rangle_{n \in \mathbf{N}} \to 0$ for $\mathfrak{T}_s(\boldsymbol{E}, \boldsymbol{E}')$, $\langle g x_n \rangle_{n \in \mathbf{N}} \to 0$. But as $\boldsymbol{E}$ is metrizable this is enough to show that g is continuous, i.e. that $g \in \boldsymbol{E}'$. By Grothendieck's criterion, $\boldsymbol{E}'$ must be complete.

Remark This result is of course true for all Fréchet spaces $\boldsymbol{E}$.

A1F James' theorem Let $\boldsymbol{E}$ be a complete locally convex Hausdorff linear topological space. Suppose that $B \subseteq \boldsymbol{E}$ is closed for $\mathfrak{T}_s(\boldsymbol{E}, \boldsymbol{E}')$ and has the property that every $f \in \boldsymbol{E}'$ is bounded and attains its bounds on B. Then B is compact for $\mathfrak{T}_s(\boldsymbol{E}, \boldsymbol{E}')$.

Proof PRYCE W.C.

A1G Krein's theorem Let E be a complete locally convex Hausdorff linear topological space. Suppose that $A \subseteq E$ is relatively compact for $\mathfrak{T}_s(E, E')$. Then so is its convex hull $\Gamma(A)$.

Proof Let B be the closure of $\Gamma(A)$ for $\mathfrak{T}_s(E, E')$. Let $f \in E'$. Then because $\bar{A}$, the $\mathfrak{T}_s(E, E')$-closure of A, is $\mathfrak{T}_s(E, E')$-compact, f is bounded and attains its bounds on $\bar{A}$. But clearly the bounds of f on $\bar{A}$ are the same as its bounds on B, and $\bar{A} \subseteq B$. Thus f is bounded and attains its bounds on B. By James' theorem [A1F], B is weakly compact, i.e. $\Gamma(A)$ is relatively weakly compact.

Alternative proof KÖTHE 24.5.4.

A1H Proposition Let E and F be linear spaces in duality. Let $\langle x_n \rangle_{n \in \mathbf{N}}$ be a sequence in E such that every subsequence is summable for $\mathfrak{T}_s(E, F)$. Then:

(a) If $A \subseteq F$ is relatively countably compact for $\mathfrak{T}_s(F, E)$ (i.e. if every sequence in A has a cluster point in F for $\mathfrak{T}_s(F, E)$), then

$$\inf_{n \in \mathbf{N}} \sup_{y \in A} |\langle x_n, y \rangle| = 0.$$

(b) In fact, $\qquad \lim_{n \to \infty} \sup_{y \in A} |\langle x_n, y \rangle| = 0.$

(c) $\langle x_n \rangle_{n \in \mathbf{N}} \to 0$ for the Mackey topology $\mathfrak{T}_k(E, F)$.

(d) $\{x_n : n \in \mathrm{N}\}$ is bounded for the strong topology $\mathfrak{T}_b(E, F)$.

*(e) ('Orlicz' theorem') $\langle x_n \rangle_{n \in \mathbf{N}}$ is summable for the topology $\mathfrak{T}$ of uniform convergence on those sets of F which are relatively countably compact for $\mathfrak{T}_s(F, E)$, and therefore is summable for $\mathfrak{T}_k(E, F)$.

Proof (a) ? Suppose otherwise. Then

$$\epsilon = \tfrac{1}{2} \inf_{n \in \mathbf{N}} \sup_{y \in A} |\langle x_n, y \rangle| > 0.$$

For each $n \in \mathrm{N}$, choose a $y_n \in A$ such that $|\langle x_n, y_n \rangle| \geqslant \epsilon$. Let y be any cluster point of $\langle y_n \rangle_{n \in \mathbf{N}}$ for $\mathfrak{T}_s(F, E)$.

Obtain a strictly increasing sequence $\langle n(i) \rangle_{i \in \mathbf{N}}$ in N inductively, as follows. Set $n(0) = 0$. Given $n(k)$, let $m \in \mathrm{N}$ be such that for any $r > m$,

$$|\langle x_r, y \rangle| \leqslant 2^{-(k+1)}, \quad |\langle x_r, y_i \rangle| \leqslant 2^{-(k+1)} \quad \forall\ i \leqslant n(k),$$

which exists as certainly $\langle x_r \rangle_{r \in \mathbf{N}} \to 0$ for $\mathfrak{T}_s(E, F)$. Let

$$n(k+1) > \max(m, n(k))$$

be such that $\quad |\langle x_i, y_{n(k+1)} - y \rangle| \leqslant 2^{-(k+1)} \quad \forall\ i \leqslant n(k),$

which exists as y is a cluster point of $\langle y_r \rangle_{r \in \mathbf{N}}$. Continue.

We see now that $\langle n(i)\rangle_{i\in\mathbf{N}}$ is a strictly increasing sequence such that

$$|\langle x_{n(k)}, y\rangle| \leqslant 2^{-k}, \quad |\langle x_{n(k)}, y_{n(i)}\rangle| \leqslant 2^{-k}, \quad |\langle x_{n(i)}, y_{n(k)}-y\rangle| \leqslant 2^{-k},$$

whenever $0 \leqslant i < k$.

Now consider $z = \sum_{i\in\mathbf{N}} x_{n(i)}$, which exists for $\mathfrak{T}_s(F, E)$. Examine, for any $k \geqslant 1$,

$$\begin{aligned}|\langle z, y_{n(k)}-y\rangle| &= |\textstyle\sum_{i\in\mathbf{N}}\langle x_{n(i)}, y_{n(k)}-y\rangle| \\ &\geqslant |\langle x_{n(k)}, y_{n(k)}\rangle| - |\langle x_{n(k)}, y\rangle| - \textstyle\sum_{i<k}|\langle x_{n(i)}, y_{n(k)}-y\rangle| \\ &\quad - \textstyle\sum_{i>k}|\langle x_{n(i)}, y_{n(k)}\rangle| - \sum_{i>k}|\langle x_{n(i)}, y\rangle| \\ &\geqslant \epsilon - 2^{-k} - \textstyle\sum_{i<k} 2^{-k} - \sum_{i>k} 2^{-i} - \sum_{i>k} 2^{-i} \\ &= \epsilon - (k+3)\,2^{-k}.\end{aligned}$$

So, if y' is any cluster point of $\langle y_{n(k)}\rangle_{k\in\mathbf{N}}$ for $\mathfrak{T}_s(E', E)$,

$$|\langle z, y'-y\rangle| \geqslant \epsilon.$$

However, since, for any $i \in \mathbf{N}$,

$$\langle x_{n(i)}, y\rangle = \lim_{k\to\infty}\langle x_{n(i)}, y_{n(k)}\rangle = \langle x_{n(i)}, y'\rangle,$$

we must have

$$\langle z, y'\rangle = \textstyle\sum_{i\in\mathbf{N}}\langle x_{n(i)}, y\rangle = \sum_{i\in\mathbf{N}}\langle x_{n(i)}, y'\rangle = \langle z, y'\rangle,$$

which is impossible. **X**

(b) This follows at once, because we can apply (a) to any subsequence of $\langle x_n\rangle_{n\in\mathbf{N}}$.

(c) This follows trivially from (b).

(d) **?** Otherwise, let $B \subseteq F$ be a $\mathfrak{T}_s(F, E)$-bounded set such that $\sup_{n\in\mathbf{N}}\sup_{y\in B}|\langle x_n, y\rangle| = \infty$. For each $n \in \mathbf{N}$, choose a $y_n \in B$ such that

$$\alpha_n = |\langle x_n, y_n\rangle| \geqslant \sup_{y\in B}|\langle x_n, y\rangle| - 1.$$

Then $\sup_{n\in\mathbf{N}}\alpha_n = \infty$. Let $\langle\beta_n\rangle_{n\in\mathbf{N}}$ be a sequence converging to 0 in $\mathbf{R}$ such that $\langle\alpha_n\beta_n\rangle_{n\in\mathbf{N}}$ does not converge to 0. Then $\langle\beta_n y_n\rangle_{n\in\mathbf{N}} \to 0$ for $\mathfrak{T}_s(F, E)$, so $C = \{\beta_n y_n : n \in \mathbf{N}\}$ is relatively compact for $\mathfrak{T}_s(F, E)$. But

$$\sup_{y\in C}|\langle x_n, y\rangle| \geqslant |\beta_n\langle x_n, y_n\rangle| = |\alpha_n\beta_n|,$$

so by (b) above $\langle\alpha_n\beta_n\rangle_{n\in\mathbf{N}} \to 0$, which contradicts the choice of $\langle\beta_n\rangle_{n\in\mathbf{N}}$. **X**

***(e)** Let $x = \sum_{n\in\mathbf{N}} x_n$ for $\mathfrak{T}_s(E, F)$, and for each $n \in \mathbf{N}$ let $z_n = \sum_{i<n} x_i$. Let $A \subseteq F$ be relatively countably compact for $\mathfrak{T}_s(F, E)$. **?** Suppose, if possible, that

$$\epsilon = \tfrac{1}{2}\inf_{n\in\mathbf{N}}\sup_{m\geqslant n}\sup_{y\in A}|\langle x - z_n, y\rangle| > 0.$$

Let $\langle n(k)\rangle_{k\in\mathbf{N}}$ be a strictly increasing sequence in $\mathbf{N}$ such that

$$\sup_{y\in A}|\langle x-z_{n(k)}, y\rangle| > \epsilon \quad \forall\ k\in\mathbf{N},$$

and for each $k\in\mathbf{N}$ choose a $y_k\in A$ such that

$$|\langle x-z_{n(k)}, y_k\rangle| > \epsilon.$$

Now let $\langle k(i)\rangle_{i\in\mathbf{N}}$ be a strictly increasing sequence in $\mathbf{N}$ such that

$$|\langle z_{n(k(i+1))}-z_{n(k(i))}, y_{k(i)}\rangle| \geqslant \epsilon \quad \forall\ i\in\mathbf{N},$$

which exists because $\langle z_{n(r)}\rangle_{r\in\mathbf{N}} \to x$ for $\mathfrak{T}_s(\boldsymbol{E}, \boldsymbol{F})$, and set

$$w_i = z_{n(k(i+1))}-z_{n(k(i))} = \sum\nolimits_{n(k(i))\leqslant j<n(k(i+1))} x_j$$

for each $i\in\mathbf{N}$. Then every subsequence of $\langle w_i\rangle_{i\in\mathbf{N}}$ is summable, being merely a collection of terms of a subsequence of $\langle x_j\rangle_{j\in\mathbf{N}}$. So we may apply (a) above to the sequence $\langle w_i\rangle_{i\in\mathbf{N}}$ and find that

$$0 = \inf_{i\in\mathbf{N}}\sup_{y\in A}|\langle w_i, y\rangle| \geqslant \inf_{i\in\mathbf{N}}|\langle w_i, y_{k(i)}\rangle| \geqslant \epsilon,$$

which is impossible. **X**

Thus $$\inf_{n\in\mathbf{N}}\sup_{m\geqslant n}\sup_{y\in A}|\langle x-z_n, y\rangle| = 0,$$

and x is a limit of $\langle z_n\rangle_{n\in\mathbf{N}}$ for the topology of uniform convergence on A. As A is arbitrary, $\langle z_n\rangle_{n\in\mathbf{N}} \to x$ for $\mathfrak{T}$, as required.

Remark Orlicz' theorem is not used in this book. But it is very close to some of the results in Chapter 8 and it has other applications in neighbouring parts of advanced measure theory.

A1I Corollary (the 'Banach–Mackey theorem') Let $\boldsymbol{E}$ and $\boldsymbol{F}$ be linear spaces in duality. Let A be a subset of $\boldsymbol{E}$ such that $\sum_{i\in\mathbf{N}}\alpha_i x_i$ exists for $\mathfrak{T}_s(\boldsymbol{E}, \boldsymbol{F})$ whenever $\langle x_i\rangle_{i\in\mathbf{N}}$ is a sequence in A and $\langle \alpha_i\rangle_{i\in\mathbf{N}}$ is a non-negative sequence in $\mathbf{R}$ such that $\sum_{i\in\mathbf{N}}\alpha_i = 1$. (For instance, suppose that A is convex, bounded and complete for some linear space topology on $\boldsymbol{E}$ finer than $\mathfrak{T}_s(\boldsymbol{E}, \boldsymbol{F})$.) Then A is bounded for $\mathfrak{T}_b(\boldsymbol{E}, \boldsymbol{F})$.

Proof Observe that in fact $\sum_{i\in\mathbf{N}}\alpha_i x_i$ exists for $\mathfrak{T}_s(\boldsymbol{E}, \boldsymbol{F})$ whenever $\langle x_i\rangle_{i\in\mathbf{N}}$ is a sequence in A and $\langle \alpha_i\rangle_{i\in\mathbf{N}}$ is a non-negative summable sequence. So if $\langle x_i\rangle_{i\in\mathbf{N}}$ is a sequence in A, every subsequence of $\langle 2^{-i}x_i\rangle_{i\in\mathbf{N}}$ is $\mathfrak{T}_s(\boldsymbol{E}, \boldsymbol{F})$-summable. By A1Hd, $\{2^{-i}x_i : i\in\mathbf{N}\}$ is bounded for $\mathfrak{T}_b(\boldsymbol{E}, \boldsymbol{F})$. As $\langle x_i\rangle_{i\in\mathbf{N}}$ is arbitrary, A is bounded for $\mathfrak{T}_b(\boldsymbol{E}, \boldsymbol{F})$.

Alternative proof See SCHAEFER T.V.S. II.8.5 or KÖTHE 20.11.3.

***A1J Exercise** Use A1I to show (i) that if $\boldsymbol{E}$ is any locally convex linear topological space, then a $\mathfrak{T}_s(\boldsymbol{E}, \boldsymbol{E}')$-bounded set is $\mathfrak{T}_k(\boldsymbol{E}, \boldsymbol{E}')$-bounded, (ii) that if $\boldsymbol{E}$ is a complete locally convex linear topological space, then a $\mathfrak{T}_s(\boldsymbol{E}, \boldsymbol{E}')$-bounded set is $\mathfrak{T}_b(\boldsymbol{E}, \boldsymbol{E}')$-bounded.

A2 Spaces of continuous functions

Some of the most important topological Riesz spaces are spaces of continuous real-valued functions. In this section I give some definitions and state a few basic properties in the language I have used in this book. As a supplement I show how a result that is normally proved using measure theory is amenable to more primitive techniques.

A2A Definition Let X be a topological space. I shall write $\boldsymbol{C}(X)$ for the space of all continuous real-valued functions on X. $\boldsymbol{C}(X)$ is a Riesz subspace of $\mathbf{R}^X$ [1XD]. It is a uniformly complete Archimedean Riesz space [25Hb].

A2B Definition Let X be a topological space. I shall write $\boldsymbol{C}_b(X)$ for the space of all bounded continuous functions from X to $\mathbf{R}$. It is an order-dense solid linear subspace of $\boldsymbol{C}(X)$ [A2A], and a norm-closed Riesz subspace of $\boldsymbol{l}^\infty(X)$ [2XA]. Under the norm $\| \ \|_\infty$, it is an M-space with unit χX [26Ad].

A2C Definition Let X be a topological space. Then $\boldsymbol{C}_0(X)$ will be the space of those functions $x \in \boldsymbol{C}(X)$ 'convergent to 0 at ∞', i.e. such that

$$\{t : |x(t)| \geqslant \epsilon\}$$

is compact for every $\epsilon > 0$. $\boldsymbol{C}_0(X)$ is a norm-closed solid linear subspace of $\boldsymbol{C}_b(X)$, and in its own right is an M-space [26Ac].

A2D Definition Let X be a topological space. Then $\boldsymbol{K}(X)$ is the space of continuous functions 'of compact support', i.e. those functions $x \in \boldsymbol{C}(X)$ such that

$$\overline{\{t : x(t) \neq 0\}}$$

is compact in X. $\boldsymbol{K}(X)$ is a solid linear subspace of $\boldsymbol{C}_0(X)$, so it is uniformly complete [25Ni]. Also, $\boldsymbol{K}(X)$ is dense in $\boldsymbol{C}_0(X)$ for $\| \ \|_\infty$. **P** Let $x \in \boldsymbol{C}_0(X)$ and $\epsilon > 0$. Set

$$y = (x - \epsilon\chi X)^+ - (x + \epsilon\chi X)^-.$$

Then it is easy to verify that $y \in \boldsymbol{K}(X)$ and that $\|y - x\|_\infty \leqslant \epsilon$. **Q**

***A2E Lemma** Let X be a locally compact Hausdorff topological space.

(a) Suppose that $t \in X$ and that G is a neighbourhood of t. Then there is an $x \in \boldsymbol{K}(X)$ such that $x(t) = 1$ and $0 \leqslant x \leqslant \chi G$ in $\mathbf{R}^X$.

(b) If $G \subseteq X$ is open and $F \subseteq G$ is compact, then there is an $x \in \boldsymbol{K}(X)$ such that $\chi F \leqslant x \leqslant \chi G$.

Proof (a) Let H be a neighbourhood of t such that $\overline{H}$ is compact. Then $\overline{H}$ is compact and Hausdorff, therefore normal, and by Urysohn's lemma there is a continuous function $y: \overline{H} \to \mathbf{R}$ such that $y(t) = 1$ and $y(u) = 0$ for every $u \in \overline{H} \backslash \operatorname{int}(G \cap H)$. Now if we set $y(u) = 0$ for every $u \in X \backslash \overline{H}$, we see that $y \in \boldsymbol{K}(X)$ and that $x = y^+ \wedge \chi X$ satisfies the conditions.

(b) Let $A = \{y : y \in \boldsymbol{K}(X), 0 \leqslant y \leqslant \chi G\}$. For each $y \in A$, let

$$G_y = \{t : y(t) > \tfrac{1}{2}\}.$$

By (a) above, we know that $\bigcup_{y \in A} G_y = G$, and of course $\{G_y : y \in A\} \uparrow$ because $A \uparrow$. As F is compact, there is a $y \in A$ such that $G_y \supseteq F$. Now set $x = 2y \wedge \chi X$.

***A2F How to avoid the Riesz representation theorem** One of the most important applications of Riesz' theorem [73D/E] in functional analysis is to the following lemma. I think it is therefore of interest that the result can be proved without any call on measure theory.

Lemma Let X be a compact Hausdorff topological space. Let $\langle x_n \rangle_{n \in \mathbf{N}}$ be a sequence in $\boldsymbol{C} = \boldsymbol{C}(X)$ which is bounded for $\| \ \|_\infty$ and such that $\langle x_n(t) \rangle_{n \in \mathbf{N}} \to 0$ for every $t \in X$. Then $\langle x_n \rangle_{n \in \mathbf{N}} \to 0$ for $\mathfrak{T}_s(\boldsymbol{C}, \boldsymbol{C}')$.

First proof By 25G, $\boldsymbol{C}' = \boldsymbol{C}^\sim$. Now if $f \in \boldsymbol{C}^{\sim +}$, then by Riesz' theorem [72D/E] there is a Radon measure μ on X such that

$$fx = \int x \, d\mu \quad \forall \ x \in \boldsymbol{C},$$

and $\mu X < \infty$. Now $\langle x_n \rangle_{n \in \mathbf{N}}$ is bounded above and below in $\mathfrak{L}^1(\mu)$ by constant functions, and is convergent to 0 at each point, so by Lebesgue's theorem [63Md],

$$\langle f x_n \rangle_{n \in \mathbf{N}} = \langle \textstyle\int x_n \, d\mu \rangle_{n \in \mathbf{N}} \to 0.$$

As f is arbitrary, $\langle x_n \rangle_{n \in \mathbf{N}} \to 0$ for $\mathfrak{T}_s(\boldsymbol{C}, \boldsymbol{C}^\sim) = \mathfrak{T}_s(\boldsymbol{C}, \boldsymbol{C}')$.

Second proof By 25G, $C' = C^{\sim}$. Let $f \in C^{\sim +}$ and $\epsilon > 0$. For each $m \in \mathbf{N}$, consider the sequence $\langle \sup_{m \leqslant i \leqslant m+n} |x_i| \rangle_{n \in \mathbf{N}}$. This is an increasing sequence, bounded above by a constant function, so

$$\alpha_m = \sup_{n \in \mathbf{N}} f(\sup_{m \leqslant i \leqslant m+n} |x_i|) < \infty.$$

Let n be such that, setting $z_m = \sup_{m \leqslant i \leqslant m+n} |x_i|$, $f(z_m) \geqslant \alpha_m - 2^{-m}\epsilon$. Now, for any $k \geqslant m$,

$$f(|x_k| - z_m)^+ \leqslant f(\sup_{m \leqslant i \leqslant n+k} |x_i| - z_m) \leqslant \alpha_m - f(z_m) \leqslant 2^{-m}\epsilon.$$

Set $w_r = \inf_{m \leqslant r} z_m$ for every $r \in \mathbf{N}$. Then $\langle w_r \rangle_{r \in \mathbf{N}} \downarrow$. Moreover, for any $t \in X$,

$$w_r(t) \leqslant z_r(t) \leqslant \sup_{i \geqslant r} |x_i(t)| \quad \forall\ r \in \mathbf{N},$$

so $\langle w_r(t) \rangle_{r \in \mathbf{N}} \downarrow 0$. By Dini's theorem, there is an $r \in \mathbf{N}$ such that $\|w_r\|_\infty \leqslant \epsilon$, so that $f(w_r) \leqslant \epsilon \|f\|$. Now for any $i \geqslant r$,

$$|fx_i| \leqslant f(|x_i|) = f(|x_i| \wedge w_r) + f(|x_i| - w_r)^+$$

$$\leqslant f(w_r) + \Sigma_{m \leqslant r} f(|x_i| - z_m)^+$$

(as $(|x_i| - w_r)^+ = \sup_{m \leqslant r}(|x_i| - z_m)^+$)

$$\leqslant \epsilon \|f\| + \Sigma_{m \leqslant r} 2^{-m}\epsilon \leqslant \epsilon(\|f\| + 2).$$

As ϵ is arbitrary, we see that $\langle fx_i \rangle_{i \in \mathbf{N}} \to 0$. As f is arbitrary, $\langle x_i \rangle_{i \in \mathbf{N}} \to 0$ for $\mathfrak{T}_s(C, C^{\sim}) = \mathfrak{T}_s(C, C')$.

***A2G Applications** This lemma is used by KÖTHE [§ 24.5] to prove Krein's theorem. Other applications are:

(a) If X is a compact Hausdorff topological space, a subset A of $C = C(X)$ is compact for $\mathfrak{T}_s(C, C')$ iff it is bounded for $\| \ \|_\infty$ and compact for the topology $\mathfrak{T}_p$ of uniform convergence on finite subsets of X [KÖTHE 24.5.3].

(b) If X is a compact Hausdorff topological space, a set

$$A \subseteq C = C(X)$$

which is compact for $\mathfrak{T}_s(C, C')$ is totally bounded for $\mathfrak{T}_{|s|}(C, C')$ [because any sequence in A must have a subsequence which is convergent for $\mathfrak{T}_s(C, C')$ and therefore for $\mathfrak{T}_{|s|}(C, C')$], and therefore is compact for $\mathfrak{T}_{|s|}(C, C')$.

Remark Of course James' theorem [A1F] provides us with another proof of Krein's theorem not using measure theory [A1G].

References

The following list comprises only works referred to in the text. For a systematic bibliography of the literature on Riesz spaces, see LUXEMBURG & ZAANEN R.S.

Amemiya, I. [G.S.T.] 'A general spectral theory in semi-ordered linear spaces.' *J. Fac. Sci. Hokkaido Univ.* (1) **12** (1953) 111–56.

[O.L.T.S.] 'On ordered topological linear spaces'. In *Proceedings of the International Symposium on Linear Spaces*, Israel Academy of the Sciences and Humanities, Jerusalem, 1961, pp. 14–23.

Baker, K. A. 'Free vector lattices'. *Canad. J. Math.* **20** (1968) 58–66.

Bartle, R. G. *The Elements of Integration.* Wiley, 1966.

Birkhoff, G. *Lattice Theory.* American Mathematical Society, 1967.

Bourbaki, N. *Éléments de Mathématique.* Hermann (Actualités Scientifiques et Industrielles).

[I] Vol. I, *Théorie des Ensembles*, A.S.I. §1212 (chaps. 1–2) (1966); §1243 (chap. 3) (1963). Translated into English as *Theory of Sets*, Hermann & Addison-Wesley, 1968.

[V] Vol. V, *Espaces Vectoriels Topologiques*, A.S.I. §1189 (chaps. 1–2) (1966); §1229 (chaps. 3–5) (1964).

[VI] Vol. VI, *Intégration*, A.S.I. §1175 (chaps. 1–4) (1965); §1244 (chap. 5) (1967); §1281 (chap. 6) (1959); §1306 (chaps. 7–8) (1963); §1343 (chap. 9) (1969).

Chacon, R. V. & Krengel, U. 'Linear modulus of a linear operator'. *Proc. Amer. Math. Soc.* **15** (1964) 553–9.

Dieudonné, J. 'Sur les espaces de Köthe'. *J. Analyse Math.* **1** (1951) 81–115.

Dunford, N. & Schwartz, J. T. *Linear Operators I.* Interscience, 1958.

Fremlin, D. H. [A.K.S. I] 'Abstract Köthe spaces, I'. *Proc. Cambridge Philos.* **63** (1967) 653–60.

[A.K.S. II] 'Abstract Köthe spaces, II'. *Proc. Cambridge Philos. Soc.* **63** (1967) 951–6.

[A.K.S. III] 'Abstract Köthe spaces, III'. *Proc. Cambridge Philos. Soc.* **63** (1967) 957–62.

[S.S.] 'Stable subspaces of $L^1 + L^\infty$'. *Proc. Cambridge Philos. Soc.* **64** (1968) 625–43.

Gänssler, P. 'Compactness and sequential compactness in spaces of measures'. *Z. Wahrscheinlichkeitstheorie und Verw. Gebiete* **17** (1971) 124–46.

'A convergence theorem for measures in regular Hausdorff spaces'. *Math. Scand.* **29** (1971) 237–44.

Garling, D. J. H. [S.S.S.] 'On symmetric sequence spaces'. *Proc. London Math. Soc.* (3) **16** (1966) 85–106.

[I.O.] 'On ideals of operators in Hilbert space'. *Proc. London Math. Soc.* (3) **17** (1967) 115-38.

Grothendieck, A. [C.C.] 'Critères de compacité dans les espaces fonctionnels généraux'. *Amer. J. Math.* **74** (1952) 168–86.

[A.L.] 'Sur les applications linéaires faiblement compactes d'espaces du type C(*K*)'. *Canad. J. Math.* **5** (1953) 129–73.

Halmos, P. R. [M.T.] *Measure Theory*. Van Nostrand, 1950.
[N.S.T.] *Naive Set Theory*. Van Nostrand, 1960.
[L.B.A.] *Lectures on Boolean Algebras*. Van Nostrand, 1963.
Hardy, G. H., Littlewood, J. E. & Pólya, G. *Inequalities*. C.U.P., 1934.
Hewitt, E. & Stromberg, K. *Real and Abstract Analysis*. Springer, 1969.
Ionescu Tulcea, A. & Ionescu Tulcea, C. *Topics in the Theory of Lifting*. Springer, 1969.
Jameson, G. *Ordered Linear Spaces*. Springer, 1970.
Kakutani, S. 'Concrete representation of abstract L-spaces and the mean ergodic theorem'. *Ann. of Math.* **42** (1941) 523–37.
Kelley, J. L. *General Topology*. Van Nostrand, 1955.
Kelley, J. L., Namioka, I. *et al. Linear Topological Spaces*. Van Nostrand, 1963.
Köthe, G. *Topologische Lineare Räume I*. Springer, 1960. Translated into English as *Topological Vector Spaces I*, Springer, 1969.
Krengel, U. 'Über den Absolutbetrag stetiger linear Operatoren und seine Anwendung auf ergodische Zerlegunger'. *Math. Scand.* **13** (1963) 151–87.
Loomis, L. H. *An Introduction to Harmonic Analysis*. Van Nostrand, 1953.
Luxemburg, W. A. J. 'Is every integral normal?' *Bull. Amer. Math. Soc.* **73** (1967) 685–8.
Luxemburg, W. A. J. & Zaanen, A. C. [B.F.S.] 'Notes on Banach function spaces'. *Nederl. Acad. Wetensch. Proc.* (*A*): notes VI–VII, **66** (1963) 655–81; notes VIII–XIII, **67** (1964) 104–19, 360–76, 493–543; notes XIV–XVI, **68** (1965) 229–48, 415–46, 646–67. Also printed in *Indag. Math.* **25–7** (1963–5).
[R.S.] *Riesz Spaces I*. North-Holland, 1971.
Maharam, D. 'On homogeneous measure algebras'. *Proc. Nat. Acad. Sci. U.S.A.* **28** (1942) 108–11.
McShane, E. J. 'Order-preserving maps and integration processes.' *Ann. of Math. Studies* **31**, Princeton U.P., 1953.
Meyer, P. A. *Probability and Potentials*. Blaisdell, 1966.
Moore, L. C. & Reber, J. C. 'Mackey topologies which are locally convex Riesz topologies'. *Duke Math. J.* **39** (1972) 105–19.
Munroe, M. E. *Introduction to Measure and Integration*. Addison-Wesley, 1953.
Nakano, H. [C.A.C.R.] 'Über die Charakterisierung des allgemeinen C-Raumes'. *Proc. Imperial Acad. Japan* **17** (1941) 301–7. Reprinted in NAKANO S.O.L.S.
[M.S.O.L.S.] *Modulared Semi-Ordered Linear Spaces*. Tokyo, 1950. Reprinted in an abridged form as NAKANO L.L.
[L.T.] 'Linear topologies on semi-ordered linear spaces'. *J. Fac. Sci. Hokkaido Univ.* (I) **12** (1953) 87–104.
[S.O.L.S.] *Semi-Ordered Linear Spaces*. Maruzen, Tokyo, 1955.
[L.L.] *Linear Lattices*. Wayne State University Press, 1966.
Peressini, A. L. *Ordered Topological Vector Spaces*. Harper & Row, 1967.
Phuong-Các, N. 'Generalized Köthe function spaces, I'. *Proc. Cambridge Philos. Soc.* **65** (1969) 601–11.
Pryce, J. D. [W. C.] 'Weak compactness in locally convex spaces'. *Proc. Amer. Math. Soc.* **17** (1966) 148–55.
[U.S.] 'An unpleasant set in a non-locally-convex vector lattice'. *Proc. Edinburgh Math. Soc.* (2) **18** (1973) 229–33.
Riesz, F. 'Sur les operations fonctionnelles linéaires'. *C. R. Acad. Sci. Paris* **149** (1909) 974–9.
Robertson, A. P. & Robertson, W. *Topological Vector Spaces*. C.U.P., 1964.
Royden, H. L. *Real Analysis*. Macmillan, 1963.

REFERENCES

Schaefer, H. H. [T.V.S.] *Topological Vector Spaces*. Springer, 1966. (Paperback edition 1971.)

[W.C.] 'Weak convergence of measures'. *Math. Ann.* **193** (1971) 57–64.

Schwartz, L. *Radon Measures*. To appear.

Seever, G. L. 'Measures on *F*-spaces'. *Trans. Amer. Math. Soc.* **133** (1968) 267–80.

Semadeni, Z. *Banach Spaces of Continuous Functions I*. Polish Scientific Publishers, 1971.

Sikorski, R. *Boolean Algebras*. Springer, 1964.

Taylor, A. E. *Introduction to Functional Analysis*. Wiley, 1958.

Topsøe, F. [T.M.] *Topology and Measure*. Springer, 1970.

[C.S.M.] 'Compactness in spaces of measures'. *Studia Math.* **36** (1970) 195–212.

Vulikh, B. Z. *Introduction to the Theory of Partially Ordered Spaces*. Wolters-Noordhoff, 1967.

Wells, B. B. 'Weak compactness of measures'. *Proc. Amer. Math. Soc.* **20** (1969) 124–30.

Widom, H. *Lectures on Measure and Integration*. Van Nostrand Reinhold, 1969.

Williamson, J. H. *Lebesgue Integration*. Holt, Rinehart & Winston, 1962.

Wright, J. D. M. 'Stone-algebra-valued measures and integrals'. *Proc. London Math. Soc.* (3) **19** (1969) 107–22.

'The measure extension problem for vector lattices'. *Ann. Inst. Fourier (Grenoble)* **21** (1971) 65–85.

'Vector lattice measures on locally compact spaces'. *Math. Z.* **120** (1971) 193–203.

[E.T.] 'An extension theorem'. To appear in *J. London Math. Soc.*

Zaanen, A. C. *Linear Analysis*. North-Holland, 1953.

Index of special symbols

References are to section numbers; those in **bold** type indicate definitions, those in *italic* passing references

INDEX OF SPECIAL SYMBOLS

Index

References are to section numbers; those in **bold** type indicate definitions, those in *italic* passing references